# Current Topics in Microbiology and Immunology

247

Editors

Springer
Berlin
Heidelberg
New York
Barcelona
Hong Kong
London
Milan
Paris
Singapore
Tokyo

# Immunobiology of Bacterial CpG-DNA

Edited by H. Wagner

With 34 Figures and 12 Tables

Springer

Professor Dr. HERMANN WAGNER, Ph.D.
Institut für Medizinische Mikrobiologie,
Immunologie und Hygiene
der Technischen Universität München
Trogerstr. 9
D-81675 München
GERMANY
E-mail: h.wagner@lrz.tum.de

*Cover Illustration:* Schematic representation of immune cells stimulated by bacterial CpG-DNA. Antigen presenting cells (APCs) become directly stimulated to express co-stimulatory molecules and to synthesize cytokines and chemokines (transition to professional APCs). In addition haemopoietic growth factors become induced and haemopoiesis ensues. While CpG-DNA is mitogenic to B cells, upon B cell receptor (BCR) crosslinking CpG-DNA co-stimulates proliferating B cells. Susceptibility of NK cells to IL-12 becomes co-stimulated by CpG-DNA. T cells develop sensitivity to G-rich DNA upon T cell receptor (TCR) crosslinking.

ISSN 0070-217X
ISBN 3-540-66400-9 Springer-Verlag Berlin Heidelberg New York


Library of Congress Catalog Card Number 15-12910
Printed in Germany



Cover Design: *design & production GmbH*, Heidelberg
Typesetting: Scientific Publishing Services (P) Ltd, Madras
Production Editor: Angélique Gcouta
Printed on acid-free paper SPIN: 10706763 27/3020GC 5 4 3 2 1 0

# Preface

When asked whether they were willing to contribute a chapter to this volume on the immunobiology of bacterial CpG-DNA, all colleagues approached who are working in this emerging new field responded favorably. Subsequently, and within the time required limit, they supplied fine chapters covering their respective areas of expertise.

Immune stimulatory CpG-motifs in bacterial DNA have recently attracted a great deal of interest throughout the academic community and the industry. For various reasons it is astounding that bacterial genomic DNA, so far considered only to be a genetic blueprint, may turn out to be a virulence factor, as is the case with endotoxin. If so, its role as a stimulus in septic shock needs to be evaluated. It also comes as a surprise that CpG-motifs within bacterial DNA target immature antigen-presenting dendritic cells (DCs) to transit to professional antigen-presenting cells (APCs). By virtue of this DC activating property, CpG-DNA may act as 'natural' adjuvants able to render proteinaceous antigens and T cell epitopes immunogenic. If so, the immunobiology of CpG-DNA explains its powerful adjuvanticity, which often surpasses the gold standard of complete Freund's adjuvants (CFA). Furthermore, immune recognition of CpG-motifs may represent an interesting example of the principle by which pattern recognition receptors in DCs detect subtle molecular characteristics that distinguish pathogens from our own cells. As a considerable body of information has become available – thanks to the continued efforts of much research – CpG-DNA-driven direct activation and costimulation of immune cells has become particularly useful to those who want to address details of the biological mechanisms and their relation to structure and function.

This volume attempted to cover as widely as possible current research on the immunobiology of pC-DNA. It encompasses the role of CpG-motifs in "naked DNA" used for vaccination protocols, its role in activating APCs in such a way to condition polarized Thl immune responses, and the signaling pathways involved for use as an adjuvant. Typically, in a newly developing research field, one leans back to wait for the first results from human clinical trials to learn whether the data obtained in animal model systems can be translated to humans.

The response and deep interest I experienced while presenting our experimental results during visits to many institutes, in different countries, on several continents encouraged me to undertake the task of editing the present volume. I wish to thank all the contributors, the editors of *Current Topics in Microbiology and Immunology* and, in particular, Springer-Verlag for their help in getting this volume to press.

Munich, November 1999 HERMANN WAGNER

# List of Contents

# List of Contributors

(Their addresses can be found at the beginning of their respective chapters.)

# Mechanism of Action of CpG DNA

A.M. Krieg, G. Hartmann, and A.-K. Yi



## 1 Introduction

In recent years, the position of the innate immune system in regulating nearly all immune responses has become well established. A central tenet in understanding the function of the innate immune system is the concept that it is triggered by pattern recognition receptors (PRRs), which bind microbial structures that are not present in host tissues. Examples of PRRs that have become well accepted are the lipopolysaccharide (LPS) receptors, such as CD14, and mannose-binding protein.

---

University of Iowa, Department of Internal Medicine, 540 EMRB, Iowa City, IA 52242, USA
E-mail: arthur-krieg@uiowa.edu

Likewise, complement proteins have a primitive ability to bind and be activated by microbial structures. Viral RNAs are thought to adopt certain structural conformations that are bound by PRRs and lead to the activation of the double-stranded-RNA-dependent protein kinase PKR (Kumar et al. 1997). This mechanism is thought to be responsible for immune activation by the polynucleotide (rI, rC) (Talmadge et al. 1985; Wiltrout et al. 1985). Evidence that bacterial DNA (bDNA) may also possess some structural feature that activates innate immune defenses was first provided by Tokunaga et al. who reported its surprising antitumor activity and ability to activate natural killer (NK) cells (Tokunaga et al. 1984; Yamamoto et al. 1988). Further studies by these investigators led to the proposal that this immune-stimulatory activity of bDNA resided in certain self-complimentary, palindromic sequences that contained CpG dinucleotides (Kuramoto et al. 1992). Interestingly, methylation of the CpGs was initially thought to have no influence on the immune stimulatory activities of DNA (Kuramoto et al. 1992). Independent studies by Pisetsky and colleagues showed that bDNA can also activate B-cell proliferation and that these proliferative activities can also be seen with polynucleotides, such as dG, dC (Messina et al. 1991, 1993), which also activate NK cells (Tokunaga et al. 1988).

Independent of these studies, several different groups of investigators using antisense oligonucleotides, which usually possessed nuclease-resistant phosphorothioate backbones, began to report the occurrence of immune stimulatory effects, some of which were initially thought to be due to an antisense mechanism of action but others of which were clearly due to some highly sequence-specific but non-antisense mechanism of action (Krieg et al. 1989; Hatzfeld et al. 1991; Tanaka et al. 1992; Branda et al. 1993; McIntyre et al. 1993; Mojcik et al. 1993; Pisetsky and Reich 1993; Fischer et al. 1994; Koike et al. 1995). Initially, there seemed to be no common sequence element or motif among all of these different immunostimulatory effects. However, after several years of investigating the immunostimulatory effects of different oligonucleotides and DNAs, we finally came to the realization that a rather simple sequence motif based on an unmethylated CpG dinucleotide could, in particular base contexts, link together all of the reported immunostimulatory activities (Fig. 1; Krieg et al. 1995; Krieg 1998).

It makes tremendous teleological sense for the immune system to have evolved an innate defense triggered by these unmethylated CpG motifs. While CpGs are unmethylated and usually fairly abundant in bDNA, they are methylated and highly suppressed (only about one-quarter as frequent as would be predicted if base utilization were random) in vertebrate DNA (Bird 1987). Moreover, the base context of CpG dinucleotides in the human genome is not random; in our genome, CpGs are most frequently preceded by a C or followed by a G, which is an unfavorable context for immune stimulation and which can even have neutralizing effects, blocking immune stimulation by optimal stimulatory CpG motifs (Fig. 1; Krieg et al. 1998b).

It does not seem particularly logical that the recognition of CpG DNA evolved as a general defense strategy against bacteria, since most of these normally live outside our cells and their DNA would, therefore, be shielded from detection. The

$X_1X_2\underline{CG}Y_1Y_2$

$X_1$ = purine
$X_2$ = purine or T
Y = pyrimidine

**CpG motifs can be species-specific:**
**ODN #1826 – 5' TCCATGA$\underline{CG}$TTCCTGA$\underline{CG}$TT 3'**
**(activates mouse but not human cells)**
**ODN #2006 – 5' T$\underline{CG}$T$\underline{CG}$TTTTGT$\underline{CG}$TTTTGT$\underline{CG}$TT 3'**
**(activates human and mouse cells)**

**Fig. 1.** Formula for a CpG motif. Not all DNA-containing, unmethylated CpG dinucleotides are immune stimulatory. If the CpG is preceded by a C or followed by a G, or if the CpGs are arranged in direct repeats, then they may have inhibitory rather than stimulatory effects on leukocyte activation by CpG DNA (KRIEG et al. 1998b). In general, mouse cells respond to a rather wide variety of CpG motifs, but human cells are optimally stimulated by a narrower subset. A particularly strong mouse motif is TGACGTT, while a particularly strong human motif is TGTCGTT. The figure shows examples of two phosphorothioate oligodeoxynucleotide (ODN) sequences that have immune stimulatory effects in mice, only one of which is stimulatory to human cells. Not all CpG motifs have the same range of immune activities – some are best at activating natural killer (NK) cells, with relatively little adjuvant effect, while others are potent adjuvants for antibody responses but have relatively little NK-stimulating activity. The exact range of immune effects that a given ODN will have depends on the CpG motif, the number and spacing of motifs in the ODN, the presence of other sequence motifs (such as poly-G sequences) and, very importantly, the backbone of the ODN. ODNs with nuclease-resistant phosphorothioate backbones can have dramatically increased potency if they have optimal or near-optimal CpG motifs but can have essentially no CpG activity if the motif is suboptimal, even though the same sequence may be stimulatory if made with a phosphodiester backbone (BALLAS et al. 1996; KRIEG et al. 1996). In general, phosphorothioate ODNs have lower potentials for inducing NK activation than do chimeric ODNs (BALLAS et al. 1996). The most potent backbones for stimulating B cells are, in order of potency, phosphorodithioate, phosphorothioate, phosphodiester, methylphosphonate, and 2′-O-methylribonucleotide (ZHAO et al. 1996a; KRIEG et al. 1996). Modification of the 5 position on the cytosine ring can also affect the level of immune stimulation of an ODN (BOGGS et al. 1997)

CD14 pathway and detection of endotoxins is probably a far more important defense against gram-negative extracellular bacteria. Instead, we propose that the CpG-DNA innate defense has evolved and is specialized for protection against pathogens that replicate inside host cells, such as intracellular bacteria, viruses, and retroviruses. This would predict that the CpG pathway should trigger appropriate protective defenses, which would be primarily T-helper 1 cells (Th1) although, of course, this does not exclude production of antibodies. Indeed, CpG DNA appears to be the most potent Th1-like adjuvant described (CHU et al. 1997; LIPFORD et al. 1997b; ROMAN et al. 1997; WEINER et al. 1997; DAVIS et al. 1998; H. Davis, personal communication). In a recent comparison of the immune responses generated by 19 different vaccine adjuvants against 3 different antigens, CpG DNA gave by far the strongest Th1-like T-cell responses (P. Livingston, personal communication).

If the hypothesis that innate immune recognition of CpG motifs contributes to defense against infection is correct, then it would be expected that viruses and retroviruses may have evolved ways to overcome this defense. Indeed, genomic analysis of small DNA viruses and retroviruses shows that they have remarkably repressed levels of CpG dinucleotides which, in some cases, are <10% of the expected level and which appear to represent strong evolutionary selection (Shpaer and Mullins 1990; Karlin et al. 1994; Krieg 1996). Adenoviruses appear to have developed a different way of overcoming the CpG defenses by dramatically skewing the flanking bases of their CpG motifs so that the immune-stimulatory motifs are under-represented by approximately 30:1 compared with the neutralizing motifs (Krieg et al. 1998b).

The purpose of this chapter is to review what is known about the molecular and cellular mechanisms of action of immune-stimulatory CpG DNA. This chapter will focus on studies of B cells and macrophages since comparatively little has yet been reported on other cell types.

## 2 Molecular Mechanisms of Action of CpG DNA

### 2.1 Cell-Surface Receptors and CpG DNA

It has long been appreciated that cell membranes have a variety of DNA-binding proteins on their surfaces (Lerner et al. 1971; Aggrarwal et al. 1975; Loke et al. 1989; Yakubov et al. 1989; Bennett 1993). To date, none of these membrane DNA-binding proteins has actually been shown to mediate DNA uptake into cells, with the exception of a protein expressed on kidney epithelial cells that has been shown to function as a voltage-gated channel (Hanss et al. 1998a). However, oligodeoxynucleotide (ODN) uptake at low concentrations (below 1μM) appears consistent with receptor-mediated endocytosis while, at higher concentrations, fluid-phase endocytosis appears to be more important (Beltinger et al. 1995).

Of course, a cell-surface CpG receptor could potentially transduce a signal into the cell without requiring uptake of the DNA. However, there is no evidence of any CpG-sequence specificity in the cell-membrane binding of DNA. Yamamoto et al. found that there was no difference in the spleen cell binding of $^{32}$P-labeled ODN, which had stimulatory or non-stimulatory palindromes (Yamamoto et al. 1994). Likewise, we found no difference in the binding of CpG or non-CpG ODN to B cells or macrophages (Krieg et al. 1995).

Nonetheless, human B cells were reported to be stimulated by immobilized ODN, implying activation through a cell-surface receptor (Liang et al. 1996). In these experiments, the ODNs were synthesized with a lysine at the 5′ end, coupled to cyanogen bromide-activated sepharose 4B beads, and then cultured with B cells for 4 days before performing $^{3}$H-thymidine-incorporation assays. Aside from concerns over whether the ODNs actually remain immobilized under these culture conditions, Manzel and MacFarlane have recently found that ODNs coupled to sepharose beads in this manner are in fact taken up by cells during incubation in tissue culture, a phenomenon that could be responsible for the immune-stimulatory effects (Manzel and MacFarlane 1999). Under these same culture conditions,

CpG ODNs were not stimulatory if they were linked to latex or magnetic or gold beads, suggesting that cell uptake may indeed be required. This conclusion is in accord with results of our own studies, in which CpG ODNs were immobilized on Teflon fibers or avidin-coated plates, in which case stimulation was not seen using short-term-stimulation assays after 4h (KRIEG et al. 1995). The conclusion that ODN stimulation does not involve cell-surface signaling is further supported by the finding that lipofection of ODN into spleen cells enhances their immune-stimulatory effects (YAMAMOTO et al. 1994). Furthermore, while unmodified ODN usually must be at least 18 bases long to induce NK activity (YAMAMOTO et al. 1994), this length requirement is reduced to just 6 bases if the ODN are instead transfected into the cells (SONEHARA et al. 1996).

ODNs containing G-rich sequences, such as three or four Gs in a row ("poly-G ODNs"), have higher binding to cell membranes and have also been reported to have increased binding to the scavenger receptor on macrophages (HUGHES et al. 1994; KIMURA et al. 1994). The cell-surface integrin CD11B/CD18 (MAC-1) has also been reported to bind ODN in a non-sequence-specific fashion and mediate uptake and the generation of reactive oxygen species (ROSs; BENIMETSKAYA et al. 1997). However, mice genetically deficient in either one of these molecules appear to have essentially normal ODN uptake into cells (S. Crooke, personal communication). Furthermore, cells that do not express either molecule, such as B cells, are strongly activated by CpG DNA, indicating that neither receptor is required.

## 2.2 Cellular Uptake and Intracellular Localization of CpG ODN

It appears that all cell types are capable of taking up DNA spontaneously in culture, although certain cell types have higher rates of uptake than others (KRIEG et al. 1991; ZHAO et al. 1996b; HARTMANN et al. 1998). It is important that investigators performing such experiments be aware of an artifact arising from the fact that apoptotic and dead cells have increased ODN uptake (ZHAO et al. 1993; GIACHETTI and CHIN 1996). In principle, mechanisms that could mediate ODN uptake include receptor-mediated endocytosis, adsorptive endocytosis, pinocytosis, potocytosis, and phagocytosis. It is clear that ODNs do not simply diffuse across cell membranes; cellular uptake is energy- and temperature-dependent and saturable (BELTINGER et al. 1995; KRIEG 1995). It is possible that several of these mechanisms contribute to ODN uptake. However, potocytosis seems unlikely to play a major role in ODN uptake because, although potocytosis has been reported to be dependent upon protein kinase C and cell-surface glycosyl phosphatidyl inositol-linked proteins (SMART et al. 1994), ODN uptake is not (A.M. Krieg, unpublished data). Following binding of ODNs, they are initially located in endosomes, which then become progressively acidified (BENNETT et al. 1985; TONKINSON and STEIN 1994; YI et al. 1998a). Most of the DNA remains in the endosomes but, based on studies with confocal microscopy, a small fraction appears to reach the nucleus (KRIEG et al. 1993; ZHAO et al. 1996b; ZHAO et al. 1993). The proportion of DNA that enters the nucleus can be increased by conjugation of the ODN to compounds, such as cholesterol, or by transfection of the ODN into cells using cationic lipids (Krieg et al. 1993; Wagner et al. 1993; Bennett 1998). If the DNA is directly micro-injected into the cytoplasm of a cell, it is localized to the nucleus within a few seconds (Chin et al. 1990; Leonetti et al. 1991). With the exception noted above for ODNs containing poly-G sequences, cellular uptake and intracellular localization of ODNs are non-sequence-specific (Hanss et al. 1998b; A.M. Krieg, unpublished data).

## 2.3 Requirement for Endosomal Acidification/Maturation for CpG-Induced Immune Stimulation

Since DNA is internalized into cells via endocytosis into acidified vesicles, it was of interest to determine whether drugs that interfere with endosomal acidification and maturation alter the immune-stimulatory effects of CpG DNA. Indeed, we recently reported that monensin, chloroquine, and bafilomycin A, which interfere with endosomal acidification, completely block leukocyte activation by CpG DNA but not by other stimulatory agents, including phorbol 12-myristate 13-acetate, anti-immunoglobulin M (IgM), anti-CD40, or LPS (Yi et al. 1998a). Other compounds related to chloroquine, such as quinacrine, are even more potent inhibitors of immune activation by CpG DNA (MacFarlane and Manzel 1998). These compounds do not bind ODN nor do they prevent ODN uptake by cells (MacFarlane and Manzel 1998). However, it remains possible that these agents alter the intracellular localization of ODN. The ability of chloroquine and bafilomycin A to specifically inhibit CpG-induced activation of macrophages and dendritic cells (DCs) has been reported by Häcker et al. (Häcker et al. 1998). At present, the mechanism of these inhibitory effects is unclear. One interpretation is that endosomal acidification may be essential for the release of ODN from the endosomes into the cytoplasm, where they may exert their stimulatory effect. However, it appears that quinacrine and chloroquine successfully block immune stimulation by CpG ODN at concentrations that do not interfere with endosomal acidification (D. MacFarlane, personal communication). Alternatively, perhaps some other aspect of endosomal function is disrupted by these compounds and specifically blocks the effects of CpG DNA but not other lymphocyte mitogens that do not function through endosomes. Further studies will clearly be required to resolve this question. The inhibitory effects of chloroquine on CpG-induced activation are already apparent after 5min, indicating that the inhibitory effect occurs at a very early stage in the activation pathway (Yi et al. 1998a).

## 2.4 What is the Intracellular CpG Receptor?

Given the high degree of sequence specificity in the effects of CpG DNA and the fact that activation requires cell uptake and is not mediated through a cell-surface receptor, it appears clear that there must be some intracellular molecule that specifically interacts with CpG DNA to transduce the stimulatory signal. In order to identify proteins that may have CpG-specific patterns of binding to ODN, we performed electrophoretic mobility shift assays using ODN with stimulatory CpG motifs, methylated CpG motifs, or no motif as probes. In these preliminary studies, we identified factors in cytoplasmic and nuclear extracts of B-cell and monocytic cell lines; these factors specifically bind unmethylated CpG motifs (R. Tuetken, B. Noll, W. Shen, A.M. Krieg, unpublished data). Efforts to purify the binding proteins are currently underway. Identification of these proteins will be extremely important in understanding the CpG signaling pathways.

## 2.5 Production of Intracellular Reactive Oxygen Species Following Stimulation by CpG DNA

Many lymphocyte mitogens alter the redox balance of the cell by inducing the generation of ROSs, such as superoxide, hydrogen peroxide, or lipid peroxides (Hothersall et al. 1997; Joneson and Bar-Sagi 1998). Regulation of the redox balance of leukocytes is extremely important, because the binding of several transcription factors to DNA can be altered depending on the oxidative state of the cell (Sen and Packer 1996; Cotgreave and Gerges 1998). In lymphocytes, the generation of ROSs has been proposed to be important in mediating stimulation or apoptosis through the B-cell antigen receptor or CD40 (Fang et al. 1995; Fang et al. 1997; Lee and Koretzky 1997). Expression of excessive amounts of intracellular ROS has been linked to apoptosis. The concentration, type, and intracellular localization of ROS are all likely to be important in determining their biologic effects.

To investigate whether leukocyte activation by CpG DNA may involve the generation of ROS, we stimulated cells in the presence of the ROS-sensitive dye dihydrorhodamine-123. These studies showed that leukocyte stimulation with CpG DNA, but not with non-CpG DNA, was associated with the rapid intracellular production of ROS, which was specifically blocked by chloroquine (Yi et al. 1996c, 1998a). Multiple cellular enzymatic pathways are capable of producing ROSs, including xanthine oxidase, reduced nicotinamide adenine dinucleotide phosphate oxidase, cyclooxygenase, cytochrome p450 enzymes, and the mitochondria. Inhibition of xanthine oxidase with allopurinol has no effect on the ROS induction by CpG DNA, nor has inhibition of cyclooxygenase with indomethacin in our preliminary studies. Studies are currently underway to identify the enzymatic source of these ROSs and to specifically identify which ROSs are produced.

## 2.6 Activation of Mitogen-Activated Protein Kinases by CpG DNA

Mitogen-activated protein kinases (MAPKs) are important mediators of many cellular activation responses. The three main MAPK pathways are the extracellular receptor kinase or ERK pathway, the p38 MAPK pathway, and the c-Jun $NH_2$-

terminal kinase or JNK pathway. The p38 and JNK pathways are activated in both B cells and macrophages as early as 7min following activation with CpG DNA (Häcker et al. 1998; Yi and Krieg 1998b). These pathways are also activated in human primary B cells exposed to human activating CpG DNA motifs but not by non-CpG DNA (G. Hartmann and A.M. Krieg, unpublished data). More recently, we identified the upstream kinases MAPK kinase kinase 3 (MKK3), MKK4, and MKK6 as being activated very rapidly after CpG-DNA treatment of the murine B-cell line WEHI-231 (A.-K. Yi and A.M. Krieg, unpublished data). CpG DNA does not appreciably induce ERK-1 or ERK-2 activity in B-cell lines (Yi and Krieg 1998b) but does cause some activation of this pathway in macrophages (K.J. Stacey and D.A. Hume, personal communication).

## 2.7 Activation of Transcription Factors by CpG DNA

The p38 and MAPK pathways have as their downstream targets several transcription factors, including activating transcription factor 2 and activator protein 1 (Gupta and Siber 1995; Bireland and Monroe 1997; DeFranco 1997). Therefore, it may not be surprising that these transcription factors become phosphorylated and show increased DNA binding and enhanced transcriptional activity in cells treated with CpG DNA (Häcker et al. 1998; Yi and Krieg 1998b).

Inhibition of p38 kinase with the specific inhibitor SB203580 inhibits CpG-induced cytokine production by macrophages and B cells, indicating the requirement for this kinase activity in mediating CpG effects (Häcker et al. 1998; Yi and Krieg 1998b). Nuclear factor κB (NFκB) has a critical role in mediating many inflammatory responses and is rapidly upregulated in leukocyte responses to many mitogens (Kistler et al. 1998; Sha 1998). The first report that ODN could induce NFκB activation was an unexpected observation in an antisense experiment in which cells treated with a sense oligonucleotide to the p65 subunit of NFκB showed enhanced NFκB activity (McIntyre et al. 1993). This sense ODN contained a CpG motif that was subsequently found to be responsible for the observed upregulation of NFκB activity (A.M. Krieg, R. Narayanan, unpublished data). Following the identification of CpG DNA, the first report linking this to the activation of NFκB was from Stacey et al. who showed that macrophages treated with CpG DNA have enhanced transcriptional activity from the human immunodeficiency virus long, terminal repeat; this activity is regulated by NFκB (Stacey et al. 1996). Supershift studies showed that this NFκB activation includes the p50 and p65 subunits but not p52 (Sparwasser et al. 1997b). NFκB activation was also seen following CpG-DNA treatment of B cells and was associated with the degradation of IκBα and IκBβ (Yi et al. 1996b; Yi and Krieg 1998a). CpG-induced NFκB activation in B cells was associated with protection against the induction of apoptosis (Yi and Krieg 1998a). Similar CpG-specific induction of transcription-factor binding is seen in human primary B cells (G. Hartmann and A.M. Krieg, unpublished data). CpG DNA also induces increased levels of mRNA for other transcriptional factors, including c-*myc*, *ets*-2, and CCAAT enhancer binding protein-β and -Δ (SWEET et al. 1998; YI and KRIEG 1998a; K.J. Stacey and D.A. Hume, personal communication).

## 2.8 Induction of Gene Expression by CpG DNA

The cell-signaling pathways that are induced by CpG DNA lead to enhanced nuclear transcription within 15–30min (YI et al. 1996b, 1998b). Although the expression of multiple early-response genes, proto-oncogenes, and cytokine genes is induced by CpG DNA, the effects show a particular pattern that is suggestive of a Th1-like immune response. CpG DNA is a particularly strong inducer of the expression of interleukin 12 (IL-12), especially as compared to other agents, such as LPS, which induce IL-12 expression from macrophages or DCs more weakly (CHACE et al. 1997; LIPFORD et al. 1997b; JAKOB et al. 1998; SPARWASSER et al. 1998). There is no detectable induction of expression of the Th2-like cytokines, IL-4, or IL-5 (KLINMAN et al. 1996) although B cells are induced to express IL-10, which appears to be partially responsible for reducing the level of IL-12 secretion that is induced by CpG DNA (REDFORD et al. 1998).

The expression of tumor necrosis factor (TNF)-α is induced very rapidly following CpG stimulation and appears to be responsible for the induction of the systemic inflammatory-response syndrome in mice that have been rendered sensitive to the effects of TNF (HALPERN et al. 1996; SPARWASSER et al. 1997a,b). Other cytokines whose expression is induced by CpG DNA include IL-6, IL-1β, IL-1RA, macrophage inflammatory protein 1β, monocyte chemoattractant protein 1, interferon (IFN)-α/β, IL-18, and IFN-γ (YI et al. 1996c; ROMAN et al. 1997; SCHWARTZ et al. 1997; ZHAO et al. 1997; SUN et al. 1998; YAMAMOTO et al. 1988). Aside from these studies in mouse cells, more recently we have also found CpG-specific induction of IL-6 and TNF in human monocytes (HARTMANN et al. 1999), and IL-12 secretion from purified human DCs (G. Hartmann and A.M. Krieg, unpublished data). In addition, human peripheral blood mononuclear cells (PBMCs) cultured with CpG DNA show induction of expression of IFN-γ (A.M. Krieg, unpublished data; YAMAMOTO et al. 1994b). Aside from the induction of these genes, CpG DNA also induces several cell-cycle-regulating proto-oncogenes and anti-apoptosis genes including, most notably, c-*myc* and *bcl*-$x_L$ (YI et al. 1996b, 1998b).

# 3 Cellular Mechanisms of Action of CpG DNA

## 3.1 B-Cell Activation by CpG DNA

All of the above pathways described as mediating CpG effects are inducible in primary B cells or B-cell lines following treatment with CpG DNA. CpG DNA is the most potent single B-cell mitogen that has been described and is capable of driving more than 95% of B cells into the cell cycle (KRIEG et al. 1995). In contrast to some other B-cell mitogens, such as 8-substituted guanines, CpG DNA appears to be equally stimulatory to both resting and activated B-cell subsets (GOODMAN 1991; KRIEG 1998). B cells enter the G1 phase of the cell cycle within a few hours and progress through the cell cycle under appropriate conditions. B-cell stimulation results in the secretion of IL-6 and IL-10 within a few hours (YI et al. 1996c; REDFORD et al. 1998). In contrast to B-cell activation by LPS, which is suppressed by the addition of IFN-γ, B-cell activation by CpG DNA synergizes with IFN-γ to promote stronger responses (YI et al. 1996a). CpG-activated B cells are induced to secrete IgM in an IL-6-dependent fashion (YI et al. 1996c).

Under certain well-described conditions, B cells will undergo apoptosis. Two experimental situations in which this has been investigated include the spontaneous apoptosis of primary B cells (ILLERA et al. 1993; MOWER et al. 1994) and the induction of apoptosis in the B-cell line WEHI-231, in which crosslinking of the B-cell receptor leads to programmed cell death (BOYD and SCHRADER 1981; FISCHER et al. 1994). The regulation of apoptosis in the WEHI-231 system is thought to involve the activity of NFκB and the expression of c-*myc* and *bcl*-$x_L$, which are thought to be protective (FISCHER et al. 1994; GOTTSCHALK et al. 1994; FANG et al. 1995; LEE et al. 1995; MERINO et al. 1995). CpG treatment is a very strong rescuer of B cells from apoptosis in these experimental models. In the case of the WEHI-231 model, CpG DNA can be added as late as 8h after the apoptosis-inducing signal of ligating-surface IgM without losing its anti-apoptotic activity (YI et al. 1996b; YI and KRIEG 1998a). This rescuing activity of CpG DNA appears to require the induction of NFκB, since protection is lost if cells are treated with inhibitors of NFκB activation (YI and KRIEG 1998a,b).

In addition to these changes, B cells activated by CpG DNA show increased expression of surface class-II major histocompatibility (MHC) molecules and the co-stimulatory molecules B7-1 and B7-2 (KRIEG et al. 1995; DAVIS et al. 1998). This suggests the possibility that CpG may enhance the antigen-presenting function of B cells directly. No other cell type is required for these activities of CpG DNA on B cells.

The responses of human B cells to CpG DNA generally appear to be very similar to those of murine cells, with the exception that the optimal motifs for stimulation are different (Fig. 1). ODNs with good human CpG motifs are very strong B-cell mitogens, but there is also a relatively non-sequence-specific stimulatory effect of the phosphorothioate backbone; this effect becomes evident at higher ODN concentrations of around 10μM (LIANG et al. 1996). Seemingly paradoxically, *Escherichia coli* DNA does not induce $^3$H-thymidine incorporation by human B cells (LIANG et al. 1996) but does trigger B cells to proliferate, as detected by the CFSE (carboxyfluorescein succinimidyl ester) assay or by the expression of cell-surface CD86 (G. Hartmann and A. Krieg, manuscript in preparation). These seemingly contradictory results may be understood if we consider that the intracellular degradation of bacterial DNA (which, unlike phosphorothioate, is highly susceptible to degradation) releases free thymidine that effectively competes with radioactive thymidine in the proliferation assay, thereby giving a false negative result (Matson and Krieg 1992). To avoid this effect in experimental assays using non-nuclease-resistant ODN, it is important to design appropriate readouts that will not be affected. In our experimental systems, phosphodiester-backbone CpG DNA shows strong and reproducible stimulatory activities on human cells that are CpG-sequence-specific even at high concentrations. In fact, many CpG motifs that will not activate human cells when present in an ODN with a phosphorothioate backbone are stimulatory when used in a phosphodiester or chimeric phosphorothioate/phosphodiester backbone (Ballas et al. 1996; G. Hartmann and A.M. Krieg, manuscript in preparation).

### 3.2 T-Cell Activation by CpG DNA

The effects of CpG DNA on T cells are much less clear. An initial report using partially purified T-cell populations (about 85% pure) reported the direct stimulation of cytokine secretion by CpG DNA in T cells (Klinman et al. 1996). However, subsequent studies by several labs have shown that, in fact, there is no direct activation of T cells by CpG DNA (Lipford et al. 1997b; Sun et al. 1998b). However, by inducing production of type-I IFNs from adherent cells, CpG DNA activates T cells to express CD69 and B7-2, although their proliferative responses to ligation of the T-cell receptor are reduced (Sun et al. 1998b). This functional inhibition resulted from the production of type-I IFN by antigen-presenting cells (Sun et al. 1998a). However, highly purified T cells that are stimulated through the T-cell receptor show synergistic proliferative responses to CpG DNA, indicating a mechanism through which CpG DNA could promote antigen-specific T-cell responses (Bendigs et al. 1999). Further studies will be required to clarify the significance of these stimulatory and inhibitory activities of CpG DNA in vivo.

### 3.3 NK-Cell Activation by CpG DNA

Early studies by Tokunaga and colleagues demonstrated that bDNA is a strong stimulus for NK lytic activity and for IFN-γ production (Tokunaga et al. 1984, Yamamoto et al. 1992). However, these effects of CpG DNA are not observed with highly purified NK cells (Ballas et al. 1996; Cowdery et al. 1996). Instead, these studies demonstrated that NK cell activation by CpG DNA was blocked by neutralizing antibodies to IL-12, TNF-α, and/or type-I IFNs. Since these cytokines are produced by adhering cells, this raised the possibility that the effect of CpG DNA on NK cells may be indirect. It is now clear that this is partially true, but it also appears that NK cells incubated with IL-12 together with CpG DNA are much more strongly stimulated than those incubated with an equivalent amount of IL-12 alone (Cowdery et al. 1996). Thus, as with T cells, NK cells appear to be co-stimulated by the combination of CpG DNA and specific activating cytokines. Interestingly, there is no such co-stimulatory activity for IL-2, which is otherwise an extremely potent activator of NK cells (Z. Ballas, G. Weiner, J. Cowdery, A.M. Krieg).

### 3.4 Monocytes, Macrophages and DC Activation by CpG DNA

These antigen-presenting cell types appear to be directly stimulated by CpG DNA in the mouse to produce a variety of cytokines. A common theme that has emerged from studies of the cytokine responses of antigen-presenting cells (APCs) to CpG DNA is the prominence of Th1-like responses. For example, skin-derived DCs are stimulated by LPS to produce large amounts of TNF-α and relatively low levels of IL-12 but are stimulated by CpG DNA to produce very large amounts of IL-12 with relatively low amounts of TNF-α (Jakob et al. 1998). APCs are also induced to produce IL-1 and IL-6 and to express increased levels of cell-surface class-II MHC and the co-stimulatory molecules B7-1 and B7-2 (Stacey et al. 1996; Lipford et al. 1997a; Sparwasser et al. 1997a,b). These effects appear to be directly mediated, since no accessory cells are required.

Some differences are observed in CpG-induced activation of human cells. The cytokine response of human monocytes to CpG DNA is delayed compared to that of LPS. As measured by intracellular cytokine staining, LPS-activated monocytes produce high levels of TNF-α and IL-6 within 4h; CpG-activated monocytes have no detectable cytokine response at this time point (Hartmann et al. 1999). Instead, the human monocyte cytokine response to CpG DNA becomes detectable at approximately 12–18h and is lower in magnitude than that which can be induced by LPS (Hartmann et al. 1999). Human DCs show much faster kinetics in their responses to CpG DNA (Hartmann).

## 4 Interaction of CpG DNA with other Cell Activation Pathways

### 4.1 Synergy versus Antagonism between CpG DNA and the B-Cell Antigen Receptor

A common biological theme in cellular responses to environmental stimuli is the interaction between different cell signaling pathways, which can synergize or interfere with one another. In the case of CpG DNA, its role as a "danger signal" that indicates the presence of infection makes it logical that the immune system should have evolved to enhance cellular activation in response to foreign antigens that are encountered in the presence of CpG DNA. Indeed, although high concentrations of CpG DNA can drive more than 95% of B cells into the cell cycle, at low concentrations of CpG DNA synergy with the B-cell receptor (BCR) is very prominent, resulting in at least a tenfold increase in the proliferation and Ig secretion of B cells that have also been activated through their antigen receptor

(KRIEG et al. 1995). Thus, although CpG DNA by itself is, of course, a non-antigen-specific signal, it preferentially promotes the induction of antigen-specific antibody responses. CpG DNA and the BCR also synergize for induction of B-cell IL-6 secretion (YI et al. 1996c). With the identification of the intracellular kinase pathways that are activated by CpG DNA, it has become possible to determine the molecular mechanism of the synergy at a more proximal level. CpG DNA and the BCR synergize for inducing MKK3, MKK4, MKK6, and JNK, but not ERK (YI and KRIEG 1998b; A.K. Yi and A.M. Krieg, unpublished data).

Although mature B cells are typically activated by crosslinking of the BCR, immature B cells may be induced to undergo apoptosis (NORVELL et al. 1995). This process is thought to be important in maintaining immune tolerance to self antigens present in the bone marrow, which will prevent autoreactive B cells from maturing, thereby reducing the risk of autoimmunity. A B-cell line with an immature phenotype, WEHI-231, has been commonly used as a model of this activation-induced apoptosis, since crosslinking of the BCR causes WEHI-231 cells to undergo rapid growth arrest followed by apoptotic death (BOYD et al. 1981; SCOTT et al. 1985). Apoptotic death has been linked to changes in c-*myc* gene expression and to a loss of NFκB activity (MCCORMACK et al. 1984; FISCHER et al. 1994; LEE et al. 1995; SCHAUER et al. 1996; WU et al. 1996). These studies have shown that agents that restore NFκB activation, such as LPS and CD40L, can prevent apoptosis of WEHI-231 cells following BCR ligation. CpG DNA, which also induces c-*myc* expression and NFκB activation, also protects WEHI-231 cells from cell death induced by BCR ligation (YI and KRIEG 1998a; YI et al. 1996b). BCR ligation in WEHI-231 cells does not immediately lead to irreversible apoptosis, since CpG DNA can still provide substantial protection even when added 8h after BCR ligation (YI et al. 1996b). The protective effect of CpG DNA is associated with increased binding of the p50/c-Rel heterodimer of NFκB as compared to the p50 homodimer, which is dominant after BCR ligation alone (YI and KRIEG 1998a).

Given its ability to interfere with apoptosis in this model of tolerance induction, it might be expected that CpG DNA would interfere with the deletion of autoreactive cells in vivo and trigger autoimmune disease. Surprisingly, this is not the case; CpG DNA does not induce lupus in normal mice nor flare lupus in genetically lupus-prone mice. In fact, the cytokines induced by CpG DNA actually reduce disease severity (GILKESON et al. 1996; MOR et al. 1997; GILKESON et al. 1998)! Thus, it appears that additional mechanisms that are not blocked by CpG DNA must protect against autoimmunity in vivo. Nonetheless, studies in the WEHI-231 system show the complexity of the signaling pathways induced by CpG DNA and the BCR and the ways in which these pathways can antagonize as well as synergize.

## 4.2 Interaction of CpG DNA with Other B-Cell Activation Pathways

Aside from the BCR, several other cell-surface molecules on B cells are known to regulate their activation. For example, B cells are stimulated by crosslinking of

CD40 or class-II MHC molecules, both of which can also synergize with activation through the BCR. Therefore, we also examined whether CpG DNA may synergize with these signaling pathways. These studies showed that CpG DNA synergized with B-cell activation through class-II MHC but not through CD40 (A.M. Krieg and G. Bishop, unpublished data).

B cells can also be affected by cytokines, such as IFN-γ. The effects of IFN-γ on B cells can be rather complex, with both positive and negative effects reported, depending on the experimental system (Leibson et al. 1984; Cowdery et al. 1992; Chace et al. 1993). For example, IFN-γ attenuates LPS-induced B-cell activation (Chace et al. 1993). In contrast, IFN-γ promotes B-cell activation by CpG DNA, as demonstrated by the fact that this activation is weaker in mice, which are genetically deficient in IFN-γ (Yi et al. 1996a). Since CpG DNA is, indirectly, a strong stimulus for NK cell production of IFN-γ, this would tend to promote antigen-specific B-cell responses. CpG DNA synergizes with LPS for the activation of B-cell proliferation and Ig secretion (A.-K. Yi and A.M. Krieg, unpublished data).

## 4.3 Interaction of CpG DNA with Monocyte Activation by LPS

LPS is a potent stimulus for monocyte cytokine production, which superficially shares many similarities with CpG DNA. Both agents are mitogenic for mouse B cells and induce high levels of monocyte TNF-α, IL-6, and other cytokines. However, differences between the biologic activities of these agents have been noted. For example, macrophages are induced to express inducible nitric oxide synthase by LPS, but CpG DNA does not exert this effect unless the macrophages have also been primed with IFN-γ (Stacey et al. 1996, 1998). Conversely, RAW264 macrophage cells are stimulated to produce IL-12 by CpG DNA but not by LPS (Anitescu et al. 1997). More marked differences have been observed in the responses of human cells to LPS and CpG. In human PBMCs treated with LPS, the monocytes show rapid induction of TNF-α and IL-6 production within a few hours, but the production of these cytokines in response to CpG DNA is delayed until about 18h (Hartmann et al. 1999). In fact, this difference in the kinetics of cytokine secretion in response to LPS and CpG DNA can be used to detect the presence of LPS in synthetic ODN or bDNA samples.

Several investigators have reported synergy between CpG DNA and LPS for induction of cytokine production in vitro and in vivo and for induction of the systemic inflammatory-response syndrome in vivo. Thus, although CpG DNA by itself is quite non-toxic, it effectively primes for the Schwartzman reaction, as demonstrated by a very high mortality in mice given a sublethal dose of endotoxin 4h after a dose of CpG DNA (Cowdery et al. 1996). This mortality is associated with synergistic induction of TNF-α, which is dramatically enhanced by administration of both CpG DNA and LPS (Cowdery et al. 1996; Hartmann et al. 1996; Sparwasser et al. 1997a).

## 5 Conclusion

Our understanding of the mechanism of action of CpG DNA has advanced remarkably far in the few years since the discovery that the immune system recognizes bDNA as "foreign" because of its content of unmethylated CpG motifs. The ability to detect CpG DNA lies both in cells of the innate immune system, such as monocytes and macrophages, but also in the acquired immune system, where it is a remarkably potent B-cell mitogen. As a result, CpG DNA not only triggers rapid innate immune responses that can provide a sort of non-specific innate protection against challenge with a broad range of infectious pathogens (Krieg et al. 1998a; Zimmerman et al. 1998; Elkins et al. 1999), it also strongly enhances the generation of antigen-specific B-cell responses (Krieg et al. 1995). In fact, CpG DNA has now emerged as an astonishingly potent and non-toxic vaccine adjuvant whose therapeutic effects are even more marked than those of the "gold standard" (complete Freund's adjuvant), but without the associated toxicity (Chu et al. 1997; Lipford et al. 1997b; Roman et al. 1997; Weiner et al. 1997; Davis et al. 1998; Moldoveanu et al. 1998; Sun et al. 1998a).

Several of the molecular signaling pathways through which the CpG DNA signal is transduced have been identified, as reviewed above. Perhaps the most important immediate question for understanding the molecular mechanism of CpG DNA is the identification of its receptor. It appears that this receptor is a cytoplasmic protein or belongs to a family of proteins, and it is hoped that, with its identification, the remaining pieces of the activation pathway will gradually be filled in.

Along with our understanding of the mechanism of action of CpG DNA has come increasing evidence of its potent therapeutic effects in a wide variety of animal models. As of this writing, CpG will imminently be entering human clinical trials, which will be the final test for whether this fascinating immune modulator can be used successfully to improve human health.

*Acknowledgements*. AMK was supported by grants from the Department of Veterans Affairs, CpG ImmunoPharmaceuticals and the National Institutes of Health. Gunther Hartmann is supported by grant Ha 2780/1-1 of the Deutsche Forschungsgemeinschaft. AKY was supported by a grant from the Lupus Foundation of America.

## References

Aggrarwal SK, Wagner RW, McAllister PK, Rosenberg B (1975) Cell-surface-associated nucleic acid in tumorigenic cells made visible with platinum-pyrimidine complexes by electron microscopy. Proc Natl Acad Sci USA 72:928–932

Anitescu M, Chace JH, Tuetken R, Yi A-K, Berg DJ, Krieg AM, Cowdery JS (1997) Interleukin-10 functions in vitro and in vivo to inhibit bacterial DNA-induced secretion of interleukin-12. J Interferon Cytokine Res 17:781–788

Ballas ZK, Rasmussen WL, Krieg AM (1996) Induction of natural killer activity in murine and human cells by CpG motifs in oligodeoxynucleotides and bacterial DNA. J Immunol 157:1840–1845

Beltinger C, Saragovi HU, Smith RM, LeSauteur L, Shah N, DeDionisio L, Christensen L, Raible A, Jarett L, Gewirtz AM (1995) Binding, uptake, and intracellular trafficking of phosphorothioate-modified oligodeoxynucleotides. J Clin Invest 95:1814–1823

Bendigs S, Salzer U, Lipford GB, Wagner H, Heeg K (1999) CpG-oligodeoxynucleotides costimulate primary T cells in the absence of APC. Eur J Immunol 29:1209–1218

Benimetskaya L, Loike JD, Dhaled Z, Loike G, Silverstein SC, Cao L, el Khoury J, Cai TQ, Stein CA (1997) Mac-1 (CD11b/CD18) is an oligodeoxynucleotide-binding protein. Nature Med 3:414–420

Bennett RM, Gabor GT, Merritt MM (1985) DNA binding to human leukocytes. Evidence for a receptor-mediated association, internalization, and degradation of DNA. J Clin Invest 76:2182–2190

Bennett RM (1993) As nature intended? The uptake of DNA and oligonucleotides by eukaryotic cells. Antisense Res Devel 3:235–241

Bennett CF (1998) Use of cationic lipid complexes for antisense oligonucleotide delivery. In: Applied Oligonucleotide Technology, CA Stein and AM Krieg, (eds), John Wiley and Sons, Inc., New York, NY, pp 129–145

Bird AP (1987) CpG islands as gene markers in the vertebrate nucleus. Trends in Genetics 3:342–347

Bireland ML, Monroe JG (1997) Biochemistry of antigen receptor signaling in mature and developing B lymphocytes. Critical Rev Immunol 17:353–385

Boggs RT, McGraw K, Condon T, Flournoy S, Villiet P, Bennett CF, Monia BP (1997) Characterization and modulation of immune stimulation by modified oligonucleotides. Antisense Nucl Acid Drug Develop 7:461–471

Boyd AW, Schrader JW (1981) The regulation of growth and differentiation of a murine B cell lymphoma. II. The inhibition of WEHI 231 by anti-immunoglobulin antibodies. J Immunol 126:2466

Branda RF, Moore AL, Mathews L, McCormack JJ, Zon G (1993) Immune stimulation by an antisense oligomer complementary to the rev gene of HIV-1. Biochem Pharmacol 45:2037–2043

Chace JH, Abed NS, Adel GL, Cowdery JS (1993) Regulation of differentiation in CD5+ and conventional B cells. Sensitivity to LPS-induced differentiation and interferon-γ-mediated inhibition of differentiation. Clin Immunol Immunopathol 68:327

Chace JH, Hooker NA, Mildenstein KL, Krieg AM, Cowdery JS (1997) Bacterial DNA-induced NK cell IFN-γ production is dependent on macrophage secretion of IL-12. Clin Immunol Immunopath 84:185–193

Chin DJ, Green GA, Zon G, Szoka FC, Straubinger RM (1990) Rapid nuclear accumulation of injected oligodeoxyribonucleotides. The New Biologist 2:1091–1100

Chu RS, Targoni OS, Krieg AM, Lehmann PV, Harding CV (1997) CpG oligodeoxynucleotides act as adjuvants that switch on Th1 immunity. J Exp Med 186:1623–1631

Cotgreave IA, Gerges RG (1998) Recent trends in glutathione biochemistry – glutathione-protein interactions: a molecular link between oxidative stress and cell proliferation? Biochem Biophys Res Comm 242:1–9

Cowdery JS, Flemming AL (1992) In vivo depletion of CD4 T cells increases B cell sensitivity to polyclonal activation: the role of interferon-γ. Clin Immunol Immunopathol 62:72

Cowdery JS, Chace JH, Yi A-K, Krieg AM (1996) Bacterial DNA induces NK cells to produce interferon-γ in vivo and increases the toxicity of lipopolysaccharides. J Immunol 156:4570–4575

Davis HL, Weeranta R, Waldschmidt TJ, Tygrett L, Schorr J, Krieg AM (1998) CpG DNA is a potent adjuvant in mice immunized with recombinant hepatitis B surface antigen. J Immunol 160:870–876

DeFranco AL (1997) The complexity of signaling pathways activated by the BCR. Current Opinion Immunol 9:296–308

Elkins KL, Rhinehart-Jones TR, Stibitz S, Conover JS, Klinman DM (1999) Bacterial DNA containing CpG motifs stimulates lymphocyte-dependent protection of mice against lethal infection with intracellular bacteria. J Immunol 162:2291–2298

Fang W, Rivard JJ, Ganser JA, LeBien TW, Nath KA, Mueller DL, Behrens TW (1995) Bcl-XL rescues WEHI-231 B lymphocytes from oxidant-mediated death following diverse apoptotic stimuli. J Immunol 155:66–75

Fang W, Nath KA, Mackey MF, Noelle RJ, Mueller DL, Behrens TW (1997) CD40 inhibits B-cell apoptosis by upregulating Bcl-xL expression and blocking oxidant accumulation. Amer J Physiol 272:C950–6

Fischer G, Kent SC, Joseph L, Green DR, Scott DW (1994) Lymphoma models for B cell activation and tolerance. X. Anti-mu-mediated growth arrest and apoptosis of murine B cell lymphomas is prevented by the stabilization of *myc*. J Exp Med 179:221–228

Giachetti C, Chin DJ (1996) Increased oligonucleotide permeability in keratinocytes of artificial skin correlates with differentiation and altered membrane function. J Invest Dermatol 107:256–262
Gilkeson GS, Ruiz P, Pippen AMM, Alexander AL, Lefkowith JB, Pisetsky DS (1996) Modulation of renal disease in autoimmune NZB/NZW mice by immunization with bacterial DNA. J Exp Med 183:1389–1397
Gilkeson GS, Conover J, Halpern M, Pisetsky DS, Feagin A, Klinman DM (1998) Effects of bacterial DNA on cytokine production by (NZB/NZW)F1 Mice. J Immunol 161:3890–3895
Goodman MC (1991) Cellular and biochemical studies of substituted guanine ribonucleoside immunostimulants. Immunopharmacology 21:51–68
Gottschalk AR, Boise LH, Thompson CB, Quintans J (1994) Identification of immunosuppressant-induced apoptosis in a murine B-cell line and its prevention by Bcl-x but not Bcl-2. Proc Natl Acad Sci USA 91:7350–7354
Gupta RK, Siber GR (1995) Adjuvants for human vaccines – current status, problems and future prospects. Vaccines 13:1263–1276
Häcker H, Mischak H, Miethke T, Liptay S, Schmid R, Sparwasser T, Heeg K, Lipford GB, Wagner H (1998) CpG-DNA-specific activation of antigen-presenting cells requires stress kinase activity and is preceded by non-specific endocytosis and endosomal maturation. EMBO J 17:6230–6240
Halpern MD, Kurlander RJ, Pisetsky DS (1996) Bacterial DNA induces murine interferon-γ production by stimulation of interleukin-12 and tumor necrosis factor-α. Cell Immunol 167:72–78
Hanss B, Leal-Pinto E, Bruggeman LA, Copeland TD, Klotman PE (1998a) Identification and characterization of a cell membrane nucleic acid channel. Proc Natl Acad Sci USA 95:1921–1926
Hanss B, Stein CA, Klotman PE (1998b) Cellular uptake and biodistribution of oligodeoxynucleotides. In: Applied oligonucleotide technology, Stein CA and Krieg AM, (eds), John Wiley and Sons, Inc., New York, NY, pp 431–448
Hartmann G, Krug A, Waller-Fontaine K, Endres S (1996) Oligodeoxynucleotides enhance lipopolysaccharide-stimulated synthesis of tumor necrosis factor: dependence on phosphorothioate modification and reversal by heparin. Mol Med 2:429–438
Hartmann G, Krug A, Bidlingmaier M, Häcker U, Eigler A, Albrecht R, Strasburger CJ, Endres S (1998) Spontaneous and cationic lipid-mediated uptake of antisense oligonucleotides in human monocytes and lymphocytes. J Pharmacol Exp Ther 285:920–928
Hartmann G, Krieg AM (1999a) CpG DNA and LPS induce distinct patterns of activation in human monocytes. Gene Therapy, 6:893–903
Hartmann G, Weiner G, Krieg AM (1999b) CpG DNA: a potent signal for growth, activation and maturation of human dendritic cells. PNAS 96:9305–9310
Hatzfeld J, Li M-L, Brown EL, Sookdeo H, Levesque J-P, O'Toole T, Gurney C, Clark SC, Hatzfeld A (1991) Release of early human hematopoietic progenitors from quiescence by antisense transforming growth factor β1 or Rb oligonucleotides. J Exp Med 174:925–929
Hothersall JS, Cunha FQ, Neild GH, Norohna-Dutra AA (1997) Induction of nitric oxide synthesis in J774 cells lowers intracellular glutathione: effect of modulated glutathione redox status on nitric oxide synthase induction. Biochem J 322:477–481
Hughes JA, Avrutskaya AV, Juliano RL (1994) Influence of base composition on membrane binding and cellular uptake of 10-mer phosphorothioate oligonucleotides in Chinese hamster ovary (CHRC5) cells. Antisense Res Develop 4:211–215
Illera VA, Perandones E, Stunz LL, Mower DA Jr, Ashman RF (1993) Apoptosis in splenic B lymphocytes: regulation by protein kinase C and IL-4. J Immunol 151:2965
Jakob T, Walker PS, Krieg AM, Udey MC, Vogel JC (1998) Activation of cutaneous dendritic cells by CpG-containing oligodeoxynucleotides: A role for dendritic cells in the augmentation of Th1 responses by immunostimulatory DNA. J Immunol 161:3042–3049
Joneson T, Bar-Sagi D (1998) A Rac1 effector site controlling mitogenesis through superoxide production. J Biol Chem 273:17991–17994
Karlin S, Doerfler W, Cardon LR (1994) Why is CpG suppressed in the genomes of virtually all small eukaryotic viruses but not in those of large eukaryotic viruses? J Virol 68:2889–2897
Kimura Y, Sonehara K, Kuramoto E, Makino T, Yamamoto S, Yamamoto T, Kataoka T, Tokunaga T (1994) Binding of oligoguanylate to scavenger receptors is required for oligonucleotides to augment NK cell activity and induce IFN. J Biochem 116:991–994
Kistler B, Rolink A, Marienfeld R, Neumann M, Wirth T (1998) Induction of nuclear factor-κB during primary B cell differentiation. J Immunol 160:2308–2317

Klinman D, Yi A-K, Beaucage SL, Conover J, Krieg AM (1996) CpG motifs expressed by bacterial DNA rapidly induce lymphocytes to secrete IL-6, IL-12 and IFN. Proc Natl Acad Sci USA 93: 2879–2883

Koike M, Ishino K, Ikuta T, Ho N-H, Kuroki T (1995) Growth enhancement of normal human keratinocytes by the antisense oligonucleotide of retinoblastoma susceptibility gene. J Biochem 116: 991–994

Krieg AM, Gause WC, Gourley MF, Steinberg AD (1989) A role for endogenous retroviral sequences in the regulation of lymphocyte activation. J Immunol 143:2448–2451

Krieg AM, Gmelig-Meyling F, Gourley MF, Kisch WJ, Chrisey LA, Steinberg AD (1991) Uptake of oligodeoxyribonucleotides by lymphoid cells is heterogeneous and inducible. Antisense Research and Development 1:161–171

Krieg A, Tonkinson J, Matson S, Zhao Q, Saxon M, Zhang L-M, Bhanja U, Yakubov L, Stein CA (1993) Modification of antisense phosphodiester oligodeoxynucleotides by a 5′ cholesteryl moiety increases cellular association and improves efficacy. Proc Natl Acad Sci USA 90:1048–1052

Krieg AK, Yi A-K, Matson S, Waldschmidt TJ, Bishop GA, Teasdale R, Koretzky G, Klinman D (1995) CpG motifs in bacterial DNA trigger direct B-cell activation. Nature 374:546–549

Krieg AM (1995) Uptake and localization of phosphodiester and chimeric oligodeoxynucleotides in normal and leukemic primary cells. In: Delivery strategies for antisense oligonucleotide therapeutics. Editor, Akhtar S. CRC Press, Inc., pp 177–190

Krieg AM (1996) An innate immune defense mechanism based on the recognition of CpG motifs in microbial DNA. J Lab Clin Med 128:128–133

Krieg AM, Matson S, Herrera C, Fisher E (1996) Oligodeoxynucleotide modifications determine the magnitude of immune stimulation by CpG motifs. Antisense Res Dev 6:133–139

Krieg AM, Love-Homan L, Yi A-K, Harty JT (1998a) CpG DNA induces sustained IL-12 expression in vivo and resistance to Listeria monocytogenes challenge. J Immunol 161:2428–2434

Krieg AM, Wu T, Weeratna R, Efler SM, Love-Homan L, Zhang L, Yang L, Yi AK, Short D, Davis H (1998b) Sequence motifs in adenoviral DNA block immune activation by stimulatory CpG motifs. Proc Natl Acad Sci USA 95:12631–12636

Krieg AM (1998) Leukocyte stimulation by oligodeoxynucleotides. In: Applied oligonucleotide technology, Stein CA and Krieg AM, (eds), John Wiley and Sons, New York, NY, pp 431–448

Kumar A, Yang YL, Flati V, Der S, Kadereit S, Deb A, Haque J, Reis L, Weissmann C, Williams BR (1997) Deficient cytokine signaling in mouse embryo fibroblasts with a targeted deletionin the PKR gene: role of IRF-1 and NF-κB. EMBO J 16:406–416

Kuramoto E, Yano O, Kimura Y, Baba M, Makino T, Yamamoto S, Yamamoto T, Kataoka T, Tokunaga T (1992) Oligonucleotide sequences required for natural killer cell activation. Jpn J Cancer Res 83:1128–1131

Lee H, Arsura M, Wu M, Duyao M, Buckler AJ, Sonenshein GE (1995) Role of *Rel*-related factors in control of c-*myc* gene transcription in receptor-mediated apoptosis of the murine B-cell WEHI-231 line. J Exp Med 181:1169

Lee JR, Koretzky GA (1997) Production of reactive oxygen intermediates following CD40 ligation correlates with c-*Jun* *N*-terminal kinase activation and IL-6 secretion in murine B lymphocytes. Eur J Immunol 28:4188–4197

Leibson HJ, Gefter M, Zlotnik A, Marrack P, Kappler JW (1984) Role of γ-interferon in antibody-producing responses. Nature 309:799

Leonetti JP, Mechti N, Degols G, Gagnor C, Lebleu B (1991) Intracellular distribution of microinjected antisense oligonucleotides. Proc Natl Acad Sci USA 88:2702–2706

Lerner RA, Meinke W, Goldstein DA (1971) Membrane-associated DNA in the cytoplasm of diploid human lymphocytes. Proc Natl Acad Sci USA 68:1212–1216

Liang H, Nishioka Y, Reich CF, Pisetsky DS, Lipsky PE (1996) Activation of human B cells by phosphorothioate oligodeoxynucleotides. J Clin Invest 98:1119–1129

Lipford GB, Sparwasser T, Bauer M, Zimmermann S, Koch E-S, Heeg K, Wagner H (1997a) Immunostimulatory DNA: sequence-dependent production of potentially harmful or useful cytokines. Eur J Immunol, 27:3420–3426

Lipford GB, Bauer M, Blank C, Reiter R, Wagner H, Heeg K (1997b) CpG-containing synthetic oligonucleotides promote B and cytotoxic T cell responses to protein antigen: a new class of vaccine adjuvants. Eur J Immunol 27:2340–2344

Loke SL, Stein CA, Zhang XH, Mori K, Nakanishi M, Subasinghe C, Cohen JS, Neckers LM (1989) Characterization of oligonucleotide transport into living cells. Proc Natl Acad Sci USA 86:3474–3478

MacFarlane DE, Manzel L (1998) Antagonism of immunostimulatory CpG-oligodeoxynucletides by quinacrine, chloroquine, and structurally related compounds. J Immunol 160:1122–1131
Manzel L, Macfarlane DE (1999) Immune stimulation by CpG-oligodeoxynucleotide requires internalization. Antisense and Nucleic Acid Drug Development 9:459–464
Matson S, Krieg AM (1992) Nonspecific suppression of 3H-thymidine incorporation by control oligonucleotides. Antisense Research and Development 2:325–330
McCormack J-E, Pepe VH, Kent RB, Dean M, Marshak-Rothstein A, Sonenshein GE (1984) Specific regulation of c-*myc* oncogene expression in a murine B-cell lymphoma. Proc Natl Acad Sci USA 81:5546
McIntyre KW, Lombard-Gillooly K, Perez JR, Kunsch C, Sarmiento UM, Larigan JD, Landreth KT, Narayanan R (1993) A sense phosphorothioate oligonucleotide directed to the initiation codon of transcription factor NF-κB p65 causes sequence-specific immune stimulation. Antisense Res Develop 3:309–322
Merino R, Grillot DAM, Simonian PL, Muthukkumar S, Fanslow WC, Bondada S, Nunez G (1995) Modulation of anti-IgM-induced B cell apoptosis by Bcl-XL and CD40 in WEHI-231 cells: dissociation from cell cycle arrest and dependence on the avidity of the antibody-IgM receptor interaction. J Immunol 155:3830
Messina JP, Gilkeson GS, Pisetsky DS (1991) Stimulation of in vitro murine lymphocyte proliferation by bacterial DNA. J Immunol 147:1759–1764
Messina JP, Gilkeson GS, Pisetsky DS (1993) The influence of DNA structure on the in vitro stimulation of murine lymphocytes by natural and synthetic polynucleotide antigens. Cell Immunol 147:148–157
Mojcik C, Gourley MF, Klinman DM, Krieg AM, Gmelig-Meyling F, Steinberg AD (1993) Administration of a phosphorothioate oligonucleotide antisense to murine endogenous retroviral MCF env causes immune effects in vivo in a sequence-specific manner. Clin Immunol Immunopath 67:130–136
Moldoveanu Z, Love-Homan L, Huang WQ, Krieg AM (1998) CpG DNA, a novel adjuvant for systemic and mucosal immunization with influenza virus. Vaccine 16:1216–1224
Mor G, Singla M, Steinberg AD, Hoffman SL, Okuda K, Klinman DM (1997) Do DNA vaccines induce autoimmune disease? Human Gene Therapy 8:293–300
Mower DA Jr, Peckham DW, Illera VA, Fishbaugh JK, Stunz LL, Ashman RF (1994) Decreased membrane phospholipid packing and decreased cell size precede DNA cleavage in mature mouse B cell apoptosis. J Immunol 152:4832
Norvell A, Mandik L, Monroe JG (1995) Engagement of the antigen-receptor on immature murine B lymphocytes results in death by apoptosis. J Immunol 154:4404–4413
Pisetsky DS, Reich CF (1993) Stimulation of murine lymphocyte proliferation by a phosphorothioate oligonucleotide with antisense activity for herpes simplex virus. Life Sciences 54:101–107
Redford TW, Yi A-K, Ward CT, Krieg AM (1998) Cyclosporine A enhances IL-12 production by CpG motifs in bacterial DNA and synthetic oligodeoxynucleotides. J Immunol 161:3930–3935
Roman M, Martin-Orozco E, Goodman JS, Nguyen M-D, Sato Y, Ronaghy A, Kornbluth RS, Richman DD, Carson DA, Raz E (1997) Immunostimulatory DNA sequences function as T helper-1-promoting adjuvants. Nature Med 3:849–854
Schauer SL, Wang Z, Sonenshein GE, Roghstein TL (1996) Maintenance of nuclear factor-κB/*Rel* and c-*myc* expression during CD40 ligand rescue of WEHI-231 early B cells from receptor-mediated apoptosis through modulation of IκB proteins. J Immunol 157:81
Schwartz D, Quinn TJ, Thorne PS, Sayeed S, Yi A-K, Krieg AM (1997) CpG motifs in bacterial DNA cause inflammation in the lower respiratory tract. J Clin Invest 100:68–73
Scott DW, Tuttle J, Livnat D, Haynes W, Cogswell J, Keng P (1985) Lymphoma models for B-cell activation and tolerance. II. Growth inhibitions by anti-mu of WEHI-231 and the selection of properties of resistants. Cell Immunol 93:124
Sen CK, Packer L (1996) Antioxidant and redox regulation of gene transcription. FASEB J 10:709–720
Sha WC (1998) Regulation of immune responses by NF-κB/*Rel* transcription factors. J Exp Med 187:143–146
Shpaer EG, Mullins JI (1990) Selection against CpG dinucleotides in lentiviral genes: a possible role of methylation in regulation of viral expression. Nucl Acids Res 18:5793–5797
Smart EJ, Foster DC, Ying Y-S, Kamen BA, Anderson RGW (1994) Protein kinase C activators inhibit receptor-mediated potocytosis by preventing internalization of caveolae. J Cell Biol 124:307–313
Sonehara K, Saito H, Kuramoto E, Yamamoto S, Yamamoto T, Tokunaga T (1996) Hexamer palindromic oligonucleoitdes with 5′-CG-3′ motif(s) induce production of interferon. J Interferon Cytokine Res 16:799–803

Sparwasser T, Miethke T, Lipford G, Borschert K, Häcker H, Heet K, Wagner H (1997a) Bacterial DNA causes septic shock. Nature 386:336–337

Sparwasser T, Miethe T, Lipford G, Erdmann A, Häcker H, Heeg K, Wagner H (1997b) Macrophages sense pathogens via DNA motifs: induction of tumor necrosis factor-α-mediated shock. Eur J Immunol 27:1671–1679

Sparwasser T, Koch E-S, Vabulas RM, Heeg K, Lipford GB, Ellwart J, Wagner H (1998) Bacterial DNA and immunostimulatory CpG oligonucleotides trigger maturation and activation of murine dendritic cells. Eur J Immunol 28:2045–2054

Stacey KJ, Sweet MJ, Hume DA (1996) Macrophages ingest and are activated by bacterial DNA. J Immunol 157:2116–2122

Sun S, Kishimoto H, Sprent J (1998a) DNA as an adjuvant: capacity of insect DNA and synthetic oligodeoxynucleotides to augment T cell responses to specific antigen. J Exp Med 187:1145–1150

Sun S, Zhang X, Tough DF, Sprent J (1998b) Type I interferon-mediated stimulation of T cells by CpG DNA. J Exp Med 188:2335–2342

Sweet MJ, Stacey KJ, Ross IL (1998) Involvement of Ets, *Rel* and Sp1-like proteins in lipopolysaccharide-mediated activation of the HIV-1 LTR in macrophages. J Inflam 48:67–83

Talmadge JE, Adams J, Phillips H, Collins M, Lenz B, Schneider M, Schlick E, Ruffmann R, Wiltrout RH, Chirigos MA (1985) Immunomodulatory effects in mice of polyinosinic-polycytidylic acid complexed with poly-L-lysine and carboxymethylcellulose. Cancer Res 45:1058–1065

Tanaka T, Chu CC, Paul WE (1992) An antisense oligonucleotide complementary to a sequence in Ic2b increases c2b germline transcripts, stimulates B cell DNA synthesis, and inhibits immunoglobulin secretion. J Exp Med 175:597–607

Tokunaga T, Yamamoto H, Shimada S, Abe H, Fukuda T, Fujisawa Y, Furutani Y, Yano O, Kataoka T, Sudo T, Makiguchi N, Suganuma T (1984) Antitumor activity of deoxyribonucleic acid fraction from mycobacterium bovis GCG. I. Isolation, physicochemical characterization, and antitumor activity. JNCI 72:955–962

Tokunaga T, Yamamoto S, Namba K (1988) A synthetic single-stranded DNA, poly(dG,dC), induces interferon-α/β and -γ, augments natural killer activity, and suppresses tumor growth. Jpn J Cancer Res 79:682–686

Tonkinson JL, Stein CA (1994) Patterns of intracellular compartmentalization, trafficking and acidification of 5′-fluorescein labeled phosphodiester and phosphorothioate oligodeoxynucleotides in HL60 cells. Nucl Acids Res 22:4268–4275

Wagner RW, Matteucci MD, Lewis JG, Gutierrez AJ, Moulds C, Froehler BC (1993) Antisense gene inhibition by oligonucleotides containing C-5 propyne pyrimidines. Science 260:1510–1513

Weiner GJ, Liu H-M, Wooldridge JE, Dahle CE, Krieg AM (1997) Immunostimulatory oligodeoxynucleotides containing the CpG motif are effective as immune adjuvants in tumor antigen immunization. Proc Natl Acad Sci USA, 94:10833

Wiltrout RH, Salup RR, Twilley TA, Talmadge JE (1985) Immunomodulation of natural killer activity by polyribonucleotides. J Biol Resp Mod 4:512–517

Wu M, Lee H, Bellas RE, Schauer SL, Arsura M, Katz D, FitzGerald MJ, Rothstein TL, Sherr DH, Sonenshein GE (1996) Inhibition of NF-κB/*Rel* induces apoptosis of murine B cells. EMBO J 15:4682–4690

Yakubov LA, Deeva EA, Zarytova VF, Ivanova EM, Ryte AS, Yurchenko LV, Vlassov VV (1989) Mechanism of oligonucleotide uptake by cells: Involvement of specific receptors? Proc Natl Acad Sci USA 86:6454–6458

Yamamoto S, Kuramoto E, Shimada S, Tokunaga T (1988) In vitro augmentation of natural killer cell activity and production of interferon-α/β and -γ with deoxyribonucleic acid fraction from mycobacterium bovis BCG. Jpn J Cancer Res 79:866–873

Yamamoto S, Yamamoto T, Kataoka T, Kuramoto E, Yano O, Tokunaga T (1992) Unique palindromic sequences in synthetic oligonucleotides are required to induce INF and augment INF-mediated natural killer activity. J Immunol 148:4072–4076

Yamamoto T, Yamamoto S, Kataoka T, Tokunaga T (1994a) Lipofection of synthetic oligodeoxyribonucleotide having a palindromic sequence of AACGTT to murine splenocytes enhances interferon production and natural killer activity. Microbiol Immunol 38:831–836

Yamamoto T, Yamamoto S, Kataoka T, Komuro K, Kohase M, Tokunaga T (1994b) Synthetic oligonucleotides with certain palindromes stimulate interferon production of human peripheral blood lymphocytes in vitro. Jpn J Cancer Res 85:775–779

Yi A-K, Chace JH, Cowdery JS, Krieg AM (1996a) IFN-γ promotes IL-6 and IgM secretion in response to CpG motifs in bacterial DNA and oligodeoxynucleotides. J Immunol 156:558–564

Yi A-K, Hornbeck P, Lafrenz DE, Krieg AM (1996b) CpG DNA rescue of murine B lymphoma cells from anti-IgM induced growth arrest and programmed cell death is associated with increased expression of c-*myc* and Bcl-$x_L$. J Immunol 157:4918–4925

Yi A-K, Klinman DM, Martin TL, Matson S, Krieg AM (1996c) Rapid immune activation by CpG motifs in bacterial DNA: Systemic induction of IL-6 transcription through an antioxidant-sensitive pathway. J Immunol 157:5394–5402

Yi A-K, Tuetken R, Redford T, Kirsch J, Krieg AM (1998a) CpG motifs in bacterial DNA activates leukocytes through the pH-dependent generation of reactive oxygen species. J Immunol 160: 4755–4761

Yi A-K, Chang M, Peckham DW, Krieg AM, Ashman RF (1998b) CpG oligodeoxyribonucleotides rescue mature spleen B cells from spontaneous apoptosis and promote cell cycle entry. J Immunol 160:5898–5906

Yi A-K, Krieg AM (1998a) CpG DNA rescue from anti-IgM induced WEHI-231 B lymphoma apoptosis via modulation of IκBα and IκBβ and sustained activation of nuclear factor-κB/c-*Rel*. J Immunol 160:1240–1245

Yi A-K, Krieg AM (1998b) Rapid induction of mitogen activated protein kinases by immune stimulatory CpG DNA. J Immunol 161:4493–4497

Zhao Q, Matson S, Herrara CJ, Fisher E, Yu H, Waggoner A, Krieg AM (1993) Comparison of cellular binding and uptake of antisense phosphodiester, phosphorothioate, and mixed phosphorothioate and methylphosphonate oligonucleotides. Antisense Research and Development 3:53–66

Zhao Q, Temsamani J, Iadarola PL, Jiang Z, Agrawal S (1996a) Effect of different chemically modified oligodeoxynucleotides on immune stimulation. Biochem Pharmac 51:173–182

Zhao Q, Song X, Waldschmidt T, Fisher E, Krieg AM (1996b) Oligonucleotide uptake in human hematopoietic cells is increased in leukemia and is related to cellular activation. Blood 88:1788–1795

Zhao Q, Temsamani J, Zhou R-Z, Agrawal S (1997) Pattern and Kinetics of cytokine production following administration of phosphorothioate oligonucleotides in mice. Antisense and Nucleic Acid Drug Develop 7:495–502

Zimmerman S, Egeter O, Hausmann S, Lipford GB, Rocken M, Wagner H, Geeg K (1998) Cutting Edge: CpG oligodeoxynucleotides trigger protective and curative Th1 responses in lethal murine leishmaniasis. J Immunol 160:3627–3630

# Oligodeoxyribonucleotides with 5′-ACGT-3′ or 5′-TCGA-3′ Sequence Induce Production of Interferons

S. YAMAMOTO[1], T. YAMAMOTO[1], and T. TOKUNAGA[2]



## 1 Introduction

More than 1 billion children in 182 countries throughout the world have been vaccinated with the Bacillus of Calmette and Guerin (BCG) against tuberculosis. BCG appears to be able to elicit protective cell-mediated immune responses in the absence of progressive disease. In addition, experimental uses of BCG and related mycobacteria have greatly contributed to the progress of modern immunology. Trials of cancer immunotherapy with BCG were carried out on a worldwide scale in the 1970s and contributed much information to the fields of basic and clinical immunology (BAST 1974). Evidence foreshadowing our finding of senseless immunostimulatory DNA was also obtained by the experimental use of BCG.

Until recently, DNA had only been considered as the blueprint of life and was thought to be immunologically uniform and essentially inert. However, recent advances in research on antisense DNA and gene vaccination changed this view; this research indicated that DNA can operate as a genetic-code-based intracellular

[1] Department of Bacterial and Blood Products, National Institute of Infectious Diseases, 4-7-1 Gakuen, Musashimurayama, Tokyo 208-0011, Japan
[2] National Institute of Infectious Diseases, 1-23-1 Toyama, Shinjuku, Tokyo 162-8640, Japan

signal. Our findings – that immunocytes of mice and humans can be stimulated in a senseless manner by DNA derived from BCG, resulting in interferon (IFN) production and tumor-growth inhibition (Tokunaga 1984; Shimada 1986; Yamamoto S 1988) – further transformed this conceptual framework (Stein 1995; Pisetsky 1996; Krieg 1996b). Using a variety of synthetic, single-stranded oligodeoxyribonucleotides (ODN), we reported that hexamer palindromic sequences with 5′-ACGT-3′ or 5′-TCGA-3′ motifs are essential for inducing IFNs (Kuramoto 1992; Tokunaga 1992). We also reported that bacterial, viral, or invertebrate DNA (but not vertebrate or plant DNA) can induce IFNs, and suggested that this mechanism may contribute to the self–non-self discrimination for foreign DNA (Yamamoto S, Yamamoto T, Shimada S, et al. 1992).

Interest in this line of study was galvanized by the demonstration of murine B-cell activation by CpG motifs in bacterial DNA (Krieg, Yi, et al. 1995). Today, immunostimulation with DNA is one of the most active topics in the research fields of host-defense mechanisms, plasmid vaccination, etiology of certain diseases, and therapeutic application. In this review, we will describe how our BCG studies led to the discovery of immunostimulatory DNA.

## 2 Antitumor Activity of DNA from BCG

In the 1970s, trials of cancer immunotherapy with BCG were carried out energetically and extensively. The mode of antitumor action of BCG was considered to be host mediated, and major effector cells were thought to be macrophages activated by lymphokines produced by BCG-sensitized lymphocytes. Intratumor administration of viable BCG was effective, but the trials were given up because the side reactions induced by viable BCG were recognized to be severe. Many investigators made efforts to isolate components of BCG or related bacteria possessing antitumor activity but not side effects. We found that BCG cytoplasm precipitated by streptomycin contained a substance strongly active against mouse and guinea pig syngeneic tumors. This fraction was purified, and a fraction possessing strong activity but few side effects was obtained. We named this fraction MY-1 but, in this review, we call it BCG-DNA.

The BCG-DNA was mostly composed of nucleic acids (98%); its protein content was only 0.15%, and the sugar content was 0.20%. The elution profile indicated that BCG-DNA was distributed over a wide range of molecular weights, and its elution peak corresponded to molecular weights of about $2–4 \times 10^5$kDa. The DNA was single stranded.

The antitumor activities of BCG-DNA were examined against various syngeneic tumors of mice or guinea pigs. Repeated intra-regional injections with BCG-DNA regressed or inhibited almost all of the various solid tumors. Repeated subcutaneous injections against leukemia were also effective (Tokunaga 1984; Shimada 1985).

We were very hesitant to conclude that DNA itself is active against tumor, because few incidences of biologic activity of DNA had been reported before, although synthetic double-stranded RNA, such as polyI:C, and synthetic single-stranded poly(dG,dC) were known to induce IFN and activate both natural killer (NK) cells and macrophages (Tokunaga 1988). In the early days of IFN research, Isaacs and his co-workers had observed that foreign DNA induced IFN, but they later questioned their observation. We further treated BCG-DNA with either RNase or DNase, purified the products by shaking with chloroform/alcohol and eluting with Sephadex G-100, and obtained the fractions named RNase-digest or DNase-digest, respectively (Tokunaga 1984). The RNase-digest contained mostly DNA (97.0%), and the DNase-digest contained 97.0% RNA and 1.5% DNA. Amazingly, RNase-digest produced IFN and showed antitumor activity, as did BCG-DNA, while DNase-digest lost its activity completely, suggesting strongly that the activity is a result of its DNA.

Direct in vitro cytotoxicity of DNA extracted from various cells had been reported by others, but BCG-DNA (1000μg/ml) showed no cytotoxicity in vitro for cells of 13 cell lines, suggesting that the action of BCG-DNA is host mediated (Shimada 1985). An intra-peritoneal (ip) injection of BCG-DNA (100μg) into mice rendered peritoneal cells cytotoxic (Shimada 1986; Yamamoto 1988). The effector activity was destroyed by anti-asialo-GM1 antiserum or carrageenan in vitro, but not by anti-Thy1.2 or carbonyl-iron uptake, suggesting that the effector cells are NK cells (Shimada 1986; Yamamoto 1988). Ip injection with DNase-digest of BCG-DNA did not bring such NK activation.

Intra-dermal (id) injection of BCG-DNA induced in situ infiltration of mononuclear cells (most of which were asialo-GM1-positive) and lasted for a week. The infiltrating cells were also positive for Ly-5 and partially positive for Thy1.2 but negative for Mac-1, Ia, mu-chain, Lyt-1, Lyt-2, L3T4, and Fc receptor II. They contained neither peroxidase nor nonspecific esterase, suggesting that the cells were NK. DNase destroyed all these effects of BCG-DNA (Kuramoto 1989, 1992).

Peripheral blood lymphocytes (PBL) from healthy human donors and 20 cancer patients were assessed for NK activity after 24h of incubation with BCG-DNA, RNase-digest, or DNase-digest. BCG-DNA or RNase-digest (but not DNase-digest) augmented the NK activity of PBL from healthy donors and all cancer patients, although the degree of augmentation varied depending upon the PBL used (Mashiba 1988).

When normal mouse spleen cells were incubated with BCG-DNA or RNase-digest (but not DNase-digest) for 20h, the cells produced both IFN and macrophage-activating factor (MAF). IFN was measured by observing inhibition of the cytopathic effects of vesicular stomatitis virus (VSV); MAF was measured by observing the extent to which normal peritoneal macrophages became cytotoxic towards tumor cells (Yamamoto 1988). The MAF activity was destroyed by anti-IFN-$\gamma$ antiserum, and the anti-virus activity was eliminated by anti-IFN-$\alpha/\beta$.

# 3 Particular Base Sequences in BCG-DNA Required for Immuno-Stimulation

## 3.1 Significance of Palindromic Sequences Having –CG– Motifs

We wondered whether or not the activity of BCG-DNA is base-sequence dependent. Based on the fact that the molecular sizes of the DNA molecules contained in BCG-DNA preparation were distributed over a broad range but peaked at 45 bases (Tokunaga 1984), we synthesized 13 kinds of 45-mer single-stranded ODNs having sequences randomly selected from the cDNA encoding BCG proteins and tested their ability to produce IFN-α and -γ and augment NK cells (Tokunaga 1992). Amazingly, 6 of the 13 ODNs activated NK cells, while the others did not. IFN-inducing activities paralleled these results, suggesting that some stereochemical structures constructed by particular sequences are required for expression of the biologic activity (Tokunaga 1992). ODNs that were biologically active in vitro also inhibited tumor growth (Kataoka 1992).

We further synthesized a variety of 30-mer single-stranded ODNs with sequences chosen from the cDNA encoding proteins of various species, including humans. ODNs having palindromic sequences of either GACGTC, AGCGCT, or AACGTT were active, whereas ODNs with ACCGGT were inactive. When a portion of the sequence of the inactive ODNs was substituted with GACGTC, AGCGCT, or AACGTT (but not ACCGGT), the ODN acquired biologic activity (Yamamoto S, Yamamoto T, Kataoka T, et al. 1992).

The 30-mer ODNs active in mice were also active in humans. IFN activity was detected in the culture fluid of PBL after 8h, and the amount of IFN reached the maximum after 18h. IFN-α was predominantly produced, and small amounts of IFN-β and IFN-γ were also found (Yamamoto T, Yamamoto S, Kataoka T, Komuro K, et al. 1994).

In order to know the minimum and essential sequences responsible for the biological activity, the palindromic sequence (GACGTC) in an active 30-mer ODN, 5′-ACCGAT *GACGTC* GCCGGT GACGGC ACCACG-3′, was replaced with each of the 63 theoretically possible 6-mer palindromic sequences, and they were tested for NK-enhancing ability (Kuramoto 1992). The analogue 30-mer ODNs that enhanced NK activity more strongly than the original ODN were 8 ODNs with one of the following palindromic sequences: AACGTT, AGCGCT, ATCGAT, CGATCG, CGTACG, CGCGCG, GCGCGC, and TCGCGA. All included one or more 5′-CG-3′ motif(s). The converse was not true; for instance, ACCGGT and CCCGGG were impotent. Palindromes composed entirely of adenines (A) and thymines (T) and those with a sequence of either Pu (purine)-Pu-Pu-Py (pyrimidine)-Py-Py or Py-Py-Py-Pu-Pu-Pu were generally unfavorable for the activity. Since stacking between Py is less stable than that between Pu or Pu-Py, stable helical structures formed by the palindromes including the 5′-CG-3′ sequence but without a Py-Py-Py sequence seemed to be favorable (Kuramoto

1992). BALLAS et al. pointed out that palindromic sequences are not always necessary for NK activation (BALLAS 1986).

## 3.2 Significance of Sequences Outside of Potent Hexamers

Sequences outside of hexamer palindromic sequences were also meaningful; studies were designed to get a comprehensive picture of the sequence requirements of ODNs. A comparison was made of the NK-augmenting activities of single-stranded 30-mer homo-oligomers, all of which possess one of the potent sequences AACGTT, CGATCG, GACGTC, or ATCGAT at the center position (KURAMOTO 1992). The homo-oligomers with no palindromic structures showed no activity. Oligo-G having a potent palindromic sequence showed the highest activity, but oligo-A and oligo-C having this sequence gave only marginal activity. No activity was seen in the oligo-G containing an impotent palindrome. These results indicate the independent and cooperative effects of palindromic and extra-palindromic sequences on the activity of ODN (KURAMOTO 1992).

The effects of the number and location of the palindromic sequences were also investigated. Among the 30-mer oligo-G nucleotides containing a different number of AACGTT palindromes, an ODN with one particular palindrome showed the strongest activity. The ODN with AACGTT at the 5′-end or 3′-end showed slightly stronger activity than the ODN with the sequence in the center, although the activity was influenced more drastically by the number of palindromes than by their location (KURAMOTO 1992).

It is unlikely that the cooperation between the palindromic and extra-palindromic sequences is due to secondary structures, such as bulges and hairpin loops possibly composed of these sequences, for several reasons. First, the thermostability of secondary structures was not correlated with the activity of ODNs, as judged by temperature-gradient gel electrophoresis. Second, there was no correlation between the activity and the preferred secondary structure of the ODN as predicted by the thermodynamic calculation. Third, the activity of a single-stranded ODN is comparable to that of its double-stranded counterpart, although the latter is less likely to form secondary structures than the former.

The effect of ODN base length on the ability of ODNs to induce IFN was also investigated. Ten kinds of 12- to 30-mer nucleotides having the AACGTT sequence were synthesized. Immunostimulatory activity was observed in ODNs of 18 bases or more in length and was proportional to the base length, with a maximum of 22–30 bases. However, the ODNs 16 bases or less in length were not active even if they possessed an AACGTT sequence (YAMAMOTO T, YAMAMOTO S, KATAOKA T and TOKUNAGA T 1994a).

The NK-enhancing and IFN-producing activities of 30-mer single-stranded oligo-G having AACGTT (ODN-1) were inhibited by a homo-oligomer of G, dextran sulfate, and polyvinyl sulfate. ODN-1 inhibited acetyl-low-density-lipoprotein binding to the scavenger receptor, suggesting that the sequences outside of the potent hexamer may play a role in the binding of ODN to cell surfaces (KIMURA 1994).

# 4 DNA from Bacteria, but not Vertebrates, Activates Immunocytes

In 1984, we noticed that DNA-rich fractions from various species of bacteria exhibited biological activities similar to those of BCG-DNA but DNA from calf thymus or salmon testis did not. Since these results were difficult to explain, we hesitated to publish them. We prepared DNA fractions from six bacterial species, calf thymus, and salmon testis, using exactly the same procedures used in the preparation of BCG-DNA and by other methods. The DNA fractions from all of the bacteria caused remarkable elevations of NK activity in mouse spleen cells and induced generation of both MAF and IFN in a concentration-dependent manner. However, the DNA fractions from calf thymus and salmon testis did not show any activity. The bacterial DNA fractions caused tumor inhibition, but the vertebrate DNA fraction did not. The activities of the bacterial DNA were completely abrogated by pre-treatment with DNase but not RNase (Yamamoto S, Yamamoto T, Shimada S, et al. 1992).

Further, we examined the increased number of DNA samples from a variety of animals and plants. As illustrated in Fig. 1, all of the DNA from four species of bacterium augmented NK activity and induced IFN, while the DNA from ten kinds of vertebrate, including three fish and five mammals, showed no such activities. The DNA from four species of plant were also inactive (Yamamoto S, Yamamoto T, Shimada S, et al. 1992).

To explain such differences of biological activities among DNAs from different sources, we set up the following six hypotheses: (i) the bacterial DNA fraction might contain toxic substances like lipopolysaccharide (LPS); (ii) the molecular size of the DNA contained in the DNA fractions might be different in bacteria and animals, a fact which might influence biologic activities; (iii) DNA from animals might be more sensitive than bacterial DNA to the DNase employed, which was derived from higher animals; if so, the bacterial DNA would persist longer in the cultures of mouse spleen cells; (iv) G + C ratios might influence the activities; (v) the majority of –CG– motifs are known to be methylated in vertebrates but not in bacteria, and the presence of methyl cytosine may obstruct the activity of –CG– motifs; (vi) the frequency of a particular base sequence essential for the biologic activity of the DNA may be different.

We examined all of the six possibilities. (i) The activities of the bacterial DNA fractions were not influenced by the presence of polymyxin B, an inhibitor of LPS; these activities were present even in spleen cells from LPS-insensitive C3H/HeJ mice, indicating that the activities were not ascribed to LPS contamination. (ii) The profiles of elution patterns obtained by agarose-gel electrophoresis were essentially the same in all of the DNA fractions from eight different sources. (iii) The absorbance of BCG-DNA and calf thymus DNA at 260nm before and after DNase treatment was the same. (iv) The (G + C) ratios of the bacterial DNAs used varied from 70.7% for *Streptomyces aureofaciens* to 27.1% for *Clostridium perfringens*,

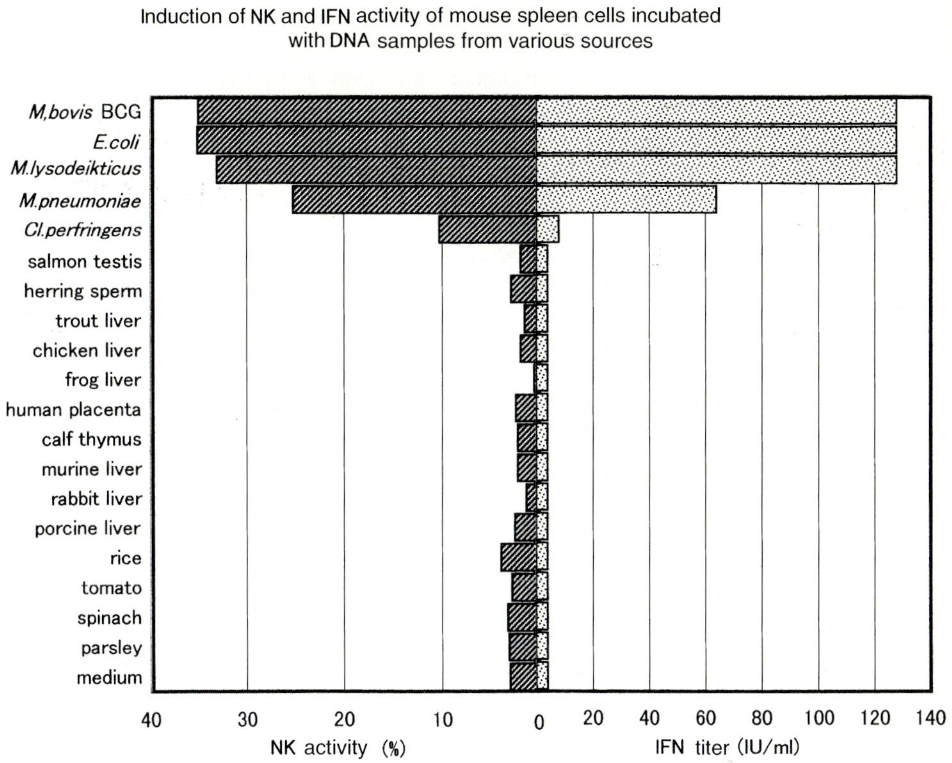


**Fig. 1.** BALB/c mouse spleen cells ($1 \times 10^7$cells/ml) were incubated with 100μg/ml of each of the DNA samples for 20h and centrifuged. The cell fractions were used for natural killer (NK) assays, and the culture supernatants were used for interferon (IFN) assays. NK cell activity was measured by a 4-h $^{51}$Cr-release assay using YAC-1 lymphoma cells as target cells. The level of IFN in the supernatant was measured in terms of its ability to inhibit the cytopathic effects of vesicular stomatitis virus

and those in calf and salmon DNAs are 50.2% and 40.2%, respectively. No correlation between the (G+C) ratio and the activity of DNA was found. (v) We previously reported that a synthetic, single-stranded poly(dG,dC) is active in vitro. Therefore, we synthesized poly(dG,dC) where all cytosines were substituted with methyl cytosines. The resulting poly (dG,methyl-dC) augmented NK activity and induced IFN as effectively as the original poly(dG,dC). Furthermore, we synthesized both a 30-mer nucleotide having 5′-AACGTT-3′ and its analogue, in which C was replaced with methyl cytosine; both ODNs could induce IFN equally. Accordingly, we thought that the fifth hypothesis was incorrect. In 1994, we found that certain hexamers, acting alone, could induce IFN when they were encapsulated in liposomes (Yamamoto T, Yamamoto S, Kataoka T and Tokunaga T 1994b). Using this liposome system, we recognized that methylation of the cytosine of AACGTT resulted in a significant decrease of the activity (Sonehara 1996). We also found that, when *Escherichia coli* DNA was incubated with CpG methylase, the resulting DNA reduced its IFN-inducing activity with incubation time.

Therefore, we now think that hypothesis (v) must be one of the reasons for the difference in activity between vertebrate and non-vertebrate DNAs. (vi) We tried to test the incidence of the nine potent palindromic sequences of cDNA reported in the GenBank DNA database. Among 17 kinds of living things registered in the database, we randomly chose one or more sequences, and the incidences of the potent palindromic sequences in all the hexamer palindromic sequences were expressed as the number per 1000 base pairs. The incidence of the potent palindromic sequences in all the DNAs from the vertebrates and plants was less than 1.0, while the incidence in most of the bacteria, viruses, and silkworms was larger than 1.0. There were some exceptions; the incidence, for instance, in the cDNA from *Mycoplasma pneumoniae* and *Cl. perfringens* was very low (0.2–0.4), but the activities of these DNAs were high. This may be due to the fact that our analysis was limited to only parts of the genomic DNA. In addition, particular 8-, 10- or 12-mer palindromic sequences show stronger immunostimulatory activities than certain hexamers (Sonehara 1996). Bird stated that CpG in bulk vertebrate DNA occurs at about one-fifth of the expected frequency (Bird 1986). We think that the different frequencies of potent sequences in DNA in vertebrate and invertebrate DNAs must be one reason for their different activities.

## 5 Immunostimulatory Sequences

As described, we reported first that hexamer palindromic sequences having –CG– motifs are essential for IFN production and NK augmentation, with some exceptional cases. Krieg et al. used the term "CpG motif" for the basic structure of B-cell stimulation (Krieg 1995); this motif consists of the CpG dinucleotide flanked by two 5′ purines (preferably a GpA dinucleotides) and two 3′ pyrimidines (preferably a TpT). Ballas et al. (Ballas 1996) examined the necessity of palindromic sequences for NK activation and observed that the presence of a hexamer palindrome is irrelevant for NK activation; for instance, GTCGTT and GACGTT are active, and GACGTC is inactive. They concluded that unmethylated CpG motifs, but not palindromic sequences, are definitely required, and two flanking bases at the 5′ and 3′ ends showed stringent requirements. Boggs et al. tested various sequences of ODNs and confirmed that the CpG motif was stimulatory for NK cells only in specific sequence contexts (Boggs 1997). Monteith et al. tested the ability of various ODN sequences to induce splenomegaly and B-cell stimulation; a 5′-AACGTT-3′ palindrome was the most effective (Monteith 1997).

Since we found previously that a hexamer palindromic ODN (5′-AACGTT-3′), acting alone, induced IFN from mouse spleen cells when added with cationic liposomes (Yamamoto 1994), 32 kinds of hexamer palindromic ODN were synthesized and tested for their IFN-inducing ability in the presence of liposomes (Sonehara 1996). The sequence–activity analysis revealed a conspicuous dependency of the IFN-inducing activity on the presence of CG motifs, and the sequences showing the highest activity were able to express as NACGTN and NTCGAN, where N represents any complementary pair of bases. Moderate activity was observed in ACGCGT and TCGCGA. The activity of CGNNCG sequences was quite low. These results agree (with some exceptions) with previous results obtained with the use of the 64 theoretically possible kinds of hexamer palindromic sequences contained in a 30-mer ODN (Sonehara 1996). An exception was that CGATCG showed only marginal activity, although it was among the most potent sequences in the previous experiment. We think that the results obtained with liposome-encapsulated ODN definitely reveal the sequence–activity relationship of IFN-inducing ODN, because possible effects of the flanking sequence, which is important for efficient uptake of ODN (Kimura 1994), can be excluded, and a possible influence of secondary and tertiary structures on the activity is minimal for short, hexameric ODNs. In addition, the base-substitution study is free of the complication in which elimination of a palindrome gives rise to another palindrome. To obtain more precise information, each of the bases of AACGTT except CG were replaced with a distinct base and encapsulated in a liposome. Original palindromic sequences showed the highest activity, and the activity was markedly reduced by any single base substitution except substitution of G for the first A (that is, GACGTT; Sonehara 1996). A high activity of GACGTT was also reported in other laboratories (Ballas 1996; Boggs 1997). We noticed, however, that some incomplete palindromic sequences, such as AACGTT and CACGTT, showed reduced but significant activities (Sonehara 1996). Nevertheless, we found that one 30-mer ODN (C13, which encodes a regulator protein of IFN-β and has the sequence 5′-GGGCAT CGGTC GAAGTG AAAGTG AAAGTGA-3′) showed extremely strong activity. This ODN contains no hexamer palindromic sequence but does contain TCGA.

It is also true that the formula, Pu-Pu-CG-Py-Py, proposed by Krieg et al. (Krieg 1996b) and Klinman et al. (Klinman 1996), which has been used widely by many investigators, has many exceptions. At present, therefore, the general term "immunostimulatory sequence (ISS) of DNA" (Sato 1996; Romann 1997; Tighe 1998) or "CpG dinucleotides in particular base contexts" (Sato 1996; Boggs 1997; Yi and Tuetken 1998) may be appropriate. Therefore, we will use "ISS DNA" in this review.

After the observation that ISS DNA induces mouse B cells to proliferate and secrete immunoglobulin (Krieg 1995), it was reported that the same ODNs induced the rapid and coordinated secretion of interleukin 6 (IL-6), IL-12, and IFN-γ but not IL-2, IL-3, IL-4, IL-5, or IL-10 (Klinman 1996). It was also reported that bacterial DNAs stimulate macrophages/monocytes to secrete IL-12 and tumor necrosis factor α (TNF-α), resulting in IFN-γ production from NK cells (Halpern 1996), and that IFN-γ promotes IL-6 and immunoglobulin M (IgM) secretion responses to ISS-DNA (Yi and Chase 1996). Bacterial DNA induces macrophage IL-12 production which, in turn, stimulates NK cell IFN-γ production (Chase 1997). IL-10 inhibits IL-12 secretion induced by bacterial DNA (Anitesccu 1997). Nitric oxide production from macrophages in response to bacterial DNA is absolutely dependent on IFN-γ priming (Sweet 1998). ISS DNAs trigger maturation and activation of mouse dendritic cells (Sparwasser 1998).

Interestingly, in the field of plasmid DNA immunization, preferential induction of T-helper 1 (Th1) response and inhibition of specific IgE formation by plasmid DNAs were reported (Raz 1996). Sato et al. pointed out that the immunogenicity of plasmid DNA required short ISSs that contain a CpG in a particular base context (Sato 1996). Roman et al. showed that particular ODNs potently stimulate immune responses to administered antigens; they suppress IgE synthesis but promote IgG and IFN-γ production and initiate IFN-γ, IFN-α, IFN-β, IL-12, and IL-18 production, resulting in fostering of Th1 responses and enhancement of cell-mediated immunity (Roman 1997).

Intracelluair mechanisms of immunostimulatory DNA are still unclear. We showed that scavenger receptors of mouse macrophages are the binding sites on cell surfaces in the particular case of G homo-oligomers having –AACGTT– (Kimura 1994). In addition, we observed that bone-marrow cells from scavenger-receptor knockout mice could be stimulated with ISS ODN. However, it was reported that Mac-1 protein is an ODN-binding protein on polymorphnuclear cells, although the base sequence of ODN is unclear (Benimetsaya 1997).

It seems clear that ISS ODNs enter into cells and react with the cytoplasmic and nuclear apparatus. In our experiments, a high level of IFN was produced by a hexamer ODN alone when entrapped in liposome (Yamamoto T, Yamamoto S, Kataoka T, Tokunaga T 1994b), suggesting that the ODN worked inside the cells. The necessity of cellular uptake of the DNA was proven in B cells (Krieg 1995) and macrophages (Yi and Krieg 1998).

In the early days of our research, we speculated that a potent, palindromic sequence may bind to a certain negative transcriptional factor like IkB, or a suppressive factor released from the IFN regulatory gene, resulting in activation of the IFN gene (Yamamoto S, Yamamoto T, Kataoka T, et al. 1992). Although many transcriptional factors, such as c-*myc* and AP-1, were known to recognize short palindromic sequences, no transcriptional factor regulating IFN-gene expression was known to recognize active ODN sequences. Then we proposed an alternative candidate: the cyclic adenosine monophosphate (cAMP) response-element (CRE)-binding protein, which recognizes the CRE (GACGTC; Sonehara 1996). Our preliminary data showed that IFN induction by ISS ODN is inhibited by forskolin, a cAMP agonist (Sonehara 1996).

Yi et al. reported that ISS-DNA-induced IL-6 production from B cells and monocytes is mediated through the generation of reactive oxygen species (ROSs) (Yi, Tuetken et al. 1998). The ROS burst is linked to the activation of nuclear factor κB (NFκB), which induces leukocyte gene transcription and cytokine secretion. It was also stated that ISS DNAs taken up by macrophages activate a signaling cascade leading to activation of NF-κB (Yi and Krieg 1998). The inhibitors of NFκB, IκB-α and IκB-β, are degraded rapidly in cells treated with ISS ODN.

It was found that ISS ODNs inhibit apoptosis of B cells (MacFarlane 1997; Yi, Chang et al. 1998) via modulation of IκB-α and IκB-β and sustained activation of NFκB/c-Rel. Further studies on the identification and characterization of the sequence-dependent recognition mechanism of ODN within the cell are expected.

It is generally known that vaccines consisting of viable, attenuated bacterium or virus can induce strong and persistent immunity but produce relatively severe side effects. However, killed vaccines are relatively safe and induce antibody responses but induce cell-mediated immunity with difficulty. Efforts have been made to find effective and safe immunoadjuvants for devising vaccines consisting of components, recombinant subunits, or synthetic peptides. However, only aluminum salts and liposomes have been approved for use in humans. Recently, immunization with antigen-encoding plasmid DNA was found to induce immune responses to the encoded foreign proteins. At present, numerous numbers of plasmid-encoding immunogens of bacteria, viruses, parasites, and tumors are under investigation (GREGORIADIS 1998; TIGHE 1998). Many of them could induce immunity against infection, and clinical trials for treating or protecting against carcinoma and infection with herpes, influenza, hepatitis B, and human immunodeficiency viruses are in progress.

In 1996, Sato et al. reported that plasmid vectors expressing large amounts of gene product do not necessarily induce immune responses to the encoded antigens but require short ISS DNA (SATO 1996). Their effective plasmid contained two repeats of the palindromic CpG hexamer 5′-AACGTT-3′. Also, ISS DNA was reported to stimulate Th1 responses (ROMAN 1997). After these, many reports showed that the ISS within the plasmid backbone stimulates immune responses to the antigen encoded by the plasmid (CARSON 1997; CHU 1997; KLINMAN 1997; LIPFORD, BEAURE et al. 1997; DENNIS 1998; KRIEG 1998). ISS DNA stimulates immune responses to administered antigens, including β-galactosidase and HBs antigen (LECLERC 1997), the idiotype of surface IgM of a B lymphoma (WEINER 1997), intracelluar pathogens, such as *Listeria monocytogenes* (KRIEG, LOVE – HOMAN et al. 1998), and formalin-inactivated influenza virus (MOLDOVEANU 1998).

One of the greatest advantages of ISS DNA as an immunoadjuvant is that ISS DNA tends to stimulate the generation of Th1-biased immune responses to antigens that are encoded by the plasmid or are administered. ISS DNA suppresses IgM synthesis but promotes IFN-α, -β, and -γ, IL-12, and IL-18, all of which foster Th1 responses and enhance cell-mediated immunity.

Since ISS DNA is known to induce both IL-12 and TNF-α, attempts were made to find DNA sequences which differentially activate IL-12 versus TNF-α cytokine production in antigen-presenting cells (LIPFORD, SPARWASSER et al. 1997). They described a single-stranded 18-mer ODN with this characteristic and proposed it as the prototype for a useful ISS DNA that lacks harmful side effects; it contains a hexamer of GACGTT.

Krieg et al. pre-treated mice with ISS ODN and then challenged them ip with a lethal dose of *L. monocytogenes*. A 2–4 log decrease in the number of bacteria in the liver and spleen was brought about by 10μg of the ODN. We pre-treated mice with ISS ODN by the nasal route, then challenged the mice with influenza virus via the same route, but failed to protect against the infection. Tighe et al. stressed the advantages of using DNA vaccines compared with conventional protein vaccines, particularly for neonates (TIGHE 1998).

The dominance of the Th1 response generated by ISS DNA may be useful for protecting Th2-type allergic diseases (PISETSKY 1996; KLINE 1998; TIGHE 1998; ZIMMERMANN 1998). In two rodent experimental models, id gene vaccination with naked plasmid-encoding antigens reduced an ongoing IgE response to the antigen and lowered the levels of eosinophils observed in the lung after protein challenge (BROIDE 1998). Kline et al. observed that mice previously sensitized to an antigen have an established Th2 response; co-administration of ISS DNA and the antigen can redirect the immune responses to Th1 and can abolish allergic reactions on subsequent antigen challenge (KLINE 1998). Mice infected lethally with *Leishmania major*, which is known as a lethal Th2-driven disease model, were given ISS DNA having a GACGTC sequence (ZIMMERMANN 1998). When the ODN was given as late as 20 days after the lethal infection, it was curative, indicating that ISS ODN reverses an established Th2 response (ZIMMERMANN 1998). In a mouse model of asthma, airway eosinophilia, Th2 cytokine induction, IgE production, and bronchial hyper-reactivity were prevented by co-administration of ISS DNA and antigen.

Tumor immunotherapy may be one of the uses of ISS DNA. Many experimental trials have been reported (KRIEG 1998; TIGHE 1998). For instance, ISS ODNs enhance the efficacy of monoclonal-antibody therapy of murine lymphoma (WOOLDRIDGE 1997). We reported many experimental models in which intra-regional injections with ISS ODN alone strongly inhibited tumor growth (TOKUNAGA 1984; SHIMADA 1985, 1986; KATAOKA 1992; YAMAMOTO 1992). Clinical trials showed repeated intratumor injections with BCG-DNA to be remarkably effective (ISHIHARA 1988, 1990; JOHNO 1991).

It was stated that a 20-mer antisense ODN is currently in phase-III human clinical trials to treat cytomegalovirus retinitis (KRIEG 1996). The ODN used is an exceptionally potent inhibitor of virus replication and contains two GCG trinucleotides (KRIEG 1996).

Activation of the immune system with ISS DNA could have both beneficial and deleterious consequences (KRIEG 1996; PISETSKY 1996, 1997; ZIMMERMANN 1998). Schwartz et al. reported that instilling mice intra-tracheally with 20-mer ISS DNA or bacterial DNA (but not vertebrate DNA or methylated bacterial DNA) caused inflammation in the lower respiratory tract; a 50-fold increase in the concentration of TNF-$\alpha$ and a 4-fold increase of IL-6 were observed (SCHWARTZ 1997). Furthermore, DNA extracted from the sputum of patients with cystic fibrosis contained approximately 0.1–1% bacterial DNA, suggesting that ISSs of DNA may play an important pathogenic role in inflammatory lung disease. The possible role of immunostimulatory DNA in inflammation in other tissues exposed to bacteria must be considered seriously.

LPS of gram-negative bacteria has been known to cause sepsis syndrome. However, factors in addition to LPS must induce the syndrome, because bacteria that lack LPS still induce it. Cowdery et al. reported that intravenous injection with bacterial DNA or 20-mer ISS DNA (but not calf thymus DNA or 20-mer ODN without ISS) induced a rapid increase in IFN-$\gamma$ and, when followed by LPS, showed 75% mortality, while no death in mice treated with both calf thymus DNA and LPS was observed (Cowdery 1996). This mortality rate required the secretion of IFN-γ, because mice lacking the IFN-γ gene were protected. It was reported that mice given ODN having a –GACGTT– sequence suffered lethal toxic shock due to macrophage-derived TNF-α, resulting in fulminate apoptosis of liver cells (Sparwasser 1997). LPS was synergized with ISS ODN, as measured by monitoring TNF-α release.

Systemic lupus erythematosus (SLE) is an autoimmune disease of unknown pathogenesis. ISS in bacterial DNA could promote the development of SLE by inducing polyclonal B-cell activation, over-expression of immune cytokines (such as IL-6) and resistance to apoptosis, thereby potentially allowing the survival of autoreactive cells. This suggests that ISSs in bacterial DNA are possibly involved in the pathogenesis of SLE (Krieg 1995, 1996). SLE patients show deficient expression of antibodies specific for bacterial DNA (Pisetsky 1997).

Miyata et al. analyzed nine kinds of DNA clones from DNA – anti-DNA-antibody-immune complexes of 11 SLE patients and found that the DNAs were rich in hexamer palindromic ISSs (Miyata 1996). ODNs having –TTCGAA– in a cloned DNA augmented both the expression of intercellular adhesion molecule 1 on endothelial cells and the production of IL-2, IL-6, IL-8, and TNF-α, suggesting that DNA contained in the immune complex of SLE patients may exacerbate vasculitis.

Using a 10-mer palindromic DNA, TCAACGTTGA, Segal et al. reported that, when murine T cells that recognize the autoantigen myelin basic protein are exposed to either LPS or the palindromic ODN, the result is a conversion of the T cells, via an IL-12-dependent pathway, from quiescent cells into autoimmune-disease-effector Th1 cells (Segal 1997). Pre-autoimmune NZB/W mice have immunoregulatory defects that, due to immunization by bacterial DNA, allow activation of mammalian double-stranded-DNA-reactive B cells (Wloch 1997). It was found that SLE patients have a marked deficiency in the production of antibodies specific for the antigen of *Micrococcus lysodeikticus* (Pisetsky 1997).

## 6 Conclusions

Both bacterial DNA and ISS ODNs having a 5′-ACGT-3′ or 5′-TCGA-3′ motif are more advantageous IFN-α inducers than are viruses. First, the ISS ODNs are stable, and it is easy to control their quality. Second, we can use the ISS ODNs without risk of infection. Third, since the action of the ISS ODN on IFN-α-producing cells may be limited compared with the action of viral infection, it may be easier to determine the site of action of ISS ODN. Fourth, IFN-α induced by the ISS ODN can be determined without the cytopathic effects of the inducing virus.

The mechanism of IFN-α induction by the ISS ODN is still obscure at present. Since the minimum sequence requirement for ISS ODNs was reduced to 5′-ACGT-3′ or 5′-TCGA-3′, an antisense mechanism is unlikely to be involved, because sequence-specific inhibition is exerted by antisense ODNs that are strictly complementary to the target mRNA. As is well known, many transcription factors, such as c-*myc* and AP-1, recognize short palindromic sequences. Hence, the most plausible hypothesis may be that the ODNs interact with transcription factors known to regulate IFN genes. However, the transcription factors known to regulate IFN-gene expression do not recognize the highly active sequences reported by us.

Our studies have shown the structure of ISS ODN essential to IFN-α induction (Yamamoto S, 1988; Yamamoto T, Yamamoto S, Kataoka and Kohuro 1994). This study characterized the cells producing IFN-α via stimulation with the ISS ODN. From now on, our studies will deal with both the INF inducers and producers. Studies on the mechanism of IFN-α induction and production by ISS ODNs having 5′-ACGT-3′ or 5′-TCGA-3′ motifs will contribute to new developments in the field of IFN-α research. Since a large amount of DNA had never been administered to experimental animals or humans, it would be worthwhile for us to introduce clinical trials and pharmacological studies of BCG-DNA obtained in Japan in the late 1980s (Tokunaga 1999).

## References


Anitescu M, Chase JH, Tueken B, Yi A-M, Berg DJ, Krieg AM, Cowdery JS (1997) Interleukin-10 functions in vitro and in vivo to inhibit bacterial DNA-induced secretion of interleukin-12. J Interferon Cytokine Res 17:781–788

Ballas ZK, Rasmussen WL, Krieg AM (1996) Induction of NK activity in murine and human cells by CpG motifs in oligodeoxynucleotides and bacterial DNA. J Immunol 157:1840–1845

Bast RC, Zbar B, Borsos T, Rapp HJ (1974) BCG and cancer. New Engl J Med 290:1413–1458

Benimetskaya L, Loike JD, Khaled Z, Loike G, Silverman SC, Cao L, Khoury JE, Cai T-Q, Stein CA (1997) Mac-1 (CD11b/CD18) is an oligonucleotide-binding protein. Nature Medicine 3:414–420

Bird AP (1986) CpG-rich islands and the function of DNA methylation. Nature 321:209–213

Boggs RL, McGraw K, Condon T, Flournoy S, Villiet P, Bennett CF, Monia BP (1997) Characterization and modulation of immune stimulation by modified oligonucleotides. Antisense Nucleic Acid Drug Develop 7:461–471

Broide D, Schwarze J, Tighe H, Gifford T, Nguyen M-D, Malek S, Van Uden U, Martin-Orozco E, Gelfand EW, Raz E (1998) Immunostimulatory DNA sequences inhibit IL-5, eosinophilic inflammation, and airway hyperresponsiveness in mice. J Immunol 161:7054–7062

Carson DA, Raz E (1997) Oligonucleotide adjuvants for T helper1 (Th1)-specific vaccination. J Exp Med 186:1621–1622

Chase JH, Hooker NA, Mildenstein KL, Krieg AM, Cowdery JS (1997) Bacterial DNA-induces NK cell IFN-γ production is dependent on macrophage secretion of IL-12. Clin Immunol Immunother 84:185–193

Chu RA, Targori O, Krieg AM, Lehmann PV, Harding CV (1997) CpG oligodeoxynucleotides act as adjuvants that switch on T helper 1 (Th1) immunity. J Exp Med 186:1623–1631

Cowdery JS, Chase JH, Yi A-K, Krieg AM (1996) Bacterial DNA induces NK cells to produce IFN-γ in vivo and increases the toxicity of lipopolysaccharides. J Immunol 156:4570–4575

Dennis MK (1998) Therapeutic applications of CpG-containing oligodeoxynucleotides. Antisense Nucleic Acid Drug Develop 8:181–184

Gregoriadis G (1998) Genetic vaccines: Strategies for optimization. Pharmaceut Res 15:661–670

Halpern MD, Kurlander RJ, Pisetsky DS (1996) Bacterial DNA induces murine interferon-γ production by stimulation of interleukin-12 and tumor necrosis factor-α. Cell Immunol 167:72–78

Ishihara K, Ikeda S, Arao T, Matsunaka M (1988) Pilot study on local administration of a new BRM, MY-1, for the treatment of malignant skin tumors. Skin Cancer 3:273–281

Ishihara K, Ikeda S, Furue H (1990) Late phase-II clinical study on MY-1 in the treatment of malignant skin tumors. Skin Cancer 5:192–202

Johno M, Arao T, Ishihara K (1991) Adult T-cell leukemia/lymphoma and mycosis fungoides responding well to MY-1. West Jpn Dermatol. 53:1297–1306

Krieg AM (1995) CpG DNA: A pathogenic factor in systemic lupus erythematosus? J Clin Immunol 15:284–292

Krieg AM (1996a) An innate immune defense mechanism based on the recognition of CpG motifs in microbial DNA. J Lab Clin Med 128:128–133

Krieg AM (1996b) Lymphocyte activation by CpG dinucleotide motifs in prokaryotic DNA. Trends Microbiol 4:73–77

Krieg AM, Love-Homan L, Yi A-M, Harty JT (1998) CpG DNA induces sustained IL-12 expression in vivo and resistance to *Listeria monocytogenes* challenge. J Immunol 161:2428–2434

Krieg AM, Yi A-K, Matson S, Waldscmidt TJ, Bishop GA, Teasdale R, Koretzky GA, Klinman DM (1995) CpG motifs in bacterial DNA trigger direct B-cell activation. Nature 373:546–549

Krieg AM, Yi A-M, Schorr J, Davis HL (1998) The role of CpG dinucleotides in DNA vaccines. Trends Microbiol 6:23–27

Kataoka T, Yamamoto S, Yamamoto T, Kuramoto E, Kimura Y, Tokunaga T (1992) Antitumor activity of synthetic oligonucleotides with sequences from cDNA encoding proteins of *Mycobacterium bovis* BCG. Jpn J Cancer Res 83:244–247

Kimura Y, Sonehara K, Kuramoto E, Makino T, Yamamoto S, Yamamoto T, Kataoka T, Tokunaga T (1994) Binding of oligoguanylate to scavenger receptors is required for oligonucleotides to augment NK cell activity and induce IFN. J Biochem 116:991–994

Kline JN, Waldschmidt TJ, Businga TR, Lemish JE, Weinstock JV, Thorne PS, Krieg AM (1998) Modulation of airway inflammation by CpG oligodeoxynucleotides in a murine model of asthma. J Immunol 160:2555–2559

Klinman DM (1998) Therapeutic application of CpG-containing oligodeoxynucleotides. Antisense Nucleic Acid Drug Develop 8:181–184

Klinman DM, Yamshchikov G, Ishigatsubo Y (1997) Contribution of CpG motifs to the immunogenicity of DNA vaccines. J Immunol 158:3635–3639

Klinman DM, Yi A-K, Beaucaga SL, Conover J, Krieg AM (1996) CpG motifs present in bacterial DNA rapidly induce lymphocytes to secret interleukin 6, interleukin 12, and interferon γ. Proc Natl Acad Sci USA. 93:2879–2833

Kuramoto E, Toizumi S, Shimada S, Tokunaga T (1989) In situ infiltration of natural killer-like cells induced by intradermal injection of the nucleic acid fraction from BCG. Microbiol Immunol 33: 929–940

Kuramoto E, Watanabe N, Iwata D, Yano O, Shimada S, Tokunaga T (1992) Changes of host cell infiltration into Meth A fibrosarcoma tumor during the course of regression induced by infections of a BCG nucleic acid fraction. Int J Immunopharmac 14:773–782

Leclerc C, Deriaud E, Rojas M, Whalen RG (1997) The preferential induction of a Th1 immune response by DNA-based immunization is mediated by the immunostimulatory effect of plasmid DNA. Cell Immunol 179:97–106

Lipford GB, Beaure M, Blank C, Reiter R, Wagner H, Heeg K (1997) CpG-containing synthetic oligonucleotides promote B and cytotoxic T cell responses to protein antigen: a new class of vaccine adjuvants. Eur J Immunol 27:2340–2344

Lipford GB, Sparwasser T, Bauer M, Zimmermann S, Koch ES, Heeg KK, Wagner H (1997) Immunostimulatory DNA: sequence-dependent production of potentially harmful or useful cytokines. Eur J Immunol 27:3420–3426

Macfarlane DE, Manzel L, Krieg AM (1997) Unmethylated CpG-containing oligodeoxynucleotides inhibit apoptosis in WEHI 231 B lymphocytes induced by several agents: evidence for blockade of apoptosis at a distal signalling step. Immunology 91:586–593

Majima H, Nomura K (1986) Phase I clinical study of MY-1, A new biological response modifier. Jpn J Cancer Chemother 13:109–115

Mashiba H, Matsunaga K, Tomoda H, Furusawa M, Jimi S, Tokunaga T (1988) In vitro augmentation of natural killer activity of peripheral blood cells from cancer patients by a DNA fraction from *Mycobacterium bovis* BCG. Jpn J Med Sci Biol 41:197–202

Miyata M, Kanno T, Ishida T, Kobayashi H, Sato Y, Sato Y, Nishimaki T, Kasukawa R (1996) CpG motif in DNA from immune complexes of SLE patients augments expression on intercellular adhesion molecule-1 on endothelial cells. Clin Path 44:1125–1131

Moldoveanu Z, Love-Homan L, Huang WQ, Krieg AM (1998) CpG DNA, a novel immune enhancer for systemic and mucosal immunization with influenza virus. Vaccine 16:1216–1224

Monteith DK, Henry SP, Howard RB, Flounoy S, Levin AA, Benett CF, Crooke ST (1997) Immune stimulation – a class effect of phosphorothioate oligonucleotides in rodents. Anticancer Drug Design 12:421–432

Ochiai T, Suzuki T, Nakajima K, Asano T, Mukai M, Kouzu T, Okayama K, Isono K, Nishijima H, Ohkawa M (1986) Studies on lymphocyte subsets of regional lymph nodes after endoscopic injection of biological response modifiers in gastric cancer patients. Int J Immunother 4:259–265

Pisetsky DA, Drayton DM (1997) Deficient expression of antibodies specific for bacterial DNA by patients with systemic lupus erythematosis. Proc Assoc Am Physicians 109:237–244

Pisetsky DS (1996) Immune activation by bacterial DNA: A new genetic code. Immunity 5:303–310

Pisetsky DS (1997) Immunostimulatory DNA: A clear and present danger? Nature Medicine 3:829–831

Raz E, Tighe H, Sato Y, Corr M, Dudler JA, Roman M, Swain SL, Spiegelberg HL, Carson DA (1996) Preferential induction of a Th1 immune response and inhibition of specific IgE antibody formation by plasmid DNA immunization. Proc Natl Acad Sci USA. 93:5141–5145

Roman M, Orozco EM, Goodman JS, Nguyen MD, Sato Y, Ronaghy A, Kornbluth RS, Richman D, Carson DA, Raz E (1997) Immunostimulatory DNA sequences function as T helper-1 promoting adjuvants. Nature Medicine 3:849–854

Sato Y, Roman M, Tighe H, Lee D, Corr M, Nguyen M-D, Silverman GJ, Lotz M, Carson DA, Raz E (1996) Immunostimulatory DNA sequences necessary for effective intradermal gene immunization. Science 273:352–354

Schwartz DA, Quinn TJ, Thorne PS, Sayeed S, Yi A-M, Krieg AM (1997) CpG motifs in bacterial DNA cause inflammation in the lower respiratory tract. J Clin Invest 100:68–73

Segal BM, Klinman DM, Shevach EM (1997) Microbial products induce autoimmune disease by an IL-12 dependent pathway. J Immunol 158:5087–5090

Shimada S, Yano O, Inoue H, Kuramoto E, Fukuda T, Yamamoto H, Kataoka T, Tokunaga T (1985) Antitumor activity of the DNA fraction from *Mycobacterium bovis* BCG. II. Effects on various syngeneic mouse tumors. J Natl Cancer Inst 74:681–688

Shimada S, Yano O, Tokunaga T (1986) In vivo augmentation of natural killer cell activity with a deoxyribonucleic acid fraction of BCG. Jpn J Cancer Res (Gann) 77:808–816

Sonehara K, Saito H, Kuramoto E, Yamamoto S, Yamamoto T, Tokunaga T (1996) Hexamer palindromic oligonucleotides with 5′-CG-3′ motif(s) induce production of interferon. J Interferon Cytokine Res 16:799–803

Sparwasser T, Koch E-S, Vabulas RM, Heeg K, Lipford GB, Ellwart JW, Wagner H (1998) Bacterial DNA and immunostimulatory CpG oligonucleotides trigger maturation and activation of murine dendritic cells. Eur J Immunol 28:2045–2054

Sparwasser T, Miethke T, Lipford G, Borschert K, Hacker H, Heeg K, Wagner H (1997) Bacterial DNA causes septic shock. Nature 386:336–337

Sparwasser T, Miethke T, Lipford G, Erdmann A, Hacker H, Heeg K, Wagner H (1997) Macrophages sense pathogens via DNA motifs: Induction of tumor necrosis factor-α mediated shock. Eur J Immunol 27:1671–1679

Stein CA (1995) Does antisense exist? It may, but only under very special circumstances. Nature Medicine 1:1119–1121

Sweet MJ, Stacey KJ, Kakuda DK, Markovich D, Hume DA (1998) IFN-γ primes macrophage response to bacterial DNA. J Interferon Cytokine Res 18:263–271

Tighe H, Corr M, Roman M, Rax E (1998) Gene vaccination: plasmid DNA is more than just a blueprint. Immunol Today 19:89–97

Tokunaga T, Yamamoto H, Shimada S, Abe H, Fukuda T, Fujisawa Y, Furutani Y, Yano O, Kataoka T, Sudo T, Makiguchi N, Suganuma T (1984) Antitumor activity of deoxyribonucleic acid fraction from *Mycobacterium bovis* BCG. I. Isolation, physicochemical characterization and antitumor activity. J Natl Cancer Inst 72:955–962

Tokunaga T, Yamamoto S, Namba K (1988) A synthetic single-stranded DNA, poly(dG,dC), induces interferon-α/β and -γ, augments natural killer activity and suppresses tumor growth. Jpn J Cancer Res (Gann) 79:682–686

Tokunaga T, Yano O, Kuramoto E, Kimura Y, Yamamoto T, Kataoka T, Yamamoto S (1992) Synthetic oligonucleotides with particular base sequences from cDNA encoding proteins of *Mycobacterium bovis* BCG induce interferons and activate natural killer cells. Microbiol Immunnol 36:55–66

Tokunaga T, Yamamoto T, Yamamoto S (1999) How BCG led to the discovery of immunostimulatory DNA. Jpn J Infec Dis 52:1–11

Weiner GJ, Liu HM, Wooldrodge EJ, Dahle CE, Krieg AM (1997) Immunostimulatory oligonucleotides containing the CpG motif are effective as immune adjuvants in tumor antigen immunization. Proc Natl Acd Sci U.S.A. 94:10833–10837

Wloch MK, Alexander AL, Pippen AMM, Pisetsky DS, Gilkeson GS (1997) Molecular properties of anti-DNA induced in pre-autoimmune NZB/W mice by immunization with bacterial DNA. J Immunol 158:4500–4506

Wooldridge JE, Ballas Z, Krieg AM, Weiner GJ (1997) Immunostimulatory oligodeoxynucleotides containing CpG motifs enhance the efficacy of monoclonal antibody therapy of lymphoma. Blood 89:2994–2998

Yamamoto S, Kuramoto E, Shimada S, Tokunaga T (1988) In vitro augmentation of natural killer cell activity with a deoxyribonucleic acid fraction from *Mycobacterium bovi*s BCG. Jpn J Cancer Res (Gann) 79:866 873

Yamamoto S, Yamamoto T, Kataoka T, Kuramoto E, Yano O, Tokunaga T (1992) Unique palindromic sequences in synthetic oligonucleotides are required to induce IFN and augment IFN-mediated natural killer activity. J Immunol 148:4072–4076

Yamamoto S, Yamamoto T, Shimada S, Kuramoto E, Yano O, Kataoka T, Tokunaga T (1992) DNA from bacteria, but not from vertebrates, induces interferons, activates natural killer cells and inhibits tumor growth. Microbiol Immunol 36:983–997

Yamamoto T, Yamamoto S, Kataoka T, Komuro K, Kohase M, Tokunaga T (1994) Synthetic oligonucleotides with certain palindromes stimulate interferon production of human peripheral blood lymphocytes in vitro. Jpn J Cancer Res 85:775–779

Yamamoto T, Yamamoto S, Kataoka T, Tokunaga T (1994a) Ability of oligonucleotides with certain palindromes to induce interferon production and augment natural killer cell activity is associated with their base length. Antisense Res Develop 4:119–122

Yamamoto T, Yamamoto S, Kataoka T, Tokunaga T (1994b) Lipofection of synthetic oligodeoxyribonucleotide having a palindromic sequence of AACGTT to murine splenocytes enhances interferon production and natural killer activity. Microbiol Immunol 38:831–836

Yi A-K, Chang M, Peckham DW, Krieg AM, Ashman RF (1998) CpG oligodeoxyribonucleotides rescue mature spleen B cells from spontaneous apoptosis and promote cell cycle entry. J Immunol 160: 5895–5906

Yi A-K, Chase JH, Cowdery JS, Krieg AM (1996) IFN-$\gamma$ promotes IL-6 and IgM secretion in response to CpG motifs in bacterial DNA and oligodeoxynucleotides. J Immunol 157:558–564

Yi A-K, Klinman DM, Martin TL, Matson S, Krieg AM (1996) Rapid immune activation by CpG motifs in bacterial DNA. J Immunol 157:5394–5402

Yi A-K, Krieg AM (1998) CpG DNA rescue from anti-IgM induced WEHI-231 B lymphoma apoptosis via modulation of IkB$\alpha$ and IkB$\beta$ and sustained activation of nuclear factor-k B/cRel. J Immunol 160:1240–1245

Yi A-K, Tuetken R, Redford T, Waldschmidt M, Kirsch J, Krieg AM (1998) CpG motifs in bacterial DNA activate leukocytes through the pH-dependent generation of reactive oxygen species. J Immunol 160:4755–4761

Zimmermann S, Egeter O, Hausmann S, Lipford GB, Rocken M, Wagner H, Heeg K (1998) CpG oligodeoxynucleotides trigger protective and curative Th1 response in lethal murine leishmaniasis. J Immunol 160:3627–3630

# Macrophage Activation by Immunostimulatory DNA

K.J. Stacey, D.P. Sester, M.J. Sweet, and D.A. Hume



## 1 Introduction

Activation of macrophages/dendritic cells appears to be of crucial importance in vivo in immunostimulation by foreign DNA. Immune cell activation by bacterial

Centre for Molecular and Cellular Biology and Departments of Biochemistry and Microbiology University of Queensland, Brisbane, Qld 4072, Australia
E-mail: K.Stacey@cmcb.uq.edu.au

DNA was first observed using mixed spleen cell cultures (SHIMADA et al. 1986; YAMAMOTO et al. 1992) and B cells (MESSINA et al. 1991), and the role of unmethylated CpG motifs in this activation was elucidated in B cells (KRIEG et al. 1995). This work showed that oligonucleotides containing an unmethylated CG dinucleotide in a certain sequence context (ACGT but not CCGG) could mimic the activity of bacterial DNA. A combination of methylation of the vertebrate genome and suppression of the frequency of activating sequences may prevent immune activation by self DNA. In mixed spleen cell culture, bacterial DNA or stimulatory oligonucleotides (here termed "CpG DNA") promote production of a range of cytokines, including interferons-α/β and -γ (IFN-α/β, IFN-γ), interleukins-6 and -12 (IL-6, IL-12) and tumour necrosis factor-α (TNF-α). A role for macrophages in this activation was suggested by the finding that the production of IFN-γ by the non-adherent fraction of spleen cells was dependent on IL-12 and TNF-α production by the adherent fraction (HALPERN et al. 1996). Direct activation of macrophages by bacterial DNA and oligonucleotides was first shown by activation of nuclear factor (NF)-κB and induction of TNF-α mRNA and human immunodeficiency virus (HIV)-1 long terminal repeat (LTR) promoter activity (STACEY et al. 1996). Since then, it has become apparent that activation of macrophages and other antigen-presenting cells (APCs) is critical in the production of cytokines and the subsequent development of T-helper 1 (Th1)-biased T-cell lineages in response to CpG DNA. Responses of macrophages and dendritic cells will be addressed together in this review, as they can be considered to be different differentiative states of the same cell lineage (PALUCKA et al. 1998). This review will focus on the mechanism of macrophage activation, from DNA uptake to altered gene expression, and the CpG DNA-induced changes in macrophage function in vitro and in vivo.

## 2 Cellular DNA Uptake

### 2.1 Endocytic Uptake of DNA

DNA and oligonucleotides are internalised by mammalian cells, and most authors now seem to favour a requirement for DNA uptake in activation by CpG DNA, although an interacting role for signalling from a cell-surface receptor has not been excluded. In our experiments, even quite large DNA molecules, such as plasmids, can be taken up into a macrophage cell line, and some plasmids reach the nucleus in an intact state able to be expressed (STACEY et al. 1996). Uptake of naked DNA has been known for some time from studies of the fate of adenoviral DNA in culture (GRONEBERG et al. 1975). Recently, attention has been focussed on oligonucleotide uptake (reviewed in BENNETT 1993) and, at low oligonucleotide concentrations, the process seems to involve receptor-mediated endocytosis via clathrin-coated pits (BELTINGER et al. 1995). Internalised oligonucleotides were found in endosomes and dense lysosome-like vesicles, with some escape to the cytoplasm and nucleus (BELTINGER et al. 1995). Phosphorothioate oligonucleotides taken up by the pre-myelocytic leukaemic line HL60 (TONKINSON and STEIN 1994) and the mouse macrophage cell line J774 (HÄCKER et al. 1998) were largely located within acidified vacuoles. Uptake of oligonucleotides seems to have little or no sequence specificity (KRIEG et al. 1995), and distinction between activating and non-activating oligonucleotides is proposed to take place at a stage later than cell-surface-receptor recognition. Receptor-mediated endocytosis, of course, requires a cell-surface receptor with affinity for DNA, and a recent paper gives evidence that the integrin Mac-1 (CD11b/CD18) may mediate oligonucleotide uptake (BENIMETSKAYA et al. 1997). Mac-1 is expressed largely on polymorphonuclear leukocytes and macrophages. In our own studies, it seems unlikely that Mac-1 is responsible for DNA uptake in macrophages, as its cell-surface concentration does not alter during DNA stimulation as would be expected for a receptor involved in endocytosis (Beasley and Hume, unpublished observations). Several studies have addressed DNA cell-surface binding (reviewed in Bennett 1993), with one crosslinking study showing DNA interacting with at least five unidentified proteins on K562 human leukaemia cells (BELTINGER et al. 1995).

## 2.2 Clearance of DNA In Vivo

Macrophages are a major site of clearance of circulating DNA in vivo and, hence, may contribute more to the overall response than other cell types. Intravenously injected DNA was rapidly cleared from the bloodstream (90% cleared within 20min) and localised to the liver and spleen, suggesting a role for macrophages in clearance (CHUSED 1972). Single-stranded DNA taken up by the liver was primarily bound by Kupffer cells (liver macrophages) (EMLEN 1988). With the injection of phophorothioate-stabilised oligonucleotides, a wider tissue distribution was found, with a large component of clearance by the kidney; however, localisation to the liver and spleen was still apparent (SANDS et al. 1994). Interestingly, DNA need not be injected to gain access to the immune system, as uptake and even integration of foreign DNA into the genome of spleen cells has been demonstrated in an experiment where mice were fed M13 DNA (SCHUBBERT et al. 1997)! In vitro studies showed a difference in the ability of immune cells to take up oligonucleotide, with macrophages the most efficient, followed by B cells and then T cells, which had relatively low-efficiency uptake (ZHAO et al. 1996).

## 2.3 DNA Uptake as a Requirement for Activation?

Evidence that uptake of foreign DNA is required for subsequent immune activation comes from several approaches. Two groups using immobilised oligonucleotides reported that internalisation of CpG DNA is necessary for the proliferative response of B cells, although data awaits publication (KRIEG et al. 1995; L. Manzel and

D. Macfarlane, personal communication). Such experiments have obvious difficulties in both the estimation of relative concentrations of free and immobilised oligonucleotides seen by cells and the problem of cleavage of oligonucleotides from the solid support. One published study on human B cells showed that immobilised oligonucleotides were active (LIANG et al. 1996), but the observed proliferative response was specific for some phosphorothioate oligonucleotides and not bacterial DNA and was not demonstrated to be CpG specific. This B-cell activation (LIANG et al. 1996) did not seem to be similar to the well-characterised mouse response and may, in fact, be a cell-surface-mediated phenomenon or the covalent coupling of oligonucleotide to sepharose may have allowed entry of oligonucleotides into cells, as found by other workers (L. Manzel and D. Macfarlane, personal communication). No direct studies on the requirement for internalisation have been performed on macrophages, but it is likely that the situation will be analogous to that for B cells. Findings that the potency of oligonucleotides is increased by lipofection into spleen cells (YAMAMOTO et al. 1994) and that lipofection makes possible a response of cells to six-base-long oligonucleotides (SONEHARA et al. 1996) has been cited as evidence that internalisation is required for activation; however, the data are ambiguous due to the possibility of synergy between oligonucleotides and lipofection reagents.

## 2.4 Endosomal Acidification Required for CpG DNA Responses

Less direct evidence of a requirement for internalisation has been obtained using compounds that prevent acidification of endosomes and block the action of DNA. Bafilomycin, chloroquine and monensin prevented J774 macrophage cell and spleen cell TNF-α, IL-12 and IL-6 production (HÄCKER et al. 1998; YI et al. 1998b). The effect was specific for the DNA response and did not prevent activation by phorbol 12-myristate 13-acetate + ionomycin or lipopolysaccharide (LPS). Chloroquine and related compounds also inhibited CpG DNA responses of B cells and unfractionated human peripheral blood monocytes (MACFARLANE and MANZEL 1998). Chloroquine and bafilomycin A did not block uptake of oligonucleotide or vesicular localisation (HÄCKER et al. 1998). These compounds act in different ways – bafilomycin A blocks the endosomal hydrogen-ion pump, and chloroquine is a strong base that localises to the vesicle. However, recent findings that chloroquine inhibits the action of CpG DNA at concentrations that do not prevent endosomal acidification (L. Manzel and D Macfarlane, personal communication) call into question the assumption that the effect of these inhibitors is necessarily related to change in endosomal pH. Further work is required to establish exactly how these inhibitors are acting. One possibility is that they inhibit release of oligonucleotides from the receptor mediating their uptake, and prevent them from reaching the cytoplasm. Although DNA "escape" from the vacuole into the cytoplasm has been suggested as a requirement for activation, a CpG-specific receptor could be located in the endosome. Escape from the vacuole, if required, is likely to be a regulated rather than an "accidental" event, as responses to exogenously added DNA are extremely rapid and reliable (see below).

## 3 Macrophage/Dendritic Cell Responses to CpG DNA

### 3.1 Signalling

The receptors for CpG DNA and initiation of signalling remain uncharacterised. Although it was proposed early on that CpG-containing DNA may induce transcription of cytokine genes by direct interaction with transcription factors (Cowdery et al. 1996), no evidence of this has emerged. Instead, a number of studies show activation of established intracellular signalling molecules. All signalling and gene inductions that have been studied are prevented by the inhibitors of endosomal acidification. This includes reactive oxygen species (ROS) production, NF-κB activation (Yi et al. 1998b) and mitogen-activated protein kinase (MAPKinase) pathways (Häcker et al. 1998; Yi and Krieg 1998). Some of these responses occur within 15min of exposure to CpG DNA, showing that the inhibitors, such as chloroquine and bafilomycin A, interfere with a very early signalling step.

#### 3.1.1 Nuclear Factor κB

In macrophages, transcription factor NF-κB is activated by bacterial DNA and CpG-containing oligonucleotides but not by methylated or vertebrate DNA (Stacey et al. 1996; Sparwasser et al. 1997b; Yi et al. 1998b). NF-κB is bound to an inhibitory subunit, IκB, in the cytoplasm of cells until activation, when IκB is phophorylated and degraded, allowing NF-κB to be translocated to the nucleus (Baldwin 1996). Activation of NF-κB is frequently initiated in situations of cellular stress, and it is implicated in induction of a wide range of inflammatory genes, including those involved in the CpG DNA response, such as TNF-α (Drouet et al. 1991), inducible nitric oxide synthase (iNOS) (Xie et al. 1994) and IL-12 (Yoshimoto et al. 1997). The use of inhibitors of IκB phophorylation or degradation suggests that active NF-κB is required for macrophage TNF-α production in response to CpG DNA (Yi et al. 1998b). The upstream signalling involved in NF-κB activation in response to CpG DNA remains to be established but probably involves a redox-sensitive step (see below).

#### 3.1.2 Reactive Oxygen Species

Production of ROS has been implicated in signalling pathways leading to NF-κB activation (Baeuerle and Henkel 1994). CpG DNA induction of ROS was detected in the J774 macrophage cell line (Yi et al. 1998b) and B cells (Yi et al. 1996) by an increase in fluorescence of oxidation-sensitive intracellular dihydrorhodamine123. The ROS production was prevented by inhibitors of endosomal acidification (Yi et al. 1998b). The levels of ROS produced are much lower than those resulting from a respiratory burst response producing radicals for cytotoxic purposes, but they may have a role in signalling. The literature linking ROS with NF-κB activation relies heavily on the use of inhibitors, such as pyrrolidine dithiocarbamate (PDTC) and other antioxidants (SCHRECK et al. 1992; BAEUERLE and HENKEL 1994). Antioxidant compounds are potentially capable of altering the redox states of many receptors and signalling molecules, and inhibition does not necessarily indicate a direct role for ROS as second messengers. PDTC has been shown to inhibit TNF-α mRNA induction in RAW264 macrophages by CpG DNA and submaximal but not maximal concentrations of LPS (K. Stacey, unpublished). PDTC also inhibited IL-6 production in response to CpG DNA in B cells, while protein kinase C and protein kinase A inhibitors had no effect (YI et al. 1996). Caution should be exercised in interpreting this as evidence for direct involvement of ROS in signalling. Some workers have presented evidence against ROS as second messengers involved in NF-κB activation (BRENNAN and O'NEIL 1995; SUZUKI et al. 1995), although the redox state of certain signalling molecules upstream of NF-κB must be important in allowing activation.

### 3.1.3 Mitogen-Activated Protein Kinases

Many different stimuli (growth factors, cytokines, environmental stress) activate members of the MAPKinase superfamily, although the pattern of activation varies among stimuli. MAPKinases directly phosphorylate and activate certain transcription factors. The MAPKinases c-Jun *N*-terminal kinase (JNK) and p38 were activated by CpG DNA in macrophage cell lines, a B-cell line and in bone-marrow cells differentiated towards a dendritic cell phenotype with granulocyte/macrophage colony-stimulating factor (GM-CSF) (HÄCKER et al. 1998; YI and KRIEG 1998). This involves presumably sequential activation of the kinase JNK kinase 1 followed by the MAPKinase JNK, which is responsible for phosphorylation of the activator protein 1 (AP-1)-family transcription factor Jun within 15min of CpG DNA exposure (HÄCKER et al. 1998; YI and KRIEG 1998). MAPKinase p38 and its target transcription factor, activating transcription factor 2 (ATF-2), were also activated by CpG DNA (HÄCKER et al. 1998; YI and KRIEG 1998). This pathway is implicated in the CpG DNA-induced macrophage expression of IL-12 and TNF-α, as a specific p38 inhibitor reduced production of these cytokines (HÄCKER et al. 1998).

MAPKinases p44 and p42 (Erk-1 and -2) are downstream of ras in growth-factor-signalling pathways and are activated by the macrophage growth factor colony-stimulating factor-1 (CSF-1) and LPS (GEPPERT et al. 1994; LIU et al. 1994; BUSCHER et al. 1995; FOWLES et al. 1998). Although one report found little activation of Erk-1 and -2 by CpG DNA in J774 macrophages (YI and KRIEG 1998), we found that CpG DNA induced tyrosine phosphorylation of these MAPKinases in bone-marrow-derived macrophages (D. Sester, unpublished). Whilst CSF-1 and LPS gave rapid phosphorylation of Erk-1 and -2 within 5min, the response to DNA was slightly delayed, perhaps reflecting the difference between signalling from cell-surface receptors for LPS and CSF-1 and the requirement for internalisation of DNA. The common effects of LPS, CSF-1 and CpG DNA in activation of Erk-1 and -2 may relate to their shared anti-apoptotic activity (Sect. 3.6). An involvement of Erk-1 and -2 in anti-apoptotic responses (JARVIS et al. 1997; WANG et al. 1998) may explain why their activation by CpG DNA was detected in factor-dependent bone-marrow-derived macrophages starved of their growth factor, but not in factor-independent J774 macrophages.

### 3.1.4 Transcription Factors

In addition to activation of NF-κB, Jun and ATF-2, mentioned above, CpG DNA and LPS both induce macrophage expression of mRNA for c-*myc* (Introna et al. 1986; Myers et al. 1995; Yi et al. 1998b) and Ets-2 (Sweet et al. 1998b). Ets-2 and c-*myc* are also up-regulated in response to tyrosine kinase and ras signalling induced by CSF-1 (Xu et al. 1993; Stacey et al. 1995; Fowles et al. 1998). In addition to its induction, Ets-2 is activated upon phosphorylation by the MAP-Kinases Erk-1 and Erk-2 in response to LPS and CSF-1 (Fowles et al. 1998; Sweet et al. 1998b). Although it has not been demonstrated, a similar phosphorylation of Ets-2 is likely to follow the CpG DNA-mediated activation of Erk-1 and -2 (Sect. 3.1.3). The induction of c-*myc* and Ets-2 via MAPKinase-dependent routes in response to LPS, CSF-1 and CpG DNA may be related to anti-apoptotic effects of all three compounds (Sect. 3.6).

LPS and CpG DNA also both induce mRNA for CCAAT/enhancer-binding protein (C/EBP)-β and -δ transcription factors. These factors bind to *cis*-acting elements of many myeloid-specific and inducible genes, and C/EBP-β was first identified as NF-IL6, a factor involved in the expression of the IL-6 gene (Akira et al. 1990). Generation of a knockout mouse showed that NF-IL6 is essential for normal bacterial and tumour cell killing by macrophages (Tanaka et al. 1995).

## 3.2 Macrophage/Dendritic Cell Gene and Cytokine Induction by DNA

### 3.2.1 Interleukin 12

One of the CpG-induced cytokines about which there is great interest is IL-12, given its role in early the induction of IFN-γ in natural killer (NK) cells (Chace et al. 1997) and the promotion of Th1-type immunity (Scott and Trinchieri 1997). Macrophages and dendritic cells are believed to be the primary source of IL-12 in response to CpG DNA, and IL-12 induction has now been detected in a variety of cells exposed to CpG DNA, including splenic macrophages (Chace et al. 1997), J774 (Lipford et al. 1997) and ANA-1 murine macrophage cell lines, bone-marrow-derived dendritic cells (Häcker et al. 1998) and fetal-skin-derived dendritic cells (Jakob et al. 1998).

### 3.2.2 Tumor Necrosis Factor α

Together with IL-12, TNF-α may play a role in the induction of IFN-γ in spleen cell cultures exposed to CpG DNA (Halpern et al. 1996). Apart from its well-publicised role in pathological states, such as toxic shock, TNF-α is an important part of the normal inflammatory response. TNF-α mRNA and protein are rapidly produced by CpG DNA in bone-marrow-derived macrophages (STACEY et al. 1996; SWEET et al. 1998a), bone-marrow-derived dendritic cells (HÄCKER et al. 1998), peritoneal macrophages (SPARWASSER et al. 1997b) and the murine macrophage cell lines RAW264 (STACEY et al. 1996; SWEET et al. 1998a), J774 (LIPFORD et al. 1997) and ANA-1 (SPARWASSER et al. 1997b). Strongly activating oligonucleotides induce both IL-12 and TNF-α efficiently. One group has described a phosphorothioate oligonucleotide that induced IL-12 and not TNF-α (LIPFORD et al. 1997), and this would implicate more than one mechanism in the detection of foreign DNA. However, we believe this finding may be a reflection of different genes having different threshold concentrations of DNA for activation. In our experiments, a phosphodiester version of this same oligonucleotide can induce TNF-α but is less potent than some other oligonucleotides (Sweet, unpublished observations). The dose-response curves for oligonucleotides are relatively steep, achieving maximal effect over a tenfold concentration range. Therefore, if different genes have different thresholds for activation, then at an appropriate concentration a weakly activating oligonucleotide could stimulate IL-12 and not TNF-α expression. The difference between our results and those of LIPFORD et al. (LIPFORD et al. 1997) warrant further investigation of dose responses to such oligonucleotides.

### 3.2.3 Other CpG DNA-Induced Cytokines/Genes

Macrophages (LIPFORD et al. 1997), bone-marrow-derived dendritic cells (SPARWASSER et al. 1998) and B cells (YI et al. 1996) produce IL-6 in response to bacterial DNA. IL-6 is a pleiotropic cytokine important in resistance to infection and involved in the acute phase response in liver (AKIRA and KISHIMOTO 1996). Induction of mRNA for the inflammatory cytokine IL-1β and a protease inhibitor, plasminogen activator inhibitor-2 (PAI-2), was detected in RAW264 macrophage cells (STACEY et al. 1996), but message levels were considerably lower than those induced by LPS. iNOS and its product, nitric oxide (NO, a free radical implicated in macrophage-mediated killing of micro-organisms and tumour cells), are produced by bone-marrow-derived macrophages in response to CpG DNA, but only in the presence of IFN-γ (STACEY et al. 1996; SWEET et al. 1998a). This requirement for IFN-γ priming is distinct from LPS induction of iNOS (Sect. 3.5).

Although the earliest characterised DNA response was production of type-I IFNs by spleen cells (TOKUNAGA et al. 1988), little work has been done to establish which cell types produce IFN-α/β. One recent study implicates APC production of type-I IFN in subsequent T-cell activation, as IFN-I-receptor$^{-/-}$ T cells failed to show normal stimulation with CpG DNA-activated APCs (SUN et al. 1998). Another study found production of IFN-α and -β mRNA by human macrophages transfected with DNA (ROMAN et al. 1997). Confirmation of IFN-α/β production in the absence of the lipid transfection reagent and analysis of production by other cell types is yet to be presented. Similarly, IL-18 mRNA has been detected, after transfection, by the same group (ROMAN et al. 1997). IL-18 cooperates with IL-12 in the promotion of IFN-γ production and the development of Th1-type immunity (Takeda et al. 1998).

### 3.2.4 Responses of Human Monocytes

There is little published information on activation of human cells of any type in response to CpG DNA, a gap in the literature which needs to be filled before therapeutic applications can be designed or optimised. Some responses of human monocytes and macrophages have been observed. Using human peripheral blood mononuclear cells, bacterial DNA and CpG oligonucleotides were found to induce the cytokines IL-6 and TNF-α and intercellular adhesion molecule 1 (ICAM-1) in the monocyte fraction (Hartmann and Krieg 1999). However, the time course of expression was delayed compared with the rapid response to LPS, with TNF-α and IL-6 only being produced 18–24h after stimulation. In addition, relatively high levels of *Escherichia coli* DNA were required for this response. The delayed induction of normally rapidly induced genes suggests that this may be a differentiative response of monocytes to CpG DNA. Another group has found that CpG DNA specifically inhibits the adhesion of human monocyte-derived macrophages in vitro (D. Macfarlane, personal communication). One difficulty in the comparisons of human and mouse work is that mouse studies generally involve relatively mature macrophages, such as peritoneal or bone-marrow-derived macrophages, as opposed to the circulating monocytes most commonly available from human subjects. In addition, most human cell lines available are relatively immature. The use of monocyte-derived macrophages or other more mature human macrophages may well show similar DNA sensitivity to mouse macrophages. The induction of IL-18 and IFN-α and -β mRNA in response to transfected DNA discussed above (Sect. 3.2.3) was performed in human macrophages purified from peripheral blood mononuclear cells by fibronectin adherence and occurred within 3h.

## 3.3 Negative Regulation of Macrophage Responses

Macrophage production of IL-12 in response to DNA leads to NK-cell IFN-γ production, which in turn enhances the macrophage response to DNA, leading to increased production of TNF-α, NO (Sweet et al. 1998a) and, most likely, IL-12 itself. This gives the possibility of a self-amplifying loop, which must be kept under control in vivo if toxic shock is to be avoided. One cytokine which may play a role in this control is IL-10, which has been shown to inhibit macrophage IL-12 production in response to DNA (Anitescu et al. 1997). Induction of IL-10 has been detected in splenocytes following bacterial DNA treatment and is probably B-cell derived (Anitescu et al. 1997; Redford et al. 1998). Addition of anti-IL-10 antibodies increased spleen cell IL-12 production in response to CpG DNA (Redford et al. 1998) and, thus, IL-10 may be one of the cytokines acting to keep the response to foreign DNA under control.

## 3.4 Maturation of Cell Function

CpG DNA induces both acute responses, like induction of TNF-α mRNA expression in macrophages, which peaks at 1h (Stacey et al. 1996), and slower differentiative responses, in particular the maturation of APC function (Jakob et al. 1998; Sparwasser et al. 1998). Like LPS, CpG DNA treatment for 18h caused maturation of fetal-skin-derived dendritic cells to exhibit decreased adhesion, increased expression of major histocompatibility complex (MHC) class-II and co-stimulatory molecules CD86 and CD40 (Jakob et al. 1998). Compared with the LPS response, CpG-treated cells made more IL-12 and less IL-6 and TNF-α. CpG-DNA-induced maturation of professional APC function has also been observed in bone-marrow-derived dendritic cells cultivated in GM-CSF (Sparwasser et al. 1998).

## 3.5 Comparison of DNA and LPS Responses

The range of macrophage gene products induced by bacterial DNA is similar to an LPS response (Stacey et al. 1996), but the relative potency of the two products varies for different genes. Despite a similarity of action, an increasing number of studies are finding important differences in the actions of LPS and CpG DNA. Given that LPS is a powerful pyrogen and inducer of toxic shock and is unsuitable for use as an adjuvant in humans, establishment of a different mode of action for CpG DNA is a prerequisite to its use in human therapy.

One simple piece of evidence that LPS and DNA have substantially different actions in vivo is that the two compounds synergise in the induction of TNF-α (Sparwasser et al. 1997b). This may require the cooperation of a mixture of cell types or may be post-transcriptional, as we are unable to show synergy at the level of TNF-α mRNA expression in macrophage culture (K. Stacey, unpublished observations). CpG DNA and LPS also synergised in the induction of IL-6 in human monocytes (Hartmann and Krieg 1999). This synergy in the induction of a gene within one cell precludes the activation of identical signalling pathways or transcription factors.

The CpG DNA response is separable from the LPS response by the use of LPS non-responder C3H/HeJ mice, which seem to retain all normal responses to DNA (Shimada et al. 1986; Cowdery et al. 1996; Sparwasser et al. 1997b). The defect in the C3H/HeJ mouse has now been identified as a putative LPS receptor, Toll-like receptor 4 (Poltorak et al. 1998), a molecule which is clearly unnecessary for the DNA response. A difference in early signalling is also shown by the inhibition of CpG DNA responses but not LPS responses by chloroquine and bafilomycin A, which may be interfering with normal endosomal pathways (Sect. 2.4; Häcker et al. 1998; Yi et al. 1998b).

Signalling in response to LPS and DNA converges with the activation of MAPKinases (Häcker et al. 1998; Yi and Krieg 1998), activation by both agents of transcription factors NF-κB (Stacey et al. 1996; Sparwasser et al. 1997b) and

AP-1 (HÄCKER et al. 1998; YI and KRIEG 1998) and induction of the transcription factors C/EBP-β, C/EBP-δ and Ets-2 (SWEET et al. 1998b). However, there are clearly undiscovered differences in transcription-factor activation, as there are some separable functions of LPS and DNA in gene induction. In a comparison of the responses of two macrophage cell lines to LPS and DNA, one line produced IL-12 in response to both compounds, but RAW264 cells were responsive only to DNA (ANITESCU et al. 1997), although other LPS responses are perfectly intact in this cell line. As noted above, a difference we have detected in LPS and DNA signalling is that LPS, acting alone, can induce expression of iNOS and NO production by bone-marrow-derived macrophages and RAW264 cells, whilst CpG DNA requires IFN-γ priming for this response (STACEY et al. 1996; SWEET et al. 1998a; K. Stacey, unpublished). IFN-γ induces Stat1α transcription factor, which is required for iNOS-promoter activity. Induction of iNOS by LPS alone requires autocrine production of type-I IFNs, which subsequently induces Stat1α phosphorylation (GAO et al. 1998). We are currently investigating whether or not CpG DNA can mediate a similar activation of Stat1α.

The above examples highlight some differences between CpG DNA and LPS actions, but there are many similarities. The concept that prior exposure to sub-stimulatory amounts of LPS induces tolerance to higher doses is well established. In experiments we have performed on activation of the HIV-1 LTR in RAW264 macrophages, overnight pretreatment with 0.1ng/ml LPS markedly reduced the response to subsequently added CpG DNA or LPS (S. Cronau and D. Hume, unpublished observations). The reverse is also true, with prior exposure to CpG DNA attenuating the response to both stimuli. Presumably, both agents are able to down-modulate a signalling molecule involved in the activation of the HIV LTR.

## 3.6 Effects on Cell Survival and Growth

Bone-marrow-derived macrophages proliferate and differentiate in the presence of the growth factor CSF-1, and it is also required for survival in culture. In the absence of CSF-1 or another survival stimulus for 24h, the cells begin to commit to an apoptotic pathway. Addition of either LPS or CpG DNA prevents apoptosis (D. Sester, unpublished). CpG DNA also promotes survival of B cells (YI et al. 1998a) but, in that case, DNA and LPS are both mitogenic stimuli. In macrophages, DNA and LPS are not mitogenic stimuli, and they inhibit growth in response to CSF-1. The observed anti-proliferative and perhaps anti-apoptotic effects of CpG DNA are associated with downregulation of the CSF-1 receptor, which occurs between 10min and 60min after exposure to CpG DNA (S. Beasley and D. Hume, unpublished observations).

# 4 Responses to CpG DNA In Vivo

## 4.1 DNA and LPS in Toxic Shock

Macrophages are central to the production of cytokines in response to LPS, leading to toxic shock. Although it has been suggested that bacterial DNA may cause toxic shock through induction of TNF-α (Sparwasser et al. 1997a,b), demonstration of a toxic effect required treatment of mice with D-galactosamine, which sensitises liver cells to TNF-α-mediated apoptosis (Gutierrez-Ramos and Bluethmann 1997). In this model of sepsis, TNF-α is the major mediator of shock, whereas, in non-sensitised mice, ICAM-1 expression may be more important (Gutierrez-Ramos and Bluethmann 1997), and a more complex array of cytokines may be involved. In fact, studies now show that, unlike LPS, a single dose of up to 500μg/mouse of phosphorothioate CpG oligonucleotide alone does not cause toxic shock, but two large doses (500μg/mouse) given within 1 week can cause a fatal sepsis-like condition (A. Krieg, personal communication). This may be mediated by a priming effect of cytokines induced by the first dose, since IFN-γ and IL-12 knockout mice are resistant to this effect (A. Krieg, personal communication). Although elevated circulating IFN-γ has only been detected for 24h after administration of bacterial DNA (Cowdery et al. 1996), IL-12 can remain high for a week (Krieg et al. 1998). Administration of bacterial DNA 4h prior to LPS greatly sensitises the mice to LPS-mediated shock, probably through production of IFN-γ (Cowdery et al. 1996). As noted in Sect. 3.5, when added at the same time, LPS and DNA synergised induction of TNF-α in vivo (Sparwasser et al. 1997b), and DNA could play a synergistic role in promoting both Gram-negative and Gram-positive shock. Nevertheless, the difference in toxicity between LPS and DNA administered alone is apparent in vivo, which is encouraging for the therapeutic applications of CpG DNA.

## 4.2 Generation of Th1 Responses

Responses to foreign DNA sequences seem to drive development of Th1-type immune responses in studies using oligonucleotides as adjuvants (Chu et al. 1997; Davis et al. 1998) and in DNA vaccinations (Sato et al. 1996; Klinman et al. 1997; Leclerc et al. 1997; Tighe et al. 1998; Stacey and Blackwell 1999). CpG DNA-induced early production of IL-12 by macrophages and subsequent induction of IFN-γ in NK cells during first exposure to antigen appear to polarise subsequent development of T cells to the Th1 lineage. This response may well also involve CpG DNA induction of type-I IFNs and IL-18 from macrophages/APCs (Roman et al. 1997). Thus, macrophage/APC recognition of DNA may be seen as responsible for Th1 development, with DNA as an adjuvant, and may contribute to Th1 development in normal infections. The Th1 bias has applications not only in vaccination but also in therapy for allergy, which is a Th2-dominated process (Goodman et al. 1998; Spiegelberg et al. 1998).

## 4.3 Activation of Macrophages/APCs in DNA Vaccination

Presentation of antigen by professional bone-marrow-derived APCs is required for cytotoxic T-lymphocyte generation in DNA vaccination (CORR et al. 1996). Activation of APCs by CpG sequences in the plasmid backbone is required for effective immunisation and development of Th1 responses in DNA vaccination (SATO et al. 1996; KLINMAN et al. 1997; LECLERC et al. 1997; TIGHE et al. 1998). However, the finding that addition of only one or two AACGTT sequences to a plasmid was critical for vaccination success (SATO et al. 1996) is very surprising given the number of potential activating sequences in any plasmid. Analysis of the capacity of mouse skin, grafted from sites that had been vaccinated by gene gun, to generate immune responses in recipient and donor mice showed that cells migrating from the site within 12h after vaccination are responsible for the generation of immulological memory (KLINMAN et al. 1998). Condon et al. (CONDON et al. 1996) showed that dendritic cells are directly transfected in gene-gun-mediated DNA vaccination and migrate from the skin to draining lymph nodes. These studies have not claimed that DNA alone induces APC migration, as at least some of the APCs involved were directly transfected by plasmid-coated gold particles (CONDON et al. 1996), but this is quite likely to be the case. Intradermal injection of CpG DNA caused activation of APCs as measured by increased MHC class-II, CD86 and IL-12 expression amongst epidermal cells 12h after injection (JAKOB et al. 1998). Study of the maturation of APCs within skin may be challenging, as they are likely to have migrated by the 12-h time point (KLINMAN et al. 1998), and they may have to be followed to the draining lymph node.

## 4.4 A Role for DNA Activation in Infection?

The involvement of unmethylated CpG sequences with macrophage/APC stimulation in DNA vaccination and with active oligonucleotides is well accepted, but the role of DNA as an immune stimulant during natural infection is difficult to assess. Some bacteria obviously contain other stimulatory molecules, such as LPS, which may have a quantitatively greater effect on the immune system than the foreign DNA to which the organism becomes exposed. However, detection of foreign DNA may be important in synergising with other activating molecules or in detection of bacterial strains that have avoided other means of surveillance. Another suggestion (PISETSKY 1996) is that the detection of foreign DNA may be of most importance in viral infections, since viruses do not contain the range of lipid and sugar molecules that generally identify bacteria and pathogens to the innate immune system.

In an infection, macrophages engulf and kill bacteria, with subsequent degradation of cellular components. Fragments of DNA would either have to escape to the cytoplasm or nucleus, or actually be detected within the phagosome in order to allow immune response. Alternatively, DNA escaping from the cell could stimulate other cells. One study looked at the fate of DNA from *E. coli* phagocytosed by macrophages and found that much of the genomic DNA was degraded into short pieces and released into the medium, whereas plasmid DNA remained intact (Rozenberg-Arska et al. 1984). Whether or not this DNA will stimulate the phagocytosing cell is difficult to assess, but it remains clear that exogenously added free CpG DNA is stimulatory. In order for the DNA of an infecting virus to be detected, it would be anticipated that the putative CpG DNA receptor would have to be located in the cytoplasm or nucleus rather than in the endosomes. Real assessment of the role of DNA as an immunostimulant in an infection awaits elucidation of the activating DNA receptor and generation of knockout mice or other approaches to interfering with its actions.

## 5 Summary

Macrophage/dendritic cells and B cells remain the only cell types where direct responses to CpG DNA are well established. The role of macrophages in vivo in DNA clearance and the potent cytokine induction in macrophages and dendritic cells places them in the central role in the in vivo response to foreign DNA. Although responses to DNA are unlikely to evolve and be retained if they are not significant in the immune response to infection, the relative contributions of DNA and other stimulators of the innate immune recognition of foreign organisms is difficult to assess. Although CpG DNA and LPS have similar actions, significant differences are emerging that make the use of DNA as a therapeutic immunostimulatory molecule feasible. The macrophage response to DNA generates cytokines favouring the development of Th1-type immunity, and active oligonucleotides now show promise as Th1-promoting adjuvants and as allergy treatments.

## References

Akira S, Isshiki H, Sugita T, Tanabe O, Kinoshita S, Nishio Y, Nakajima T, Hirano T, Kishimoto T (1990) A nuclear factor for IL-6 expression (NF-IL6) is a member of the C/EBP family. EMBO J 9:1897–1906

Akira S, Kishimoto T (1996) Role of interleukin-6 in macrophage function. Curr Opin Hematol 3:87–93

Anitescu M, Chace JH, Tuetken R, Yi A-K, Berg D, Krieg AM, Cowdery JS (1997) Interleukin-10 functions in vitro and in vivo to inhibit bacterial DNA-induced secretion of interleukin-12. J Interferon Cytokine Res 17:781–788

Baeuerle P, Henkel T (1994) Function and activation of NF-κB in the immune system. Annu Rev Immunol 12:141–179

Baldwin AS, Jr (1996) The NF-κB and IκB proteins: New discoveries and insights. Annu Rev Immunol 14:649–681

Beltinger C, Saragovi HU, Smith RM, LeSauteur L, Shah N, DeDionisio L, Christensen L, Raible A, Jarett L, Gewirtz AM (1995) Binding, uptake, and intracellular trafficking of phosphorothioate-modified oligodeoxynucleotides. J Clin Invest 95:1814–1823

Benimetskaya L, Loike JD, Khaled Z, Loike G, Silverstein SC, Cao L, Khoury JE, Cai T-Q, Stein CA (1997) Mac-1 (CD11b/CD18) is an oligodeoxynucleotide-binding protein. Nature Med 3:414–420

Bennett RM (1993) As nature intended? The uptake of DNA and oligonucleotides by eukaryotic cells. Antisense Res Dev 3:235–241

Brennan P, O'Neil L (1995) Effects of oxidants and antioxidants on NF-κB activation in three different cell lines: evidence against a universal hypothesis involving oxygen radicals. Biochem Biophys Acta 1260:1670–1675

Buscher D, Hipskind RA, Krautwald S, Reimann T, Baccarini M (1995) Ras-dependent and -independent pathways target the mitogen-activated protein kinase network in macrophages. Mol Cell Biol 15:466–475

Chace JH, Hooker NA, Mildenstein KL, Krieg AM, Cowdery JS (1997) Bacterial DNA-induced NK cell IFN-γ production is dependent on macrophage secretion of IL-12. Clin Immunol Immunopathol 84:185–193

Chu RS, Targoni OS, Krieg AM, Lehmann PV, Harding CV (1997) CpG oligodeoxynucleotides act as adjuvants that switch on T helper 1 (Th1) immunity. J Exp Med 186:1623–1631

Condon C, Watkins SC, Celluzzi CM, Thompson K, Falo LD (1996) DNA-based immunization by in vivo transfection of dendritic cells. Nature Med 2:1122–1128

Corr M, Lee DJ, Carson DA, Tighe H (1996) Gene vaccination with naked plasmid DNA: Mechanism of CTL priming. J Exp Med 184:1555–1560

Cowdery JS, Chace JH, Yi A-K, Krieg AM (1996) Bacterial DNA induces NK cells to produce IFN-γ in vivo and increases the toxicity of lipopolysaccharides. J Immunol 156:4570–4575

Davis HL, Weeranta R, Waldshmidt TJ, Tygrett L, Schorr J, Krieg AM (1998) CpG DNA is a potent enhancer of specific immunity in mice immunized with recombinant hepatitis B surface antigen. J Immunol 160:870–876

Drouet C, Shakov AN, Jongeneel CV (1991) Enhancers and transcription factors controlling the inducibility of the tumor necrosis factor-α promoter in primary macrophages. J Immunol 147: 1694–1700

Fowles LF, Martin ML, Nelsen L, Stacey KJ, Redd D, Clark YM, Nagamine Y, McMahon M, Hume DA, Ostrowski MC (1998) Persistent activation of mitogen-activated protein kinases p42 and p44 and ets-2 phosphorylation in response to colony-stimulating factor 1/c-fms signalling. Mol Cell Biol 18:5148–5156

Gao JJ, Filla MB, Fultz MJ, Vogel SN, Russell SW, Murphy WJ (1998) Autocrine/paracrine IFN-αβ mediates the lipopolysaccharide-induced activation of transcription factor Stat1α in mouse macrophages: Pivotal role of Stat1α in induction of the inducible nitric oxide synthase gene. J Immunol 161:4803–4810

Geppert TD, Whitehurst CE, Thompson P, Beutler B (1994) Lipopolysaccharide signals activation of tumor necrosis factor biosynthesis through the ras/raf-1/MEK/MAPK pathway. Mol Med 1:93–103

Goodman JS, Van Uden JH, Kobayashi H, Broide D, Raz E (1998) DNA immunotherapeutics: new potential treatment modalities for allergic disease. Int Arch Allergy Immunol 116:177–87

Groneberg J, Brown DT, Doerfler W (1975) Uptake and fate of the DNA of adenovirus type 2 in KB cells. Virology 64:115–131

Gutierrez-Ramos JC, Bluethmann H (1997) Molecules and mechanisms operating in septic shock: lessons from knockout mice. Immunol Today 18:329–334

Häcker H, Mischak H, Miethke T, Liptay S, Schmid R, Sparwasser T, Heeg K, Lipford GB, Wagner H (1998) CpG-DNA-specific activation of antigen-presenting cells requires stress kinase activity and is preceded by non-specific endocytosis and endosomal maturation. EMBO J 17:6230–6240

Halpern MD, Kurlander RJ, Pisetsky DS (1996) Bacterial DNA induces murine interferon-γ production by stimulation of interleukin-12 and tumour necrosis factor-α. Cell Immunol 167:72–78

Hartmann G, Krieg A (1999) CpG DNA and LPS induce distinct patterns of activation in human monocytes. Gene Ther (in press)

Introna M, Hamilton TA, Kaufman RE, Adams DO, Bast RC Jr (1986) Treatment of murine peritoneal macrophages with bacterial lipopolysaccharide alters expression of c-*fos* and c-*myc* oncogenes. J Immunol 137:2711–2715

Jakob T, Walker PS, Krieg AM, Udey MC, Vogel JC (1998) Activation of cutaneous dendritic cells by CpG-containing oligodeoxynucleotides: a role for dendritic cells in the augmentation of Th1 responses by immunostimulatory DNA. J Immunol 161:3042–3049

Jarvis WD, Fornari FA Jr, Auer KL, Freemerman AJ, Szabo E, Birrer MJ, Johnson CR, Barbour SE, Dent P, Grant S (1997) Coordinate regulation of stress- and mitogen-activated protein kinases in the apoptotic actions of ceramide and sphingosine. Mol Pharmacol 52:935–947

Klinman DM, Yamshchikov G, Ishigatsubo Y (1997) Contribution of CpG motifs to the immunogenicity of DNA vaccines. J Immunol 158:3635–3639

Klinman DM, Sechler JMG, Conover J, Gu M, Rosenberg A (1998) Contribution of cells at the site of DNA vaccination to the generation of antigen-specific immunity and memory. J Immunol 160: 2388–2392

Krieg AM, Love-Homan L, Yi AK, Harty JT (1998) CpG DNA induces sustained IL-12 expression in vivo and resistance to Listeria monocytogenes challenge. J Immunol 161:2428–2434

Krieg AM, Yi A-K, Matson S, Waldschmidt TJ, Bishop GA, Teasdale R, Koretzky GA, Klinman DM (1995) CpG motifs in bacterial DNA trigger direct B-cell activation. Nature 374:546–549

Leclerc C, Deriaud E, Rojas M, Whalen RG (1997) The preferential induction of a Th1 immune response by DNA-based immunization is mediated by the immunostimulatory effect of plasmid DNA. Cell Immunol 179:97–106

Liang H, Nishioka Y, Reich CF, Pisetsky DS, Lipsky PE (1996) Activation of human B cells by phosphorothioate oligodeoxynucleotides. J Clin Invest 98:1119–1129

Lipford GB, Sparwasser T, Bauer M, Zimmerman S, Koch E-S, Heeg K, Wagner H (1997) Immunostimulatory DNA: sequence-dependent production of potentially harmful or useful cytokines. Eur J Immunol 27:3420–3426

Liu MK, Herrera-Velit P, Brownsey RW, Reiner NE (1994) CD14-dependent activation of protein kinase C and mitogen-activated protein kinases (p42 and p44) in human monocytes treated with bacterial lipopolysaccharide. J Immunol 153:2642–2652

Macfarlane DE, Manzel L (1998) Antagonism of immunostimulatory CpG-oligodeoxynucleotides by quinacrine, chloroquine, and structurally related compounds. J Immunol 160:1122–1131

Messina JP, Gilkeson GS, Pisetsky DA (1991) Stimulation of in vitro murine lymphocyte proliferation by bacterial DNA. J Immunol 147:1759–1764

Myers MJ, Ghildyal N, Schook LB (1995) Endotoxin and interferon-gamma differentially regulate the transcriptional levels of proto-oncogenes and cytokine genes during the differentiation of colony-stimulating factor type-1-derived macrophages. Immunology 85:318–324

Palucka KA, Taquet N, Sanchez-Chapuis F, Gluckman JC (1998) Dendritic cells as the terminal stage of monocyte differentiation. J Immunol 160:4587–4595

Pisetsky DS (1996) The immunologic properties of DNA. J Immunol 156:421–423

Poltorak A, He X, Smirnova I, Liu MY, Huffel CV, Du X, Birdwell D, Alejos E, Silva M, Galanos C, Freudenberg M, Ricciardi-Castagnoli P, Layton B, Beutler B (1998) Defective LPS signaling in C3H/HeJ and C57BL/10ScCr mice: mutations in Tlr4 gene. Science 282:2085–2088

Redford TW, Yi AK, Ward CT, Krieg AM (1998) Cyclosporin A enhances IL-12 production by CpG motifs in bacterial DNA and synthetic oligodeoxynucleotides. J Immunol 161:3930–3935

Roman M, Martin-Orozco E, Goodman JS, Nguyen M-D, Sato Y, Ronaghy A, Kornbluth RS, Richman DD, Carson DA, Raz E (1997) Immunostimulatory DNA sequences function as T helper-1 promoting adjuvants. Nature Med 3:849–854

Rozenberg-Arska M, van Strijp JAG, Hoekstra WPM, Verhoef J (1984) Effect of human polymorphonuclear and mononuclear leukocytes on chromosomal and plasmid DNA of *Escherichia coli*. J Clin Invest 73:1254–1262

Sands H, Gorey-Feret LJ, Cocuzza AJ, Hobbs FW, Chidester D, Trainor GL (1994) Biodistribution and metabolism of internally 3H-labelled oligonucleotides. I. Comparison of a phosphodiester and a phosphorothioate. Mol Pharmacol 45:932–943

Sato Y, Roman M, Tighe H, Lee D, Corr M, Nguyen M-D, Silverman GJ, Lotz M, Carson DA, Raz E (1996) Immunostimulatory DNA sequences necessary for effective intradermal gene immunization. Science 273:352–354

Schreck R, Meier B, Männel DN, Dröge W, Baeuerle PA (1992) Dithiocarbamates as potent inhibitors of nuclear factor κB activation in intact cells. J Exp Med 175:1181–1194

Schubbert R, Renz D, Schmitz B, Doerfler W (1997) Foreign (M13) DNA ingested by mice reaches peripheral leukocytes, spleen, and liver via the intestinal wall mucosa and can be covalently linked to mouse DNA. Proc Natl Acad Sci USA 94:961–966

Scott P, Trinchieri G (1997) IL-12 as an adjuvant for cell-mediated immunity. Semin Immunol 9:285–291

Shimada S, Yano O, Tokunaga T (1986) In vivo augmentation of natural killer cell activity with a deoxyribonucleic acid fraction of BCG. Jpn J Cancer Res 77:808–816

Sonehara K, Saito H, Kuramoto E, Yamamoto S, Yamamoto T, Tokunaga T (1996) Hexamer palindromic oligonucleotides with 5′-CG-3′ motif(s) induce production of interferon. J Interferon Cytokine Res 16:799–803

Sparwasser T, Miethke T, Lipford G, Borschert K, Häcker H, Heeg K, Wagner H (1997a) Bacterial DNA causes septic shock. Nature 386:336–337

Sparwasser T, Miethke T, Lipford G, Erdmann A, Häcker H, Heeg K, Wagner H (1997b) Macrophages sense pathogens via DNA motifs: induction of tumor necrosis factor. Eur J Immunol 27:1671–1679

Sparwasser T, Koch ES, Vabulas RM, Heeg K, Lipford GB, Ellwart JW, Wagner H (1998) Bacterial DNA and immunostimulatory CpG oligonucleotides trigger maturation and activation of murine dendritic cells. Eur J Immunol 28:2045–2054

Spiegelberg HL, Tighe H, Roman M, Broide D, Raz E (1998) Inhibition of IgE formation and allergic inflammation by allergen gene immunization and by CpG motif immunostimulatory oligodeoxynucleotides. Allergy 53:93–97

Stacey KJ, Blackwell JM (1999) Immunostimulatory DNA as an adjuvant in vaccination against Leishmania major. Infect Immun (in press)

Stacey KJ, Fowles LF, Colman MS, Ostrowski MC, Hume DA (1995) Regulation of urokinase-type plasminogen activator gene transcription by macrophage colony stimulating factor. Mol Cell Biol 15:3430–3441

Stacey KJ, Sweet M, Hume DA (1996) Macrophages ingest and are activated by bacterial DNA. J Immunol 157:2116–2122

Sun S, Zhang X, Tough DF, Sprent J (1998) Type I interferon-mediated stimulation of T cells by CpG DNA. J Exp Med 188:2335–2342

Suzuki Y, Mizuno M, Packer L (1995) Transient overexpression of catalase does not inhibit TNF- or PMA- induced NF-κB activation. Biochem Biophys Res Comm 210:537–541

Sweet MJ, Stacey KJ, Kakuda DK, Markovich D, Hume DA (1998a) IFN-γ primes macrophage responses to bacterial DNA. J Interferon Cytokine Res 18:263–271

Sweet MJ, Stacey KJ, Ross IL, Ostrowski MC, Hume DA (1998b) Involvement of Ets, rel and Sp1-like proteins in lipopolysaccharide-mediated activation of the HIV-1 LTR in macrophages. J Inflammation 48:67–83

Takeda K, Tsutsui H, Yoshimoto T, Adachi O, Yoshida N, Kishimoto T, Okamura H, Nakanishi K, Akira S (1998) Defective NK cell activity and Th1 response in IL-18-deficient mice. Immunity 8: 383–390

Tanaka T, Akira S, Yoshida K, Umemoto M, Yoneda Y, Shirafuji N, Fujiwara H, Suematsu S, Yoshida N, Kishimoto T (1995) Targeted disruption of the NF-IL6 gene discloses its essential role in bacteria killing and tumour cytotoxicity by macrophages. Cell 80:353–361

Tighe H, Corr M, Roman M, Raz E (1998) Gene vaccination: plasmid DNA is more than just a blueprint. Immunol Today 19:89–97

Tokunaga T, Yamamoto S, Namba K (1988) A synthetic single-stranded DNA, poly(dG,dC), induces interferon-α/β and -γ, augments natural killer activity, and suppresses tumor growth. Jpn J Cancer Res 79:682–686

Tonkinson JL, Stein CA (1994) Patterns of intracellular compartmentalization, trafficking and acidification of 5′-fluorescein labeled phosphodiester and phosphorothioate oligodeoxynucleotides in HL60 cells. Nucl Acids Res 22:4268–4275

Wang X, Martindale JL, Liu Y, Holbrook NJ (1998) The cellular response to oxidative stress: influences of mitogen-activated protein kinase signalling pathways on cell survival. Biochem J 333:291–300

Xie QW, Kashibara Y, Nathan C (1994) Role of transcription factor NF-κB/Rel in induction of nitric oxide synthase. J Biol Chem 269:4705–4708

Xu X-X, Tessner TG, Rock CO, Jackowski S (1993) Phosphatidylcholine hydrolysis and c-*myc* expression are in collaborating mitogenic pathways activated by colony-stimulating factor 1. Mol Cell Biol 13:1522–1533

Yamamoto S, Yamamoto T, Kataoka T, Kuramoto E, Yano O, Tokunaga T (1992) Unique palindromic sequences in synthetic oligonucleotides are required to induce IFN and augment IFN-mediated natural killer activity. J Immunol 148:4072–4076

Yamamoto T, Yamamoto S, Kataoka T, Tokunaga T (1994) Lipofection of synthetic oligodeoxynucleotide having palindromic sequence of AACGTT to murine splenocytes enhances interferon production and natural killer activity. Microbiol Immunol 38:831–836

Yi A-K, Krieg AM (1998) Cutting Edge: Rapid induction of mitogen-activated protein kinases by immune stimulatory CpG DNA. J Immunol 161:4493–4497

Yi A-K, Klinman DM, Martin TL, Matson S, Krieg AM (1996) Rapid immune activation by CpG motifs in bacterial DNA. Systemic induction of IL-6 transcription through an antioxidant-sensitive pathway. J Immunol 157:5394–5402

Yi A-K, Chang M, Peckham DW, Krieg AM, Ashman RF (1998a) CpG oligodeoxyribonucleotides rescue mature spleen B cells from spontaneous apoptosis and promote cell cycle entry. J Immunol 160:5898–5906

Yi A-K, Tuetken R, Redford T, Waldshmidt M, Kirsch J, Krieg AM (1998b) CpG motifs in bacterial DNA activate leukocytes through the pH-dependent generation of reactive oxygen species. J Immunol 160:4755–4761

Yoshimoto T, Nagase H, Ishida T, Inoue J, Nariuchi H (1997) Induction of interleukin-12 p40 transcript by CD40 ligation via activation of nuclear factor-κB. Eur J Immunol 27:3461–3470

Zhao Q, Song X, Waldschmidt T, Fisher E, Krieg AM (1996) Oligonucleotide uptake in human hematopoietic cells in increased in leukemia and is related to cellular activation. Blood 88:1788–1795

# Consequences of Bacterial CpG DNA-Driven Activation of Antigen-Presenting Cells

T. SPARWASSER and G.B. LIPFORD



## 1 Introduction

Infectious pathogens that break through surface defences and reach underlying tissue or blood encounter an array of interior host defences. These fall basically into two classes: innate and adaptive defences. The innate system is the phylogenically older of the two branches and uses germline-encoded pattern-recognition receptors (PRR) to identify microbial invaders (PUGIN et al. 1994). An example is CD14, a receptor for bacterial cell-wall components (CWCs), such as lipopolysaccharides (LPS) and lipoteichoic acids. CD14 recognises carbohydrate constituents commonly expressed on pathogens (ESPEVIK et al. 1993; ZHANG et al. 1994; ULEVITCH and TOBIAS 1995; CLEVELAND et al. 1996) and signals "danger" to myeloid-lineage cells (MATZINGER 1994).

The reliance on carbohydrate recognition alone by innate defences would have the disadvantage of genetic inflexibility, since escape mutants lacking the target structures might arise. Because of this limitation, alternative mechanisms may exist

Institute of Medical Microbiology, Immunology and Hygiene, Technical University Munich, Trogerstr. 9, D-81675 Munich, Germany that activate protective innate defences. The notion that bacterial DNA might be one candidate ligand is appealing for several reasons. Although considered immunologically inert for decades, bacterial DNA displays properties required for a perfect "danger signal", including structural diversity, species specificity and the ability to cross-link receptors (Pisetsky 1996). Importantly, bacterial and vertebrate DNAs differ in their relative abundances of CpG dinucleotides, termed CpG suppression, within vertebrate DNA. Furthermore, in contrast to bacterial DNA, vertebrate DNA displays a high degree of cytosine methylation (Doerfler 1991; Hergersberg 1991; Bird 1992). These unexplained structural differences (Doerfler 1991) may be viewed as the genetic code for recognition of bacterial DNA, independent of surface antigens, such as LPS (Pisetsky 1996). Specific conserved, static DNA sequence structures appear to be a key enabling the vertebrate immune system to distinguish danger from non-danger.

The first report detailing the immune-stimulating properties of bacterial DNA was by Tokunaga and co-workers, who succeeded in attributing the tumoricidal effects of *Mycobacterium bovis* bacille Calmette-Guerin (BCG) to natural killer (NK) cell activation by mycobacterial DNA (Tokunaga et al. 1984). Subsequently, the effects of bacterial DNA on cells of the adaptive immune system were observed. Pisetsky and co-workers demonstrated an LPS-like proliferation of murine B lymphocytes on in vitro stimulation with DNA from *Escherichia coli* (Messina et al. 1991, 1993). Consequently, unexpected immune-stimulatory effects of synthetic oligodeoxynucleotides (ODNs) used as antisense or control ODNs were reported; some ODNs also triggered polyclonal proliferation of murine and human B cells in a sequence-specific manner (Helene and Toulme 1990; Tanaka et al. 1992; McIntyre et al. 1993; Pisetsky and Reich 1993; Pisetsky and Reich 1994; Branda et al. 1996a; Liang et al. 1996). In 1995, Krieg et al. defined bacterial DNA and synthetic, unmethylated CpG-DNA sequences as potent B-cell mitogens (Krieg et al. 1995).

The discovery of bacterial immune-stimulating DNA sequences has two major implications: bacterial DNA is not inert but activates immune cells; thus, immune cells can sense pathogens via pathogen DNA. This review will focus both on the recognition of bacterial CpG DNA by cells of the innate immune system and on the pathophysiologic and possible therapeutic consequences associated with this pattern recognition.

## 2 Activation of Macrophages

Tokunaga and co-workers discovered that a nucleic-acid-rich fraction from BCG induces NK cell lytic activity and interferon-γ (IFN-γ), -α and -β secretion by murine spleen cells. The immune-stimulating capacity of bacterial DNA was DNase sensitive but RNase resistant and could not be mimicked by vertebrate DNA (S.E. Yamamoto, this issue). Comparable biologic activity was displayed by synthetic 30–45-mer single-stranded (ss) ODNs; these ODNs contained certain palindromic hexamer motifs, such as AACGTT and GACGTC, and were derived from cDNA encoding for a BCG protein (Tokunaga et al. 1992; Yamamoto et al. 1992). In addition, similar sequences caused B-cell proliferation, and it was concluded that immunostimulatory bacterial "CpG-motifs" are characterised by a more liberal formula that does not require palindromic sequences but contains the motif 5′-purine-purine-CG-pyrimidine-pyrimidine-3′ (Krieg et al. 1995). If the central CpG motif was methylated or mutated, the immune-stimulating properties were lost. The efficacy of the CpG motif is influenced by flanking sequences (Pisetsky and Reich 1998). Interestingly, sequence rules that govern recognition of bacterial DNA by murine immune cells differ from the rules for human cells (Liang et al. 1996; Bauer et al. 1999). While CpG DNA directly activated murine and human B cells, further studies revealed that the originally observed biological effects of bacterial DNA on NK cells were dependent on the presence of adherent spleen cells, pointing to a direct or indirect contribution of myeloid cells (Halpern et al. 1996; Chace et al. 1997).

In 1996, the first direct proof of the susceptibility of macrophages towards CpG DNA was published. Bacterial plasmid DNA was shown to activate the transcription factor nuclear factor κB (NFκB) and to induce tumour necrosis factor α (TNF-α) mRNA in murine bone-marrow-derived macrophages and in the macrophage cell line RAW 264. In contrast to LPS-inducible genes like inducible nitric oxide synthase (iNOS), induction of iNOS mRNA in RAW cells using CpG DNA as stimulus required IFN-γ pretreatment, suggesting differences in the signal pathways (Stacey et al. 1996). In this study, initial experiments to elucidate the molecular mechanism of CpG DNA were performed. Plasmid DNA containing CpG sequences was shown to be taken up from the medium and to code for a reporter protein. While genomic DNA is known to be degraded within the lysosomes after receptor-mediated uptake, plasmid DNA apparently partly survives and reaches the nucleus (Bennett et al. 1985). The question of whether recognition of bacterial DNA requires indirect interaction (after binding to cytoplasmic proteins) or direct interaction with transcription factors or if active DNA fragments bind e.g. endosomal receptors, thus initiating downstream signaling events, remained unanswered.

Subsequently, it was demonstrated that primary murine macrophages were activated in vitro and in vivo by genomic DNA from gram-positive (*Staphylococcus aureus*, *Streptococcus faecalis*, *Micrococcus lysodeikticus*) and gram-negative (*E. coli*) bacteria. In addition, synthetic oligonucleotides containing unmethylated CpG motifs were also effective in activating peritoneal macrophages and the macrophage cell line ANA-1. In vitro, bacterial DNA or CpG ODNs triggered nuclear translocation of NFκB, accumulation of pro-inflammatory cytokine mRNAs and subsequent release of substantial amounts of TNF-α, interleukin 1β (IL-1β) and IL-6, which are known to be involved in the pathogenesis of septic shock (Sparwasser et al. 1997a,b and unpublished data). Furthermore, cytokines [such as granulocyte–macrophage colony-stimulating factor (GM-CSF), IL-12 and IL-10] and chemokines [such as macrophage inflammatory protein 1α (MIP-1α), MIP-1β, MIP-2 and monocyte chemoattractant protein 1 (MCP-1)] could be induced in macrophage cell cultures (Lipford et al. 1997b and unpublished data). These observations could also be extended to human peripheral blood monocytes that produce IL-12, IFN-α, IL-6 and TNF-α upon stimulation with bacterial DNA or synthetic ODNs (Roman et al. 1997; Bauer et al. 1999). Activation of murine macrophages and human peripheral monocytes could also be demonstrated by upregulation of surface markers, such as CD86, CD40, and major histocompatibility complex (MHC) class-II molecules (Bauer et al. 1999 and our own unpublished data). In comparative studies, similarities and dissimilarities between LPS- and DNA-induced signal transduction pathways are evident (Häcker et al. 1998; Yi and Krieg 1998). However, the parallels between LPS and the structurally unrelated CpG DNA as strong inflammatory stimuli for macrophages are remarkable. Even though we have still a poor understanding of how CpG DNA mediates acute sequence-dependent cell activation, some questions about intracellular signaling are being unravelled (A. Krieg and H. Häcker, this issue).

## 3 Activation and Maturation of Dendritic Cells

Dendritic cells (DCs) originate from a common bone-marrow precursor cell giving rise to macrophages, granulocytes and DCs. The macrophage and myeloid DC lineages are not as strictly separated as was hypothesised in the past, e.g. because blood monocytes can differentiate under certain conditions into either DCs or macrophages (Szabolcs et al. 1996). However, there are several essential differences between DCs and macrophages. DCs, as professional antigen-presenting cells (APCs), are most efficient in the activation of resting T cells and have the unique capacity to activate naive T cells. DCs exist in two functional states: immature antigen-capturing DCs and mature antigen-presenting DCs (Steinman 1991).

Immature DCs are abundant in peripheral non-lymphoid tissues, such as skin, and can be rapidly recruited to sites of local infection (Fig. 1). Immature DCs display a high endocytotic activity and high numbers of surface Fc receptors but low numbers of surface MHC class-II and co-stimulatory molecules. In order to initiate primary immune responses, immature DCs need to become activated and to differentiate into mature DCs able to present antigen in the MHC class-II and class-I contexts. This maturation/activation signal can be delivered by microbial products, such as LPS (Banchereau and Steinman 1998). Interestingly, bacterial CpG DNA can substitute for this classic danger signal. Murine bone-marrow-derived DCs (BMDDCs) cultured in vitro in the presence of GM-CSF could be activated by CpG DNA, in a CD40 ligand-independent manner, to upregulate MHC class-II molecules and express co-stimulatory molecules, such as CD86 and CD40 (Sparwasser et al. 1998). Low concentrations of CpG ODNs and bacterial genomic DNA efficiently induced secretion of pro-inflammatory cytokines, such as TNF-α, IL-12 and IL-6. The functional maturation of BMDDCs could be demonstrated in vitro both in mixed lymphocyte reactions and in staphylococcal

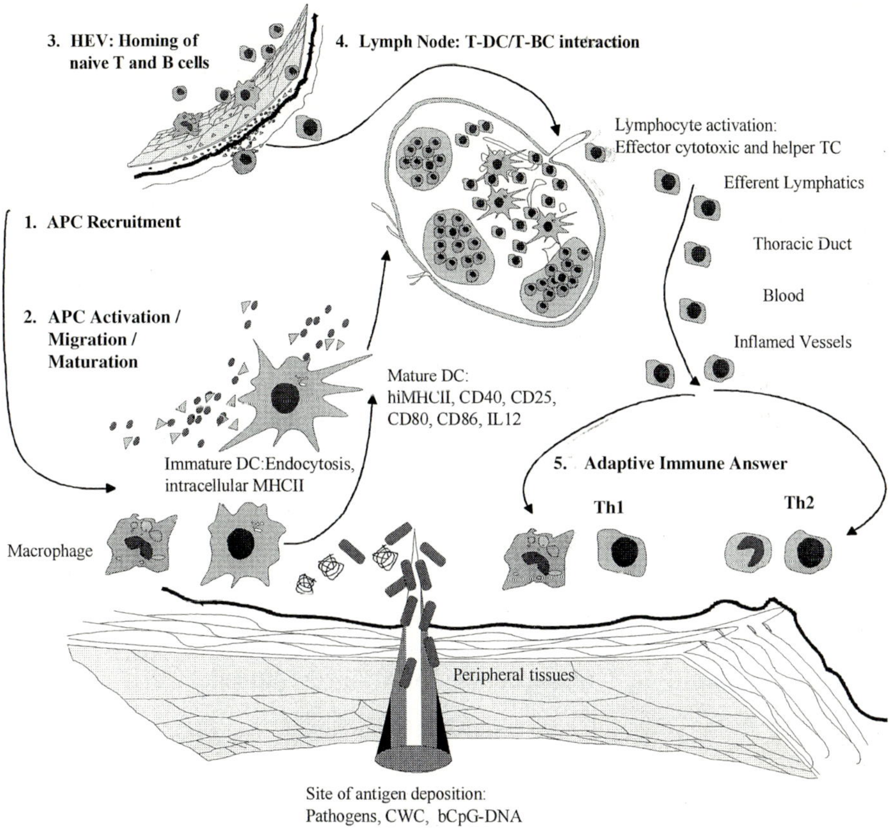


**Fig. 1.** Antigen-presenting cells (APCs) as decisive interfaces between the innate and the adaptive immune systems. After recruitment of APC progenitor cells from the bone marrow (*1*) to the site of antigen challenge, pathogens and pathogen-derived stimuli, such as bacterial cell-wall components and bacterial CpG DNA, cause activation of macrophages and dendritic cells (DCs). Immature DCs, upon activation, maturate and migrate to secondary lymphoid organs, where they translocate to T-cell areas (*2*). Naive T cells enter the draining lymph nodes via high endothelial venules (*3*) and interact with mature antigen-presenting DCs. In contrast to antigen-capturing immature DCs, mature antigen-loaded DCs display high numbers of surface major histocompatibility complex class-II complexes and co-stimulatory molecules (CD80, CD86). Depending on the expression of these molecules and the local cytokine environment (interleukin 12), cellular T-helper (Th)1 or humoral Th2-oriented immune responses are triggered (*4*). Activated cytotoxic and helper T cells migrate to the site of infection, where they act, in concert with other effector cells of the adaptive and innate immune systems, to fight the microbial invaders (*5*)

enterotoxin B (SEB)-driven naive T-cell responses (Sparwasser et al. 1998). A similar report showed activation of Langerhans' cell (LC)-like murine foetal-skin-derived DCs upon stimulation with CpG ODNs and *E. coli* DNA in vitro. In vivo, a small subset of MHC-II-bearing LCs exhibited upregulation of MHC II and CD86 and intracellular accumulation of IL-12 (Jakob et al. 1998).

Matured and activated DCs migrate to secondary lymphoid organs, where they translocate to the T-cell areas (De Smedt et al. 1996; Reis e Sousa et al. 1997).

After subcutaneous injection of CpG ODNs or plasmid DNA, activation of DCs from the draining lymph nodes could be observed. Importantly, in addition to upregulation of CD80, CD86 and CD40, local IL-12 production within the T-cell areas could be detected (Akbari et al. 1999 and our own unpublished data). Naive T cells, attracted and guided by chemokine/integrin-mediated mechanisms, enter the lymph nodes via high endothelial venules to interact with antigen-loaded DCs, leading to the generation of T-helper 1 (Th1)- or Th2-type effector cells (Sallusto et al. 1998). The interaction of activated DCs with T cells in the presence of IL-12 is thought to be responsible for a Th1-biased cellular immune response (Trinchieri 1995). A recent hypothesis proposes a 'dynamic' interaction between first DCs and $CD4^+$ T cells and subsequent $CD8^+$ T-cell activation triggered by the activated DC (Ridge et al. 1998). In this model, CD40L–CD40-mediated interaction between DCs and T-helper cells is necessary to induce co-stimulatory molecules on DCs triggering cytotoxic T-cell (CTL) activation. The demonstration of CD40L-independent DC activation suggests that it is possible to bypass T-helper cell-dependent mechanisms (Jakob et al. 1998; Sparwasser et al. 1998). In vaccination studies, for example, CD4-deficient mice can mount cellular immune responses if co-injected with CpG ODNs as adjuvants (our own unpublished data).

## 4 Pathophysiologic Consequences of Innate Immune Cell Activation

Bacterial CpG DNA operationally resembles bacterial LPS, a CWC known to be essential for morbidity and mortality caused by gram-negative bacteria. The question arises whether bacterial DNA, either liberated by lysosomal degradation within macrophages or released extracellularly from dead micro-organisms, can also trigger pathology.

### 4.1 Role of Bacterial DNA in Toxic Shock

Macrophages are regarded as key cells in the pathogenesis of septic shock (Glauser et al. 1991; Rietschel and Wagner 1996b). They are known to be activated by bacterial CWCs to release toxic amounts of pro-inflammatory cytokines that can lead to tissue damage, multi-organ failure and death (Hack et al. 1997). Cells of the innate system may be able to detect bacterial pathogens independent of CD14, the receptor for CWCs, such as LPS. This question is intriguing for several reasons. In clinical settings, about 50–60% of cases of septic shock are caused by gram-negative and about 35–40% by gram-positive bacteria (Glauser et al. 1991; Brandtzaeg 1996). While the physiological significance of the CWCs of gram-negative bacteria (such as LPS, or endotoxin) is well established, the role of lipoteichoic acids and peptidoglycans in triggering gram-positive septic shock is less clear (Wakabayashi et al. 1991; De Kimpe et al. 1995; Kusunoki et al. 1995; Rietschel et al. 1996a). Some gram-positive bacteria produce exotoxins, like SEB, that can cause superantigen-mediated toxic shock syndrome and sensitise for sub-lethal doses of LPS (Marrack and Kappler 1990; Miethke et al. 1992; Blank et al. 1997). The majority of gram-positive shock cases are superantigen independent, implying that unknown components derived from gram-positive bacteria have a role in pathophysiology (Glauser et al. 1991). Interestingly, while the inflammatory response towards whole gram-negative or gram-positive bacteria in acute cytokine-release syndrome is comparable, purified LPS from gram-negative bacteria is more than 1000-fold more potent in activating macrophages than purified gram-positive CWCs (Kusunoki et al. 1995 and references within). Importantly, LPS-non-responder mice are susceptible to septic shock triggered by both heat-killed gram-positive and gram-negative bacteria (Galanos and Freudenberg 1991). This observation suggests that additional signaling pathways are operative in septic shock, and one might claim that LPS-independent mechanisms trigger the activation of macrophages.

To investigate a possible role of bacterial DNA in septic shock, CpG ODNs and DNA from gram-positive and gram-negative bacteria were tested in a mouse shock model. These stimuli activated murine macrophages and macrophage cell lines in a manner similar to LPS. Bacterial CpG DNA triggered cytokine release by APCs, culminating in vivo in acute systemic release of pro-inflammatory cytokines and leading to TNF-$\alpha$-mediated lethal toxic shock in D-galactosamine (D-GalN)-sensitised mice (Sparwasser et al. 1997a,b). Of note, LPS and active CpG ODNs synergized in vivo in their ability to trigger TNF-$\alpha$ release. This synergy may be mediated through the common activation of NF$\kappa$B, and one might claim that there are unrecognised correlates in the gram-positive setting. In addition to the direct toxicity observed, bacterial DNA also sensitises mice (in a Shwartzman-like fashion via IFN-$\gamma$) to subsequent toxic shock from sub-lethal LPS doses (Cowdery et al. 1996 and our own unpublished results). This effect may be important in chronic infections.

## 4.2 Bacterial CpG DNA and Inflammation

Unquestionably, systemic release of cytokines leading to septic shock associated with high lethality is the worst outcome of bacterial infections. However, it will be necessary to document whether bacterial DNA liberated in the course of infection is physiologically relevant in causing local tissue damage and inflammation. Schwartz et al. demonstrated that bacterial DNA, instilled intratracheally, causes significant inflammation in the lower respiratory tracts of mice. Increased numbers of mononuclear cells and increased levels of pro-inflammatory cytokines (TNF-$\alpha$, IL-6, MIP-2) could be detected in whole-lung lavage after application of *E. coli* DNA or CpG ODNs, whereas LPS instillation caused a negligible inflammatory response (Schwartz et al. 1997). Furthermore, up to 1% of DNA extracted from the sputum of cystic fibrosis (CF) patients suffering from increased and abnormal mucus production and severe, recurrent bacterial infections was found to be of bacterial origin. Interestingly, these total DNA isolates induced acute inflammation in the mouse model similar to inflammation caused by *E. coli* DNA or CpG ODNs. One may speculate that, due to the local accumulation of bacterial DNA, tissue damage and inflammation persist in these patients even after antibiotic treatment (Schwartz et al. 1997).

## 5 Possible Therapeutic Applications

Bacterial immunostimulatory DNA sequences may trigger inflammatory disease or may even lead to septic shock. However, "beneficial" physiological host-defence functions, such as the activation and mobilisation of immune cells and the recognition of potentially dangerous pathogens, may be of major importance.

The tumoricidal effects of immune-stimulating DNA sequences were the first immunological effects of bacterial DNA to be reported and were considered to be "beneficial" (Tokunaga et al. 1984). The enhanced lytic anti-tumor activity of NK cells induced by prokaryote DNA could be demonstrated both in vitro and in vivo and was mediated by APC-derived cytokines, IL-12 and, to a minor extent, TNF-$\alpha$ and IFNs (Halpern et al. 1996; Chace et al. 1997). NK-stimulatory activity can be reproduced in the human system in vitro and enhances the efficacy of monoclonal-antibody therapy in a mouse lymphoma model (Ballas et al. 1996; Wooldridge et al. 1997). Here, further potentially therapeutic applications of CpG ODNs mediated by the activation of APCs will be discussed.

### 5.1 Role of CpG DNA as an Adjuvant

The adaptive immune system partially depends on innate immunity to distinguish infectious non-self from non-infectious self (Janeway 1992; Fearon and Locksley 1996). Danger versus non-danger information is delivered by cells of the innate system, such as macrophages or DCs sensing pathogens via pattern recognition. If non-self antigens associated with pathogens are processed and recognised as dangerous, the APCs become activated. Once activated, APCs induce productive T-cell responses, while non-activated APCs can cause peripheral tolerance (Ehl et al. 1998). The significance of this dynamic interaction between innate and adaptive immunity might have been underestimated in the past (Fearon and Locksley 1996; Fearon 1997). In fact, the need for bacterial products as APC stimuli to elicit humoral and cellular immune responses towards purified proteinaceous antigens, referred to as the "dirty little secret of immunologists", illustrates the importance of the innate system (Janeway 1989).

To enhance immunisation efficiency, various microbial products have served as adjuvants. It is believed that adjuvants primarily target APCs. These APCs have at least two duties. First, they process and present antigen peptides to T cells via MHC molecules and, after activation, they express co-stimulatory molecules. Second, APCs secrete pro-inflammatory cytokines, thereby acquiring the ability to instigate productive T- and B-cell activation (Jenkins et al. 1991; Galvin et al. 1992; Jenkins and Johnson 1993). Polarisation of activated T-helper cells into either Th1 or Th2 subsets is the result of secondary regulatory events and depends on APC secretion of effector cytokines, such as IL-12 and IL-10 (Mosmann and Coffman 1989; Abbas et al. 1996). Activation signals can be conveyed by pattern recognition of bacterial products, i.e. via CD14 or bacterial DNA.

The classical adjuvant CFA (complete Freund's adjuvant) contains dead mycobacteria in an oil-in-water emulsion (Ke et al. 1995). This adjuvant is widely used in experimental animal protocols; however, massive local inflammatory response precludes its use in humans. Various adjuvants have been shown to differentially activate Th1 and Th2 subsets, with CFA tending to induce Th1 subsets, while IFA (incomplete Freund's adjuvant), which does not contain mycobacteria, produces Th2-like responses (Forsthuber et al. 1996). Strong macrophage activation caused by CpG DNA has been shown by different groups (Stacey et al. 1996; Chace et al. 1997; Sparwasser et al. 1997b). Two recent studies demonstrate the activation and maturation of DCs upon stimulation with CpG DNA (Jakob et al. 1998; Sparwasser et al. 1998). Because of the powerful APC-stimulatory qualities of bacterial DNA or CpG ODNs, one might ask if these substances could substitute for heat-killed mycobacteria as adjuvants.

The first studies using CpG ODNs as adjuvants were performed when only its capacity to upregulate B7 and MHC-II molecules in B cells was known. CpG ODNs were shown to enhance antibody production to tetanus toxoid (Branda et al. 1996b). Subsequently, CpG ODNs were evaluated in a protocol for vaccination against the soluble antigen ovalbumin (OVA) (Lipford et al. 1997a). Two effects were prominent. First, immunostimulatory ODNs increased specific immunoglobulin G2 (IgG2)-antibody titers by more than 100-fold, demonstrating a Th1 bias. Second, the use of CpG ODNs as adjuvants allowed the induction of primary CTL responses to either unprocessed protein antigen (OVA) or its immunodominant peptide (SIINFEKL) encapsulated in liposomes. The CTL induction was independent of B-cell activation, and B-cell bystander effects as demonstrated through the use of B-cell-deficient μMT mice. Whereas direct APC-activating effects are now generally accepted, the role of direct effects of CpG ODNs on T cells are still controversial. In vitro, CpG DNA co-stimulatory effects can be observed after T-cell-receptor engagement (Bendigs et al. 1999). In contrast, in vivo data demonstrated T-cell activation only in the presence of APCs secreting type-I IFNs, suggesting that ODNs, as adjuvants, function primarily by potentiating APC function (Sun et al. 1998b).

The Th1-biasing effect of bacterial CpG sequences used as adjuvants has been described by various other groups (Raz et al. 1996; Chu et al. 1997; Roman et al. 1997; Weiner et al. 1997; Brazolot Millan et al. 1998; Davis et al. 1998; Sun et al. 1998a). Importantly, immunisation protocols leading to Th1-type cellular immunity can be essential for protective immune responses against infections. The skewing of CpG-ODN-enhanced antibody responses towards the IgG2 isotypes indicates Th1-dependent cytokine-induced class switching (Snapper and Paul 1987; Finkelman et al. 1990). In fact, the cytokine profile induced by CpG ODNs includes mainly IL-12, IL-18 (IFN-γ- inducing factor: IGIF) and IFN-γ and, thus, favours Th1-oriented T-cell responses (Halpern et al. 1996; Klinman et al. 1996; Lipford et al. 1997b; Roman et al. 1997; McCluskie and Davis 1998; Sparwasser et al. 1998). Interestingly, findings similar to those observed with synthetic ODNs can be demonstrated in DNA vaccination studies. Briefly, these vaccines consist of a bacterial plasmid backbone and a gene insert coding for the antigen. After intramuscular or intradermal injection of the plasmid, it is thought that host cells, including DCs, take up the plasmid and, thus, express the antigen. Because of its bacterial origin, these plasmids contain immunostimulatory CpG sequences. Operationally, "naked DNA" may, thus, harbour its adjuvant while acting as a translation unit. Indeed, recent evidence suggests that the immunogenicity of naked DNA is grossly influenced by the presence or absence of immunostimulatory CpG sequences, suggesting a key role in activating transfected APCs (Akbari et al. 1999; Klinman et al. 1997; Roman et al. 1997; Sato et al. 1996). Apparently, the co-administration of synthetic CpG ODNs and plasmid DNA vaccines can reduce the efficacy of the vaccines, possibly due to competitive binding at cellular receptors (Weeratna et al. 1998). Synthetic ODNs offer important advantages as adjuvants. They can be produced on a large scale with immense purity, combining the advantages of enormous stability and inexpensive synthesis as compared with protein or plasmid-based vaccines. Compared to a CpG-sequence-containing DNA vaccine, synthetic CpG ODNs, as adjuvants, may be more effective in inducing protective immune responses (Brazolot Millan et al. 1998). In direct comparison, bacterial CpG DNA may even surpass CFA as an adjuvant (Chu et al. 1997; Sun et al. 1998a). Furthermore, since mucous membranes represent an important transmission site for many pathogens and since, to date, no adjuvant has been licensed for clinical application in mucosal vaccinations, its use as a mucosal adjuvant might be promising (McCluskie and Davis 1998; Moldoveanu et al. 1998).

Overall, recent data indicate that bacterial DNA and immunostimulatory ODNs convey an adjuvant effect, in that they activate professional APCs to express co-stimulatory molecules and secrete cytokines pivotal for the initiation of productive Th1 responses. The use of immunostimulatory DNA sequences in vaccination protocols could be an interesting therapeutic application, because vaccines are the most commonly used immunotherapeutics (Klinman 1998).

## 5.2 CpG DNA as a Therapeutic Agent in Th2-based Pathology

APCs, as "sentinels at the gate", can strongly influence the differentiation of adaptive immune cells towards either cell-mediated (Th1) or antibody-mediated (Th2) forms of immunity. Intracellular pathogens, such as viruses, certain bacteria and protozoans, tend to induce cellular immunity, including CTL activity and the production of pro-inflammatory cytokines, such as IFN-γ and TNF-α. Extracellular pathogens, e.g. helminths, induce primarily humoral immune responses characterised by vigorous IgG1 and IgE responses and IL-4/IL-5/IL-10-dominated cytokine patterns (Constant and Bottomly 1997). In order to mount cellular immune responses, IL-12 needs to be present at the time of antigen recognition, driving NK cell- and Th1 cell-derived IFN-γ production and thereby inhibiting Th2 development. In addition to the appropriate "cytokine environment", co-stimulatory signals delivered by innate immunity contribute to the decision of whether Th1- or Th2-biased adaptive responses result (Seder and Paul 1994; Constant and Bottomly 1997). Th2-based pathology can be caused by two different mechanisms: (1) inappropriate Th2 responses e.g. to environmental antigens or vaccines or (2) insufficient (humoral) immune responses against intracellular microorganisms.

Apart from vaccination, for which a Th1-directed situation mediating cellular immunity is often preferable, there are other clinical situations where a Th1 shift is of therapeutic use. Allergies and asthma are, for example, characterised by a Th2-dominated status (Finkelman 1995). CpG ODNs may reduce the risk of anaphylactic shock, an extreme Th2 response and potentially fatal side effect of vaccination by its Th1 skewing effects (reduced IgE synthesis, inhibition of Th2 cells; Carson and Raz 1997). Co-administration of CpG ODNs can turn a Th2-dominated immune response caused by IFA towards hen-egg lysozyme into a Th1-type response, as defined by cytokine profile and IgG isotypes (Chu et al. 1997). In a murine asthma model, systemic administration of CpG ODNs appears to be beneficial even in animals already sensitised to the antigen (schistosome eggs), which normally causes airway eosinophilia, Th2 cytokine production, IgE production and bronchial hyper-reactivity (Kline et al. 1998). In another mouse model of Th2-mediated pathology, CpG sequences could confer immediate and sustained protection against allergen-induced airway hyper-responsiveness (Broide et al. 1998). While the acute effects were caused by innate immune cell-derived Th1-promoting cytokines, such as IL-12 and IFNs, the lasting effects over a time period of 1 week were explained by the inhibition of the bone-marrow production of eosinophils (Broide et al. 1998). These studies suggest that not only the initiation but also ongoing Th2 responses can be prevented and cured by the therapeutic use of CpG ODNs. Moreover, this treatment can confer lasting protection against inappropriate generation of Th2 responses.

Murine leishmaniasis is a well-characterised model of Th1/2-dependent pathology, rendering Th1-biased mouse strains resistant and Th2-biased strains sensitive to the intracellular pathogen (Heinzel et al. 1989; Hsieh et al. 1995; Reiner and Locksley 1995). Th2-biased BALB/c mice treated with immunostimulatory CpG sequences can be cured of normally lethal leishmania infection (Zimmermann et al. 1998). Resistance is likely to be explained by APC-derived cytokines, such as IL-12 and IL-18, and NK cell-derived IFN-γ, which promotes Th1-type cellular immunity (Heinzel et al. 1993; Sypek et al. 1993).

It might be more than a coincidence that, in developed countries, the decrease of Th1-promoting infectious diseases, such as tuberculosis and measles, correlates with a significant increase in Th2-driven atopic disorders (Shaheen et al. 1996; Shirakawa et al. 1997). In addition to natural exposure to infectious diseases, other factors, such as dietary changes, "indoor life-style", increasing hygiene standards and Th2-promoting modern vaccinations, influence the fine-tuning of the T-cell repertoire and Th1/Th2 balance (Romagnani 1994). Because "Westernisation" continually deprives the human immune system of its evolutionary input, it may be necessary to develop vaccination strategies that not only protect against specific pathogens but also maintain the correct cytokine balance and education of the adaptive immune system (Rook and Stanford 1998).

## 6 Conclusion

APCs, bridging the innate and adaptive immune systems, recognise prokaryotic DNA as a danger signal. This recognition of bacterial DNA mediates activation of macrophages and DCs, leading to secretion of pro-inflammatory cytokines, upregulation of co-stimulatory molecules and production of effector cytokines influencing the adaptive immune system's responses. Systemic toxicity of CpG DNA, triggering a cytokine-release syndrome culminating in toxic shock and local pathology caused by inflammatory responses, may represent examples of "harmful" consequences for APC activation. These same responses, however, contribute to CpG DNA's function as a therapeutic adjuvant, activating APCs and thereby enabling protective Th1-biased immune responses.

Bacterial CpG DNA is a good candidate for a conserved pattern-recognition molecule (Krieg 1996; Pisetsky 1996; Heeg et al. 1998; Wagner 1999). The mechanism of pattern recognition via non-self DNA also appears to extend to other invertebrate DNAs that have been reported to activate immune cells (Sun et al. 1996, 1998b; Brown et al. 1998). For example, DNA from *Drosophila* causes upregulation of co-stimulatory molecules in murine B cells and activation of APCs leading to cytokine release (Sun et al. 1996, 1998b). The immunostimulatory potency of viral double-stranded RNA was recognised in the early 60s. At that time, the first suspicions about immunostimulatory prokaryotic DNA arose but was questioned later (Vilcek et al. 1969). It is also possible that the mechanism of pathogen-DNA pattern-recognition originally evolved as a way to recognise viruses lacking LPS-like surface structures, thus activating infected cells via "foreign" genetic material. The postulated CpG-PRR might have evolved to signal danger to APCs during infection with gram-positive bacteria, parasites or certain viruses.

To substantiate this proposal, there is a need to evaluate the physiological relevance of prokaryotic DNA (released either extra- or intracellularly during infection) in causing activation of APCs. Independent of whether sensing of foreign DNA represents an evolutionarily developed primordial recognition system signaling danger to cells of the innate immune system, its use as a natural adjuvant in promoting Th1 responses is promising.

## References


Abbas AK, Murphy KM, Sher A (1996) Functional diversity of helper T lymphocytes. Nature 383: 787–793

Akbari O, Panjwani N, Garcia S, Tascon R, Lowrie D, Stockinger B (1999) DNA Vaccination: Transfection and Activation of Dendritic Cells as Key Events for Immunity. J Exp Med 189:169–178

Ballas ZK, Rasmussen WL, Krieg AM (1996) Induction of NK activity in murine and human cells by CpG motifs in oligodeoxynucleotides and bacterial DNA. J Immunol 157:1840–1845

Banchereau J, Steinman RM (1998) Dendritic cells and the control of immunity. Nature 392:245–252

Bauer M, Heeg K, Wagner H, Lipford GB (1999) DNA activates human immune cells through a CpG sequence dependent manner. Immunology 97:699–705

Bendigs S, Salzer U, Lipford GB, Wagner H, Heeg K (1999) CpG-oligodeoxynucleotides costimulate primary T cells in the absence of antigen presenting cells. Eur J Immunol 29:1209–1218

Bennett RM, Gabor GT, Merritt MM (1985) DNA binding to human leukocytes. Evidence for a receptor-mediated association, internalization, and degradation of DNA. J Clin Invest 76:2182–2190

Bird A (1992) The essentials of DNA methylation. Cell 70:5–8

Blank C, Luz A, Bendigs S, Erdmann A, Wagner H, Heeg K (1997) Superantigen and endotoxin synergize in the induction of lethal shock. Eur J Immunol 27:825–833

Branda RF, Moore AL, Hong R, McCormack JJ, Zon G, Cunningham-Rundles C (1996a) B-cell proliferation and differentiation in common variable immunodeficiency patients produced by an antisense oligomer to the rev gene of HIV-1. Clin Immunol Immunopathol 79:115–121

Branda RF, Moore AL, Lafayette AR, Mathews L, Hong R, Zon G, Brown T, McCormack JJ (1996b) Amplification of antibody production by phosphorothioate oligodeoxynucleotides. J Lab Clin Med 128:329–338

Brandtzaeg P (1996) Significance and pathogenesis of septic shock. In Pathology of septic shock. E.T. Rietschel and H. Wagner, eds (Berlin: Springer), pp. 15–37

Brazolot Millan CL, Weeratna R, Krieg AM, Siegrist CA, Davis HL (1998) CpG DNA can induce strong Th1 humoral and cell-mediated immune responses against hepatitis B surface antigen in young mice. Proc Natl Acad Sci USA 95:15553–15558

Broide D, Schwarze J, Tighe H, Gifford T, Nguyen MD, Malek S, Van Uden J, Martin-Orozco E, Gelfand EW, Raz E (1998) Immunostimulatory DNA sequences inhibit IL-5, eosinophilic inflammation, and airway hyperresponsiveness in mice. J Immunol 161:7054–7062

Brown WC, Estes DM, Chantler SE, Kegerreis KA, Suarez CE (1998) DNA and a CpG oligonucleotide derived from Babesia bovis are mitogenic for bovine B cells. Infect Immun 66:5423–5432

Carson DA, Raz E (1997) Oligonucleotide adjuvants for T helper 1 (Th1)-specific vaccination. J Exp Med 186:1621–1622

Chace JH, Hooker NA, Mildenstein KL, Krieg AM, Cowdery JS (1997) Bacterial DNA-induced NK cell IFN-γ production is dependent on macrophage secretion of IL-12. Clin Immunol Immunopathol 84:185–193

Chu RS, Targoni OS, Krieg AM, Lehmann PV, Harding CV (1997) CpG oligodeoxynucleotides act as adjuvants that switch on T helper (Th1) immunity. J Exp Med 186:1623–1631

Cleveland MG, Gorham JD, Murphy TL, Tuomanen E, Murphy KM (1996) Lipoteichoic acid preparations of gram-positive bacteria induce interleukin-12 through a CD14-dependent pathway. Infect Immun 64:1906–1912

Constant SL, Bottomly K (1997) Induction of Th1 and Th2 $CD4^+$ T cell responses: the alternative approaches. Annu. Rev. Immunol 15:297–322

Cowdery JS, Chace JH, Yi AK, Krieg AM (1996) Bacterial DNA induces NK cells to produce IFN-γ In vivo and increases the toxicity of lipopolysaccharides. J Immunol 4570–4575

Davis HL, Weeranta R, Waldschmidt TJ, Tygrett L, Schorr J, Krieg AM (1998) CpG DNA is a potent enhancer of specific immunity in mice immunized with recombinant hepatitis B surface antigen. J Immunol 160:870–876

De Kimpe SJ, Kengatharan M, Thiemermann C, Vane JR (1995) The cell wall components peptidoglycan and lipoteichoic acid from *Staphylococcus aureus* act in synergy to cause shock and multiple organ failure. Proc Natl Acad Sci USA 92:10359–10363

De Smedt T, Pajak B, Muraille E, Lespagnard L, Heinen E, De Baetselier P, Urbain J, Leo O, Moser M (1996) Regulation of dendritic cell numbers and maturation by lipopolysaccharide In vivo. J Exp Med 184:1413–1424

Doerfler W (1991) Patterns of DNA methylation – evolutionary vestiges of foreign DNA inactivation as a host defense mechanism. Biol Chem Hoppe Seyler 372:557–564

Ehl S, Hombach J, Aichele P, Rulicke T, Odermatt B, Hengartner H, Zinkernagel R, Pircher H (1998) Viral and bacterial infections interfere with peripheral tolerance induction and activate CD8 + T cells to cause immunopathology. J Exp Med 187:763–774

Espevik T, Otterlei M, Skjak-Braek G, Ryan L, Wright SD, Sundan A (1993) The involvement of CD14 in stimulation of cytokine production by uronic acid polymers. Eur J Immunol 23:255–261

Fearon DT (1997) Seeking wisdom in innate immunity. Nature 388:323–324

Fearon DT, Locksley RM (1996) The instructive role of innate immunity in the acquired immune response. Science 272:50–53

Finkelman FD (1995) Relationships among antigen presentation, cytokines, immune deviation, and autoimmune disease. J Exp Med 182:279–282

Finkelman FD, Holmes J, Katona IM, Urban JF, Beckmann MP, Park LS, Schooley KA, Coffman RL, Mosmann TR, Paul WE (1990) Lymphokine control of In vivo immunoglobulin isotype selection. Annu Rev Immunol 8:303–333

Forsthuber T, Yip HC, Lehmann PV (1996) Induction of Th1 and Th2 immunity in neonatal mice. Science 271:1728–1730

Galanos C, Freudenberg MA (1991) Tumor necrosis factor α mediates lethal activity of killed gram-negative and gram-positive bacteria in D-galactosamine-treated mice. Infect Immun 59:2110–2115

Galvin F, Freeman GJ, Razi-Wolf Z, Hall W Jr, Benacerraf B, Nadler L, Reiser H (1992) Murine B7 antigen provides a sufficient costimulatory signal for antigen-specific and MHC-restricted T cell activation. J Immunol 149:3802–3808

Glauser MP, Zanetti G, Baumgartner J-D, Cohen J (1991) Septic shock:pathogenesis. The Lancet 338:732–735

Hack CE, Aarden LA, Thijs LG (1997) Role of cytokines in sepsis. Adv Immunol 66:101–95, 101–195

Halpern MD, Kurlander RJ, Pisetsky DS (1996) Bacterial DNA induces murine interferon-γ production by stimulation of interleukin-12 and tumor necrosis factor-α. Cell Immunol 167:72–78

Häcker H, Mischak H, Miethke T, Liptay S, Schmid R, Sparwasser T, Heeg K, Lipford GB, Wagner H (1998) CpG-DNA specific activation of antigen presenting cells requires stress kinase activity and is preceded by non-specific endocytosis and endosomal maturation. EMBO J 17:6230–6240

Heeg K, Sparwasser T, Lipford GB, Häcker H, Zimmermann S, Wagner H (1998) Bacterial DNA: evolutionary conserved ligands signaling infectious danger to immune cells. Eur J Clin Microbiol Infect Dis 17:464–469

Heinzel FP, Sadick MD, Holaday BJ, Coffman RL, Locksley RM (1989) Reciprocal expression of interferon-γ or interleukin 4 during the resolution or progression of murine leishmaniasis. Evidence for expansion of distinct helper T-cell subsets. J Exp Med 169:59–72

Heinzel FP, Schoenhaut DS, Rerko RM, Rosser LE, Gately MK (1993) Recombinant interleukin 12 cures mice infected with *Leishmania major*. J Exp Med 177:1505–1509

Helene C, Toulme JJ (1990) Specific regulation of gene expression by antisense, sense and antigene nucleic acids. Biochim Biophys Acta 1049:99–125

Hergersberg M (1991) Biological aspects of cytosine methylation in eukaryotic cells. Experientia 47: 1171–1185

Hsieh CS, Macatonia SE, O'Garra A, Murphy KM (1995) T cell genetic background determines default T-helper phenotype development in vitro. J Exp Med 181:713–721

Jakob T, Walker PS, Krieg AM, Udey MC, Vogel JC (1998) Activation of cutaneous dendritic cells by CpG-containing oligodeoxynucleotides: a role for dendritic cells in the augmentation of Th1 responses by immunostimulatory DNA. J Immunol 161:3042–3049

Janeway CA Jr (1989) Approaching the assymptote? Evolution and revolution in immunology. Cold Spring Harb Symp Quant Biol 54:1–13

Janeway CA Jr (1992) The immune system evolved to discriminate infectious nonself from noninfectious self. Immunol Today 13:11–16

Jenkins MK, Johnson JG (1993) Molecules involved in T-cell costimulation. Curr Opin Immunol 5: 361–367

Jenkins MK, Taylor PS, Norton SD, Urdahl KB (1991) CD28 delivers a costimulatory signal involved in antigen- specific IL-2 production by human T cells. J Immunol 147:2461–2466

Ke Y, Li Y, Kapp JA (1995) Ovalbumin injected with complete Freund's adjuvant stimulates cytolytic responses. Eur J Immunol 25:549–553

Kline JN, Waldschmidt TJ, Businga TR, Lemish JE, Weinstock JV, Thorne PS, Krieg AM (1998) Modulation of airway inflammation by CpG oligodeoxynucleotides in a murine model of asthma. J Immunol 160:2555–2559

Klinman DM (1998) Therapeutic applications of CpG-containing oligodeoxynucleotides. Antisense. Nucleic Acid Drug Dev. 8:181–184

Klinman DM, Yamshchikov G, Ishigatsubo Y (1997) Contribution of CpG motifs to the immunogenicity of DNA vaccines. J Immunol 158:3635–3639

Klinman DM, Yi AK, Beaucage SL, Conover J, Krieg AM (1996) CpG motifs present in bacteria DNA rapidly induce lymphocytes to secrete interleukin 6, interleukin 12, and interferon γ. Proc Natl Acad Sci USA 93:2879–2883

Krieg AM (1996) Lymphocyte activation by CpG dinucleotide motifs in prokaryotic DNA. Trends Microbiol 4:73–76

Krieg AM, Yi AK, Matson S, Waldschmidt TJ, Bishop GA, Teasdale R, Koretzky GA, Klinman DM (1995) CpG motifs in bacterial DNA trigger direct B-cell activation. Nature 374:546–549

Kusunoki T, Hailman E, Juan TS, Lichenstein HS, Wright SD (1995) Molecules from *Staphylococcus aureus* that bind CD14 and stimulate innate immune responses. J Exp Med 182:1673–1682

Liang H, Nishioka Y, Reich CF, Pisetsky DS, Lipsky PE (1996) Activation of human B cells by phosphorothioate oligodeoxynucleotides. J Clin Invest 98:1119–1129

Lipford GB, Bauer M, Blank C, Reiter R, Wagner H, Heeg K (1997a) CpG-containing synthetic oligonucleotides promote B and cytotoxic T cell responses to protein antigen: a new class of vaccine adjuvants. Eur J Immunol 27:2340–2344

Lipford GB, Sparwasser T, Bauer M, Zimmermann S, Koch E-S, Heeg K, Wagner H (1997b) Immunostimulatory DNA: Sequence dependent production of potentially harmful or useful cytokines. Eur J Immunol 27:3420–3426

Marrack P, Kappler J (1990) The staphylococcal enterotoxins and their relatives. Science 248:705–711

Matzinger P (1994) Tolerance, danger, and the extended family. Annu Rev Immunol 12:991–1045

McCluskie MJ, Davis HL (1998) CpG DNA is a potent enhancer of systemic and mucosal immune responses against hepatitis B surface antigen with intranasal administration to mice. J Immunol 161:4463–4466

McIntyre KW, Lombard-Gillooly K, Perez JR, Karsch C, Sarmiento UM, Larigan JD, Landreth KT, Narayanan R (1993) A sense phosphorothioate oligonucleotide directed to the initiation codon of transcription factor NFκB p65 causes sequence-specific immune stimulation. Antisense Res Dev 3:309–322

Messina JP, Gilkeson GS, Pisetsky DS (1991) Stimulation of in vitro murine lymphocyte proliferation by bacterial DNA. J Immunol 147:1759–1764

Messina JP, Gilkeson GS, Pisetsky DS (1993) The influence of DNA structure on the in vitro stimulation of murine lymphocytes by natural and synthetic polynucleotide antigens. Cell Immunol 147:148–157

Miethke T, Wahl C, Heeg K, Echtenacher B, Krammer PH, Wagner H (1992) T cell-mediated lethal shock triggered in mice by the superantigen staphylococcal enterotoxin B: critical role of tumor necrosis factor. J Exp Med 175:91–98

Moldoveanu Z, Love-Homan L, Huang WQ, Krieg AM (1998) CpG DNA, a novel immune enhancer for systemic and mucosal immunization with influenza virus. Vaccine 16:1216–1224

Mosmann TR, Coffman RL (1989) Th1 and Th2 cells: different patterns of lymphokine secretion lead to different functional properties. Annu Rev Immunol 7:145–173

Pisetsky DS (1996) Immune activation by bacterial DNA: A new genetic code. Immunity 5:303–310

Pisetsky DS, Reich C (1993) Stimulation of in vitro proliferation of murine lymphocytes by synthetic oligodeoxynucleotides. Mol Biol Rep 18:217–221

Pisetsky DS, Reich CF (1994) Stimulation of murine lymphocyte proliferation by a phosphorothioate oligonucleotide with antisense activity for herpes simplex virus. Life Sci 54:101–107

Pisetsky DS, Reich CF 3 (1998) The influence of base sequence on the immunological properties of defined oligonucleotides. Immunopharmacology 40:199–208

Pugin J, Heumann D, Tomasz A, Kravchenko V, Akamatsu Y, Nishijimi M, Glauser M-P, Tobias PS, Ulevitch RJ (1994) CD14 is a pattern recognition receptor. Immunity 1:509–516

Raz E, Tighe H, Sato Y, Corr M, Dudler JA, Roman M, Swain SL, Spiegelberg HL, Carson DA (1996) Preferential induction of a $Th_1$ immune response and inhibition of specific IgE antibody formation by plasmid DNA immunization. Proc Natl Acad Sci USA 93:5141–5145

Reiner SL, Locksley RM (1995) The regulation of immunity to *Leishmania major*. Annu Rev Immunol 13:151–177

Reis e Sousa C, Hieny S, Scharton-Kersten T, Jankovic D, Charest H, Germain RN, Sher A (1997) In vivo microbial stimulation induces rapid CD40 ligand-independent production of interleukin 12 by dendritic cells and their redistribution to T cell areas. J Exp Med 186:1919–1829

Ridge JP, DiRosa F, Matzinger P (1998) A conditioned dendritic cell can be a temporal bridge between a $CD4^+$ T-helper and a T-killer cell. Nature 393:474–477

Rietschel ET, Brade H, Holst O, Brade L, Müller-Loennies S, Mamat U, Zähringer U, Beckmann F, Seydel U, Brandenburg K, Ulmer AJ, Mattern T, Heine H, Schletter J, Loppnow H, Schönbeck U, Flad HD, Hauschildt S, Schade UF, Di Padova F, Kusumoto S, Schumann RR (1996a) Bacterial endotoxin: chemical constitution, biological recognition, host response, and immunological detoxification. In: E.T. Rietschel and H. Wagner, eds. Pathology of septic shock (Springer Berlin Heidelberg New York), pp. 39–82

Rietschel ET, Wagner H eds (1996) Pathology of septic shock (Springer Berlin Heidelberg New York)

Romagnani S (1994) Regulation of the development of type 2 T-helper cells in allergy. Curr Opin Immunol 6:838–846

Roman M, Martin-Orozco E, Goodman JS, Nguyen M-D, Sato Y, Ronaghy A, Kornbluth RS, Richman DD, Carson DA, Raz E (1997) Immunostimulatory DNA sequences function as Th1-promoting adjuvants. Nat Med 3:849–854

Rook GAW, Stanford JL (1998) Give us this day our daily germs. Immunol Today 19:113–116

Sallusto F, Lanzavecchia A, Mackay CR (1998) Chemokines and chemokine receptors in T-cell priming and Th1/Th2-mediated responses. Immunol Today 19:568–574

Sato Y, Roman M, Tighe H, Lee D, Corr M, Nguyen MD, Silverman GJ, Lotz M, Carson DA, Raz E (1996) Immunostimulatory DNA sequences necessary for effective intradermal gene immunization . Science 273:352–354

Schwartz DA, Quinn TJ, Thorne PS, Sayeed S, Yi AK, Krieg AM (1997) CpG motifs in bacterial DNA cause inflammation in the lower respiratory tract. J Clin Invest 100:68–73

Seder RA, Paul WE (1994) Acquisition of lymphokine-producing phenotype by $CD4^+$ T cells. Annu Rev Immunol 12:635–73, 635–673

Shaheen SO, Aaby P, Hall AJ, Barker DJ, Heyes CB, Shiell AW, Goudiaby A (1996) Measles and atopy in Guinea-Bissau. Lancet 347:1792–1796

Shirakawa T, Enomoto T, Shimazu S, Hopkin JM (1997) The inverse association between tuberculin responses and atopic disorder. Science 275:77–79

Snapper CM, Paul WE (1987) Interferon-γ and B cell stimulatory factor-1 reciprocally regulate Ig isotype production. Science 236:944–947

Sparwasser T, Koch E-S, Vabulas RM, Lipford GB, Heeg K, Ellwart JW, Wagner H (1998) Bacterial DNA and immunostimulatory CpG oligonucleotides trigger maturation and activation of murine dendritic cells. Eur J Immunol 28:2045–2054

Sparwasser T, Miethke T, Lipford G, Borschert K, Häcker H, Heeg K, Wagner H (1997a) Bacterial DNA causes septic shock. Nature 386:336–337

Sparwasser T, Miethke T, Lipford G, Erdmann A, Haecker H, Heeg K, Wagner H (1997b) Macrophages sense pathogens via DNA motifs: induction of tumor necrosis factor-α-mediated shock. Eur J Immunol 27:1671–1679

Stacey KJ, Sweet MJ, Hume DA (1996) Macrophages ingest and are activated by bacterial DNA. J Immunol 157:2116–2122

Steinman RM (1991) The dendritic cell system and its role in immunogenicity. Annu Rev Immunol 9: 271–296

Sun S, Cai Z, Langlade Demoyen P, Kosaka H, Brunmark A, Jackson MR, Peterson PA, Sprent J (1996) Dual function of Drosophila cells as APCs for naive $CD8^+$ T cells: implications for tumor immunotherapy. Immunity 4:555–564

Sun S, Kishimoto H, Sprent J (1998a) DNA as an adjuvant: capacity of insect DNA and synthetic oligonucleotides to augment T cell responses to specific antigen. J Exp Med 187:1145–1150

Sun S, Zhang X, Tough DF, Sprent J (1998b) Type I Interferon-mediated Stimulation of T Cells by CpG DNA. J Exp Med 188:2335–2342

Sypek JP, Chung CL, Mayor SEH, Subramanyam JM, Goldman SJ, Sieburth DS, Wolf SF, Schaub RG (1993) Resolution of cutaneous leishmaniasis: interleukin 12 initiates a protective T helper type 1 immune response. J Exp Med 177:1797–1802

Szabolcs P, Avigan D, Gezelter S, Ciocon DH, Moore MA, Steinman RM, Young JW (1996) Dendritic cells and macrophages can mature independently from a human bone marrow-derived, post-colony-forming unit intermediate. Blood 87:4520–4530

Tanaka T, Chu CC, Paul WE (1992) An antisense oligonucleotide complementary to a sequence in Ic2b increases c2b germline transcripts, stimulates B cell DNA synthesis, and inhibits immunoglobulin secretion. J Exp Med 175:597–607

Tokunaga T, Yamamoto H, Shimada S, Abe H, Fukada T, Fujisawa Y, Furutani Y, Yano O, Kataoka T, Sudo T, Makiguchi N, Suganuma T (1984) Antitumor activity of deoxyribonucleic acid fraction from *Mycobacterium bovis* BCG. I. Isolation, physicochemical characterization, and antitumor activity. J Natl Cancer Inst 72:955–962

Tokunaga T, Yano O, Kuramoto E, Kimura Y, Yamamoto T, Kataoka T, Yamamoto S (1992) Synthetic oligonucleotides with particular base sequences from the cDNA encoding proteins of Mycobacterium bovis BCG induce interferons and activate natural killer cells. Microbiol Immunol 36:55–66

Trinchieri G (1995) Interleukin-12: a proinflammatory cytokine with immunoregulatory functions that bridge innate resistance and antigen-specific adaptive immunity. Annu Rev Immunol 13:251–276

Ulevitch RJ, Tobias, P.S (1995) Receptor-dependent mechanisms of cell stimulation by bacterial endotoxin. Annu Rev Immunol 13:437–457

Vilcek J, Ng MN, Friedmann-Kien AE, Krawciw T (1969) Induction of interferon synthesis by synthetic double-stranded polynucleotides. J Virol 2:648–650

Wagner H (1999) Bacterial CpG-DNA activates immune cells to signal 'infectious danger'. Adv Immunol 73:329–368

Wakabayashi G, Gelfand JA, Jung WK, Connolly RJ, Burke JF, Dinarello CA (1991) Staphylococcus epidermidis induces complement activation, tumor necrosis factor and interleukin-1, a shock-like state and tissue injury in rabbits without endotoxemia. Comparison to *Escherichia coli*. J Clin Invest 87:1925–1935

Weeratna R, Brazolot Millan CL, Krieg AM, Davis HL (1998) Reduction of antigen expression from DNA vaccines by coadministered oligodeoxynucleotides. Antisense Nucleic Acid Drug Dev 8: 351–356

Weiner GJ, Liu HM, Wooldridge JE, Dahle CE, Krieg AM (1997) Immunostimulatory oligodeoxynucleotides containing the CpG motif are effective as immune adjuvants in tumor antigen immunization. Proc Natl Acad Sci USA 94:10833–10837

Wooldridge JE, Ballas Z, Krieg AM, Weiner GJ (1997) Immunostimulatory oligodeoxynucleotides containing CpG motifs enhance the efficacy of monoclonal antibody therapy of lymphoma. Blood 89:2994–2998

Yamamoto S, Yamamoto T, Kataoka T, Kuramoto E, Yano O, Tokunaga T (1992) Unique palindromic sequences in synthetic oligonucleotides are required to induce IFN and augment IFN-mediated natural killer activity. J Immunol 148:4072–4076

Yi AK, Krieg AM (1998) Rapid induction of mitogen-activated protein kinases by immune stimulatory CpG DNA. J Immunol 161:4493–4497

Zhang Y, Broser M, Rom WN (1994) Activation of the interleukin 6 gene by *Mycobacterium tuberculosis* or lipopolysachharide is mediated by nuclear factors NF-IL6 and NF-kappa B. Proc Natl Acad Sci USA 91:2225–2229

Zimmermann S, Egeter O, Hausmann S, Lipford GB, Röcken M, Wagner H, Heeg K (1998) CpG oligonucleotides trigger curative Th1 responses in lethal murine leishmaniasis. J Immunol 160: 3627–3630

# Signal Transduction Pathways Activated by CpG-DNA

H. Häcker



## 1 Introduction

In the last few years, it has been shown that bacterial DNA stimulates and is recognized by cells of the immune system (Wagner 1999). Cell types that have been shown to respond directly to bacterial DNA include cells of the innate immune system, such as macrophages and dendritic cells, as well as B cells. Notably, the pattern of responsive cell types strongly resembles the pattern of cells stimulated by other so-called pattern-recognition factors, such as lipopolysaccharides (LPS). In addition, the spectrum of effects induced by these biochemically different agents are remarkably similar to each other. It includes induction of a variety of soluble factors, upregulation of membrane proteins and regulation of cell proliferation and survival. Due to the profound alterations in the cells' behaviour upon stimulation

Institut für Med. Mikrobiologie, Immunologie and Hygiene, Technische Universität München, Trogerstr. 9, D-81675 Munich, Germany
E-mail: hans.haecker@lrz.tum.de

by CpG-DNA, it has been compelling to postulate changes in the transcriptional and post-transcriptional activities. How can the information contained in unmethylated CpG-motifs in bacterial DNA be translated into gene expression? One possibility is that CpG-DNA, like other common ligands, engages a specific cellular receptor that transduces the signal by signal transduction pathways from the outside to the nucleus. Recently, it was shown that classic signal transduction pathways, such as the stress-kinase pathway and the nuclear factor κB (NF-κB) activation pathway, are switched on in response to CpG-DNA. The current knowledge about these pathways in the context of their activation by CpG-DNA and the upstream requirements for signal initiation are discussed in this chapter.

## 2 The NF-κB Activation Pathway

The transcription factor NF-κB is expressed in virtually all cell types of eukaryotes. In the immune system, it is involved in the inducible expression of proteins, including soluble factors like cytokines and chemokines, membrane proteins involved in co-stimulation and adhesion (Grilli et al. 1993; Kopp and Ghosh 1995) and proteins regulating cell survival (Karin 1998). In general, activation of NF-κB leads to an enhancement of immune responses. NF-κB is composed of dimers of the Rel family of DNA-binding proteins, RelA/p65, p50, p52, c-Rel, RelB and Bcl-3 (Ghosh et al. 1998). In most cell types, NF-κB activity is tightly controlled by inhibitory proteins, the IκBs. Like the NF-κB family, IκB is a family of molecules; it includes IκBα, IκBβ, IκBγ and IκBε (Ghosh et al. 1998). Bcl-3 structurally resembles other IκB family members. However, because Bcl-3 primarily seems to stimulate NF-κB-dependent gene transcription (Bours et al. 1993), it is here included in the NF-κB family. By interaction with the pre-formed NF-κB complexes, IκBs inhibit both the translocation of NF-κB to the nucleus and DNA binding. The best-characterized members of the IκB family are IκBα and IκBβ. Cell stimulation, for example by pro-inflammatory cytokines, leads to rapid phosphorylation of two conserved serine residues within the regulatory *N*-terminal domain of IκB, followed by rapid ubiquitination and degradation of IκBα/β by the 26S proteasome. The released NF-κB dimer translocates to the cell nucleus, free to bind to NF-κB enhancers and activate gene transcription. Recently, two proteins – IκB kinase α (IKKα)/conserved helix–loop–helix ubiquitous kinase (Chuk) and IκB kinase β (IKKβ) – have been shown to contribute to a higher molecular weight complex of proteins with IκB-kinase activity (DiDonato et al. 1997; Regnier et al. 1997; Zandi et al. 1997). This IKK complex specifically phophorylates IκBα and IκBβ at the serine residues critical for ubiquitination and degradation (Karin 1998). For IKKα/Chuk, direct interaction with and activation by NIK, the NF-κB inducing kinase, has been described (Regnier et al. 1997). Tumor necrosis factor (TNF)- and interleukin 1 (IL-1)-induced activation of NF-κB seems to depend on this kinase activity (Malinin et al. 1997). With respect to IL-1-induced NF-κB activation, the proteins TNF-receptor-associated factor 6 (TRAF6), IL-1-receptor-associated kinase and myeloid differentiation marker 88 (MyD88) appear to connect the signalling pathway between IL-1 receptor and NF-κB activation with only minor gaps (Fig. 1) (Wesche et al. 1997).

Recently, the IL-1-receptor-related human toll-like receptor 2 has been shown to be involved in LPS-induced signalling (Kirschning et al. 1998; Yang et al. 1998). Using *trans*-dominant negative mutants of IKKα/Chuk, IKKβ, NIK, TRAF6 and MyD88, a critical role for these molecules in the LPS-induced activation of NF-κB has been demonstrated (Fig. 1) (Kirschning et al. 1998).

## 2.1 Plasmid DNA and CpG Oligodeoxynucleotides Activate NF-κB

Stacey and colleagues were the first to show that plasmid DNA (as a source of bacterial DNA) and synthetic oligodeoxynucleotides (ODNs) that contain a palindromic CpG-motif are inducers of NF-κB and NF-κB-dependent gene expression in macrophages (Stacey et al. 1996). They directly demonstrated enhanced NF-κB-mediated DNA-binding activity of plasmid-treated RAW264 macrophages. Using a human immunodeficiency virus long terminal repeat (HIV-LTR)-driven luciferase reporter, which had been shown to be strongly dependent on NF-κB activity, they showed that plasmid DNA, but not CG-methylated plasmid DNA, induced transcriptional activity of this promoter. Furthermore, ODNs containing a palindromic AACGTT sequence, but not ODNs with the CpG palindrome ACCGGT, were able to induce HIV-LTR-dependent gene transcription. While these data show the importance of nucleotides surrounding the central CG, inversion of CG to GC also abolishes the NF-κB-inducing activity of ODN in RAW264 cells, directly proving the CpG dependency of CpG ODN (our own unpublished observation). Additionally, activation of other genes known to be induced by LPS and dependent on NF-κB activity, such as TNF, IL-1β and the HIV-1 LTR (Drouet et al. 1991; Hiscott et al. 1993; Pombo et al. 1994), were activated by plasmid DNA (Stacey et al. 1996). Notably, the pattern of activated genes strongly resembles the one provoked by LPS, although, according to Stacey et al. (1996), LPS contamination of the various DNA stimuli could be ruled out.

Later on, it was directly shown that CpG ODN, dependent on a CpG motif, induces degradation of IκBα and IκBβ in WEHI-231 B cells (Yi and Krieg 1998b), thereby inducing translocation of p50/c-Rel (Yi and Krieg 1998b), p50/p65 and p50 homodimers (Sparwasser et al. 1997). Degradation of IκBα is preceded by phosphorylation of Ser32 and Ser36 (our unpublished observation). Concurrent with degradation of IκBα, rapid induction of reactive oxygen species (ROS) has been demonstrated to occur in J774 macrophages and WEHI-231 B cells (Yi et al. 1998b). Treatment of cells with the antioxidants *N*-acetyl-L-cysteine (NAC) or pyrrolidine dithiocarbamate (PDTC) suppressed TNF (Yi et al. 1998b) and IL-6 secretion (Yi et al. 1996), respectively, implying an important role for the redox state of the responding cell during stimulation. It is important to note that almost all inducers of NF-κB tested so far can be blocked by antioxidants (Baeuerle and

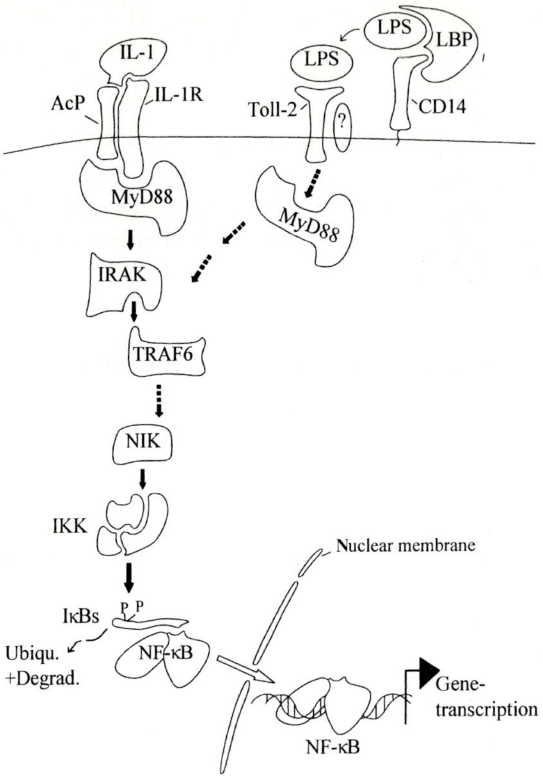


**Fig. 1.** Model of nuclear factor κB (NF-κB) activation pathways induced by interleukin 1 (IL-1) and lipopolysaccharides (LPS). *Solid arrows* indicate where direct interaction of proteins has been shown, *broken arrows* are used where direct interaction has not been demonstrated. Arguments supporting the involvement of myeloid differentiation marker 88 (*MyD88*), tumor-necrosis-factor receptor-associated factor 6 (*TRAF6*), NF-κB-inducing kinase (*NIK*) and Iκ kinase α/β in the LPS-induced NF-κB activation are exclusively based on experiments using dominant negative forms of the respective proteins. While the IL-1-receptor complex consists of two defined subunits, the situation of the LPS-receptor is more complicated and only partially elucidated. The glycosyl-phosphatidyl-inositol-linked CD14 molecule lacks a signal transduction moiety of its own and seems to be used to scavenge LPS – opsonized by LBP (LPS-binding protein) – for the actual signalling receptor. Toll-like receptor 2 (*toll-2*) has been shown to bind LPS and transduce the signal towards NF-κB (Kirschning et al. 1998; Yang et al. 1998). However, LPS-resistant mouse strains C3H/HeJ and C57BL/10ScCr both have mutations in another gene, toll-like receptor 4 (*toll-4*) (Poltorak et al. 1998). Although biochemically not demonstrated yet, involvement of *toll-4* in LPS signalling seems to be likely. Composition of the LPS receptor by different proteins – as indicated by '*?*' – is only one of many possible ways these findings might be brought in line (Ulevitch 1999). *AcP*, accessory protein; *IL-1R*, IL-1 receptor; *IRAK*, IL-1R-associated kinase; *IKK:*, Iκ kinase complex:. IKK: is a complex of about 700 kDa and consists of several proteins, including IKKα/conserved helix–loop–helix ubiquitous kinase and IKKβ. Other molecules of the complex have not been molecularly defined. All molecules in the figure are symbols only, and their shapes are not based on any biochemical data

Henkel 1994; Muroi et al. 1994; Muller et al. 1997). This could mean that induction of ROS is an important and general way of NF-κB activation. However, the critical step during signalling, which in fact depends on ROS generation, could not be unequivocally demonstrated for any of the inducers described. While there is no doubt about the principal importance of the redox states of cells, the question of the significance of its altered dynamics during cell stimulation awaits further investigation.

The demonstration of CpG-DNA-induced specific phosphorylation of IκBα at Ser32 and Ser36 strongly suggests that the IKK complex, recently found to be activated by TNF, IL-1 and probably LPS (Regnier et al. 1997; DiDonato et al. 1997; Zandi et al. 1997; Kirschning et al. 1998), is also activated by CpG-DNA, although this has not been shown formally.

It will be interesting to see at which step CpG-DNA signalling enters the known pathways of NF-κB activation and whether this pathway can be traced back to an upstream receptor. Until now, every cell type shown to be directly responsive to CpG-DNA also activated NF-κB during stimulation (Stacey et al. 1996; Yi et al. 1998b and our own unpublished observations). Therefore, this pathway seems to be a general mechanism through which CpG-DNA transduces its signal.

## 3 Mitogen-Activated Protein Kinase Pathways

Mitogen-activated protein kinases (MAPKs) are a group of serine/threonine-specific kinases that are activated in response to a variety of extracellular stimuli (Lewis et al. 1998). MAPKs are involved in the regulation of specific transcription factors and other regulatory proteins and are themselves activated by upstream kinases called MAPK kinases or MAPK/extracellular-signal-regulated kinase (ERK) kinases (MAPKKs/MEKs) and MAPKK kinases/MEK kinases (MAPKKKs/MEKKs) (Fig. 2). In vertebrates, at least three MAPK pathways can be differentiated. These pathways are able to sense and integrate incoming signals from different sources, for example membrane receptors, and transduce them into specific gene expression. Although some crosstalk between the different kinase cascades seems to exist, generally they are organized as sequentially working modules. Partially characterized scaffold proteins forming complexes with kinases belonging to a given cascade seem to contribute to the high degree of specificity (Schaeffer et al. 1998; Whitmarsh et al. 1998). According to the last MAPK in the corresponding cascade, they are referred to as the extracellular-signal-regulated ERK/MAPK pathway, the c-Jun *N*-terminal kinase (JNK)/stress-activated protein kinase (SAPK) pathway and the p38 MAPK pathway (Fig. 2). Members of both latter groups of kinases are sometimes also commonly referred to as SAPKs.

The ERK/MAPK pathway is primarily activated by growth factors and phorbol ester. Mitogenicity of growth factors largely depends on activation of the ERK pathway. Constitutive activity of members of this pathway contributes to the oncogenic transformation of cells (Roussel 1998).

The JNK and p38 MAPKs are groups of kinases that are activated by different extracellular stimuli, including pro-inflammatory cytokines, such as TNF and IL-1,

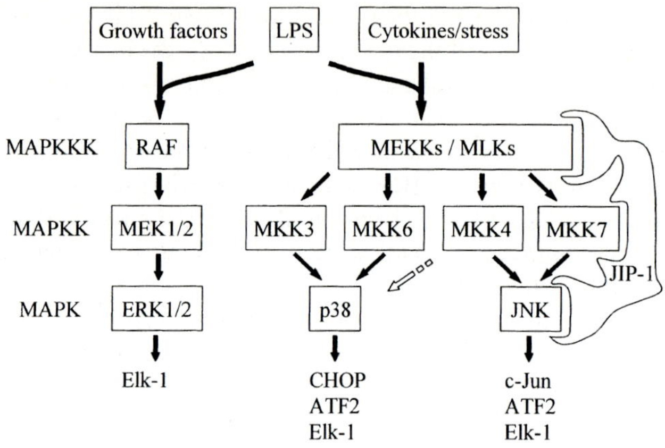


**Fig. 2.** Organisation of the mammalian mitogen-activated protein (MAP)-kinase signal-transduction pathways. The extracellular-signal-regulated kinase (ERK) pathway and the c-Jun *N*-terminal kinase (JNK)- and p38-stress-kinase pathways are illustrated schematically. Growth factors typically lead to the activation of the ERK subgroup of MAP kinases, while inflammatory cytokines and stress signals primarily activate the stress kinases JNK and p38. Lipopolysaccharides are able to activate both ERK- and stress-kinase pathways. At present, at least nine different MAP or ERK kinase kinases (MEKKs), including MEKK1–4, MAP kinase kinase kinase 5, tumor progression locus 2, transforming-growth-factor-β-activated kinase 1, apoptosis-signal-regulating kinase 1 and nuclear-factor-κB-inducing kinase, have been identified in mammals with different specificities towards MAP kinase kinases (MKKs). The MLKs (mixed-lineage kinases) are a structurally different group of enzymes. For its family members MLK-2, MLK-3 and dual leucine zipper-bearing kinase (DLK), activation of MKKs has been observed (Lewis et al. 1998). JIP-1, the JNK-interacting protein, has recently been identified as a protein interacting with multiple components of the JNK signalling pathways (Whitmarsh et al. 1998) and has been postulated to be a Ste5p-like scaffold protein, facilitating and specifying JNK signalling (Whitmarsh and Davis 1998). *MAPK*, MAP kinase; *MAPKK*, MAP kinase kinase; *MAPKKK*, MAP kinase kinase kinase

as well as LPS, ultraviolet (UV) light and stress signals (Fig. 2) (Hibi et al. 1993; Bird et al. 1994; Sluss et al. 1994; Westwick et al. 1994; Hambleton et al. 1996).

## 3.1 CpG-DNA Induces the Transcription Factor Activating Protein 1

The first evidence that MAPKs might be involved in CpG-DNA-induced signalling came from transient transfection experiments using luciferase reporter plasmids with a triplet of activating protein 1 (AP-1)-binding sites. Stimulation of transfected ANA-1 macrophages with both plasmid DNA and CpG ODN induced transcriptional activity of AP-1 in a strictly CpG-dependent manner (Hacker et al. 1998).

The transcription factor AP-1 is comprised of members of the Fos, Jun and ATF (activating factor) families, which form homodimers and heterodimers. Its activity can be regulated at different levels (Karin et al. 1997). Primarily, MAPK pathways control both the abundance of the AP-1 protein complex and its transcriptional activity. With respect to participation in AP-1 complexes, c-Fos, c-Jun and ATF-2 are especially well characterized. c-Fos, which is mainly regulated at the transcriptional level, contains two enhancer elements in its promoter region that directly confer responsiveness to MAPK pathways. These two elements are: a CRE (cyclic adenosine monophosphate-responsive element) site and a serum-responsive element (SRE). The CRE site seems to be occupied by both CREB (CRE-binding protein) and ATF proteins. The SRE element binds dimers of the serum-response factors (SRF), recruiting ternary complex factors (TCF) like Elk-1 and stress-activated protein 1. The *C*-termini of these TCFs contain clusters of phosphorylation sites that are phoshorylated and thereby activated by ERK/MAP kinases and JNK/p38 stress kinases (HIPSKIND et al. 1994; PRICE et al. 1995). c-Jun is regulated by different mechanisms. Transcriptional regulation to most stimuli seems to be controlled through a c-Jun TRE (12-O-tetradecanoyl phorbol 13-acetate-responsive element) proposed to bind c-Jun/ATF-2 heterodimers (VAN DAM et al. 1993). In addition, c-Jun activity is regulated by specific phosphorylation at two sites, Ser63 and Ser73, which enhance the transactivating potency of the AP-1 complex (SMEAL et al. 1992). This phosphorylation is achieved by JNKs (SANCHEZ et al. 1994). In contrast to c-Fos and c-Jun, ATF-2 expression is constitutive. Its activity is regulated by phosphorylation by p38 and JNK (GUPTA et al. 1995; RAINGEAUD et al. 1996). Taken together, AP-1 integrates signals from different MAPK pathways.

Gel-shift analysis in CpG-DNA-stimulated ANA-1 macrophages revealed basal AP-1-binding activity but only a slight increase during the first 4h of simulation (HACKER et al. 1998). This AP-1 complex contained c-Jun. In contrast, c-Jun phosphorylation at Ser73 and Ser63 rapidly increased within 10min and stayed constant over at least 4h. The overall amount of c-Jun did not increase significantly during that time period. As transcriptional activity of AP-1 was correlated with specific phosphorylation of c-Jun, we concluded that, during the first hours of stimulation, phosphorylation might be the principal way of AP-1 regulation by CpG-DNA. Interestingly, using WEHI-231 B cells, Yi and Krieg (YI and KRIEG 1998a) also demonstrated phosphorylation of c-Jun and an additional significant increase in AP-1-binding activity. It is possible that the different cell types explain these differences.

## 3.2 CpG-DNA Activates Stress Kinases

The kinases responsible for phosphorylation of c-Jun at Ser63 and Ser73 are the JNKs. Accordingly, JNKs and their upstream kinase, JNKK1, were found to be activated by CpG-DNA (HACKER et al. 1998). Additionally, p38, a stress kinase originally cloned as LPS-responsive kinase (HAN et al. 1994), has been found to be activated by CpG-DNA (HACKER et al. 1998; YI and KRIEG 1998a). Whether JNKK1 is also the upstream kinase of p38 in the cell types investigated so far is unknown. In general, p38 is efficiently activated by MKK3 and MKK6 (LEWIS et al. 1998). However, MKK4 has also been found to activate p38 in vitro and in vivo (SANCHEZ et al. 1994; DERIJARD et al. 1995; LIN et al. 1995) and could be involved in p38 activation (GANIATSAS et al. 1998).

Activation of JNK and p38 sufficiently explains phosphorylation of ATF-2 at Thr69/71, the critical site for transcriptional transactivation (GUPTA et al. 1995;

Livingstone et al. 1995). Interestingly, a specific inhibitor of p38, SB203580, blocked the DNA-binding activity of AP-1 in CpG-DNA-stimulated WEHI-231 cells (Yi and Krieg 1998a). Such p38 inhibition by SB203580 could affect AP-1 induction at different stages. As detailed above, c-Fos expression is regulated by a SRE site binding TCFs, which are phosphorylated and activated by different MAPKs. c-Jun is also transcriptionally regulated, involving ATF-2 and c-Jun themselves. DNA binding of ATF-2 also depends on its phosphorylation status (Abdel-Hafiz et al. 1992). It will be interesting to see which proteins are contained within this AP-1 complex and which components are affected at which stage by SB203580.

## 3.3 Downstream Effects of Stress Kinase Activation

While a panel of inducers of stress kinases in immune cells has been defined, effector functions clearly attributable to stress kinases are rare. One report demonstrated that mRNA stability of IL-2 is enhanced by JNK activity (Chen et al. 1998). JNKK1 (MKK4)-deficient thymocytes exhibit a defect in IL-2 production in response to CD3/CD28 and phorbol 12-myristate 13-acetate (PMA)/$Ca^{2+}$ ionophore stimulation (Nishina et al. 1997). However, these results are difficult to interpret, as – depending on the differentiation status of the cells – JNKK1 deficiency provokes additional phenomena, such as impairment of MKK3/6 phosphorylation (Ganiatsas et al. 1998). This implies that secondary effects must be considered in the interpretation of immune cell activation in knock-out animals. There is ample evidence that JNKs are involved in the regulation of apoptosis, although their specific role seems highly dependent on the cell type (Ip and Davis 1998) and has not been evaluated for CpG-DNA-induced signalling.

A role for p38 with respect to the release of cytokines, such as TNF, IL-6 and IFN-γ (Prichett et al. 1995; Beyaert et al. 1996; Rincon et al. 1998), during inflammatory responses in vitro and in vivo has been mainly deduced from experiments with a class of cytokine biosynthesis inhibitors called cytokine-suppressive anti-inflammatory drugs. Originally, p38 was defined by one of these specific inhibitors, SB203580 (Cuenda et al. 1995; Lee et al. 1994). As in the case of LPS-stimulated cells, CpG-DNA-induced TNF and IL-6 release is strongly suppressed by SB203580 (Hacker et al. 1998; Yi and Krieg 1998a). Remarkably, CpG-DNA-induced IL-12 release from macrophage cell lines and primary bone-marrow-derived dendritic cells was also significantly reduced (Hacker et al. 1998). However, as for many other effector functions affected by SB203580, it will be important to define the concrete target(s) downstream of p38.

## 3.4 Activation of the ERK/MAPK Pathway by CpG-DNA

Various observations strongly suggested that the ERK/MAPK pathway would also be activated by CpG-DNA. First, CpG-DNA induces strong proliferation in B cells

(MESSINA et al. 1991; KRIEG et al. 1995), obviously by stimulation of cell-cycle progression through G1 (YI et al. 1998a). This resembles the case for known mitogens, such as growth factors, which depend on ERK and MEK activity for their capacity to induce cell-cycle entry through G1 (ROUSSEL 1998). Second, LPS, which induces a very similar set of effector functions, including B-cell proliferation, induces ERK activity in different cell types (SWEET and HUME 1996).

Surprisingly, using phospho-specific antibodies against ERK1/2, Yi and Krieg reported that in WEHI-231 B cells and J774 macrophages JNK and p38, but not ERKs, are activated by CpG-DNA. Moreover, they found ERK1 to be activated by LPS in J774 to an extent comparable to that achieved by phorbol ester (PMA) (YI and KRIEG 1998a). If this holds true in more sensitive assays, it represents the first known qualitative difference between LPS and CpG-DNA-induced signalling.

Using another macrophage cell line, RAW264.7, and primary peritoneal macrophages for CpG-DNA stimulation, we observed substantial ERK2 kinase activity in in vitro kinase assays, comparable to that caused by LPS (unpublished observations). It will be interesting to see whether there exist differences in ERK activation among distinct cell types and, even more interesting, whether activation of different MAPK pathways qualitatively separates CpG-DNA from LPS.

## 4 CpG ODN-Induced Signalling Requires Non-Specific Uptake and Endosomal Acidification

### 4.1 Endosomal Uptake of CpG ODN is CpG-Independent

CpG ODN and fluorescein-isothiocyanate-labelled CpG ODN are taken up rapidly from cells and move into an endosomal compartment (Tonkinson and STEIN 1994; HACKER et al. 1998). Based on competition analyses, it has been argued that ODN uptake is receptor mediated. Several cell-surface receptors lacking sequence specificity, including CD11b/CD18 integrins and scavenger receptors, have been shown to engage single-stranded DNA (KIMURA et al. 1994; BENIMETSKAYA et al. 1997). Our own studies showed that uptake of CpG ODN could be competed for by unrelated non-CpG ODNs. Although some ODNs were more potent than others, every non-CpG ODN was essentially able to compete for CpG ODN uptake. ODNs with poly-G stretches were especially potent competitors (G. Lipford, personal communication). These ODNs originally had been found to bind to the scavenger receptor, thus promoting efficient cellular uptake (KIMURA et al. 1994). Poly-G stretches, in the context of stimulatory CpG sequences, enhance the immunogenicity of CpG ODN; increased uptake has been proposed as probable reason for this enhancement (KIMURA et al. 1994).

Importantly, both uptake and CpG-induced signalling can be competed for by non-CpG ODN. Every effector function and signalling event investigated so far can be blocked by non-CpG ODN, implying that binding of CpG DNA to a receptor represents a first, obligate step during signal induction (Hacker et al. 1998). This raises two possibilities: (1) this receptor mediates uptake *and* signalling or (2) this receptor mediates uptake, but signalling starts from a later point. The latter possibility is supported by experiments with endosomal acidification inhibitors (see below).

The observation that non-CpG ODNs compete for CpG-ODN-induced signalling raises interesting questions. Every CpG ODN without intrinsic activity will be able to act as a competitor for active CpG ODNs. Additionally, higher-molecular-weight DNAs, like bacterial DNA and plasmid DNA, contain both activating sequences and non-activating sequences and, therefore, one attractive model would be to consider DNA as a composition of stimulatory and non-stimulatory (and hence inhibitory) sequences. According to this model, vertebrate DNA is able to inhibit cytokine release of plasmid-triggered spleen cells (Wloch et al. 1998).

Recently, another possible explanation for the inhibitory effects of certain adenoviral DNAs has been proposed: the presence of specialized sequences surrounding central CGs (Krieg et al. 1998). However, more work will be required to determine whether such sequences act as effective competitors or contain signalling qualities of their own in an inhibitory manner.

## 4.2 CpG-DNA-Induced Signalling can be Blocked by Inhibitors of Endosomal Maturation/Acidification

Endosomal maturation may be defined as pH-dependent evolution of early endosomes to lysosomal compartments (Mellman et al. 1986). Compounds like bafilomycin A and chloroquine block this evolution, the former by antagonising intravesicular hydrogen pumps (Yoshimori et al. 1991), the latter by moving into acidified vesicles and acting as a neutralising base buffer (Ohkuma and Poole 1981). Neither compound significantly altered cellular uptake but modulated endosomal accumulation of ODN, possibly by altering the trafficking of lysosomal enzymes and receptors (Chapman and Munro 1994) or by affecting efflux pathways. Endosomal vesicle compartments with varying ODN efflux rates have been described (Tonkinson and Stein 1994).

Using a variety of 'read out' systems, such as cytokine production, cytokine promoter activity, AP-1 and NF-κB activation and induction of p38, it has been consistently observed that all these CpG-DNA-driven responses were sensitive to endosomal acidification inhibitors (Hacker et al. 1998; Macfarlane and Manzel 1998; Yi and Krieg 1998a; Yi et al. 1998b). Notably, LPS-induced signalling was not affected by concentrations of chloroquine and bafilomycin A that efficiently blocked CpG-DNA-induced signalling (Hacker et al. 1998; Yi and Krieg 1998a).

As described above, CpG ODNs are taken up in endosomes in a CpG-independent but receptor-mediated way. The fact that these acidification inhibitors allow uptake of CpG ODN but are still able to block signalling strongly suggests that uptake of CpG ODN and signal induction are separable events. As such, these findings imply that both CpG-independent cellular uptake and endosomal maturation/acidification are required for signalling to be initiated.

The shift to low pH could be important for different reasons. Such pH changes could either trigger dissociation of CpG-DNA from a non-specific (cell-surface) receptor or enable binding of CpG-DNA to a specific (endosomal) receptor. Alternatively, a less well-defined molecular mechanism dependent on endosomal maturation may be required for signal transduction.

## 5 The NFκB and Stress-Kinase Pathways are Activated in Parallel

As detailed above, separable signalling pathways are switched on in response to CpG-DNA. In this respect, it is important to know whether these pathways are turned on in parallel or sequentially. This is even more important in cells that – after stimulation – secrete a variety of both defined and undefined cytokines able to act in an autocrine or paracrine fashion. We therefore compared the kinetics of the NFκB and stress-kinase activation pathways, measuring the earliest defined events, i.e. IκBα degradation and JNKK1 phosphorylation, after stimulation by CpG ODN. Figure 3 shows that, using RAW264.7 cells, both pathways seem to be switched on in a parallel and not sequential way. This suggests that both pathways posses a common starting point upstream. These data argue against the possibility that activation of one pathway is the prerequisite for activation of the other. Another interesting point is the overall velocity of signal induction. CpG-ODN-induced signalling starts after about 5min and is easily detectable between 5min and 10min after stimulation (Fig. 3). LPS seems to be slightly faster (our personal observation), and signalling induced by classic transmembrane receptors, such as the IL-1R and TNF receptor p55, is significantly faster. The delay of CpG-ODN-induced signalling, in comparison to that induced by IL-1 and TNF, could be explained by the requirement for cellular uptake and endosomal maturation.

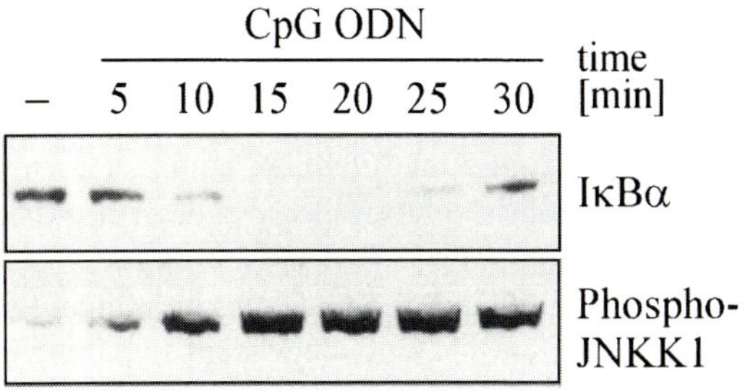


**Fig. 3.** The nuclear factor κB activation pathway and the stress-kinase pathway are induced with comparable kinetics. Shown is a Western blot analysis of RAW264.7 macrophages stimulated with CpG oligodeoxynucleotides for different time periods and analysed by antibodies to Iκ kinase Bα (*I*κBα) and the Thr223-phosphorylated form of c-Jun *N*-terminal kinase kinase 1 as indicated

## 6 Summary and Conclusion

While more and more attention has been paid to CpG-DNA with respect to its usefulness as an adjuvant, its molecular mechanism of action is less well defined. Over the last few years, at least two major signalling pathways have been shown to be utilized by CpG-DNA: the NF-κB activation pathway and the stress-kinase pathway. Direct downstream events of these pathways are induction of transcriptional activity of NF-κB and transcriptional activity of AP-1. As far as investigated, CpG-DNA uses signal transduction pathways originally described for other stimuli, such as LPS, IL-1 or TNF. Therefore, to us, the prime question is: where does CpG-DNA-induced signalling enter these known pathways? This raises questions about the existence of a CpG-DNA-sequence-specific receptor. Several points of evidence support the probability of the existence of a cellular receptor: There is a strong motif (unmethylated CpG) dependency for CpG-DNA-induced signalling. There is cell-type specificity. Dendritic cells, macrophages and B cells respond to CpG-DNA, but other cell types, such as fibroblasts and T cells, do not. In addition, classic signal-transduction pathways are rapidly switched on in a parallel manner, as is known for other receptors.

Using competing non-CpG ODNs and inhibitors of endosomal acidification, some evidence has been obtained that CpG ODNs are taken up into endosomes by a CpG-independent receptor, followed by a pH-dependent step before signalling starts. A model based on these findings is proposed in Fig. 4.

Nevertheless, other receptor-independent activities of CpG-DNA cannot yet be ruled out. Although unlikely, we should consider the possibility that CpG-DNA directly interacts with cellular nucleic acids either by direct hybridization with complementary nucleotides or by formation of DNA triplexes (Vasquez and Wilson 1998). While these possibilities have been explored by antisense technology, using a huge variety of ODNs, there is no experimental evidence that such interactions are important for the activity of CpG-DNA.

In this context, it is important to note that DNA, especially phosphothioate-stabilized ODNs with poly-G stretches, have substantial CpG-independent activities, although these activities seem not to depend on specific, antisense-like DNA–DNA interactions (Pisetsky 1996). One good example comes from experiments using ODNs on primary T cells. Co-stimulation of CD3-primed T cells with CpG ODN leads to a significant increase of IL-2 secretion and proliferation; however, these effects are CpG independent (K. Heeg, personal communication). Remarkably, these poly-G stretches seem to be inactive when transferred to double-stranded DNAs, such as plasmid DNA (Wloch et al. 1998). In contrast, to my knowledge, no immune-stimulatory effect of bacterial DNA has been described that can not be abolished by CpG-specific methylation.

Taken together, CpG-dependent and CpG-independent activities must be distinguished from one another. Among these effects, CpG-dependent signalling is better defined.

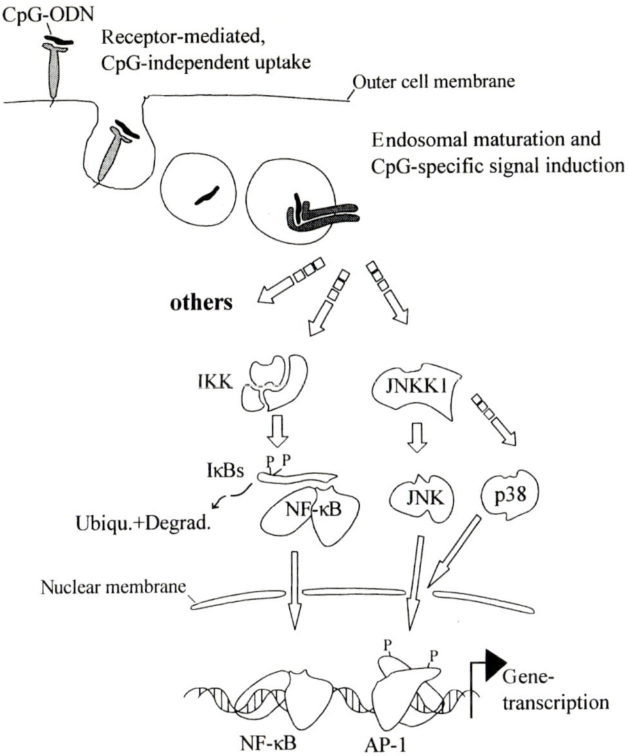


**Fig. 4.** Model of CpG-DNA-induced signal-transduction pathways. CpG oligodeoxynucleotides are taken up into endosomes by CpG-independent receptors. A pH-dependent step precedes signalling. The existence of a specific receptor in the endosomal compartment is speculative. All molecules in the figure are symbols only, and their shapes are not based on any biochemical data. *IKK:*, Iκ kinase complex:. IKK: is a complex of about 700 kDa consisting of several proteins, including IKKα/conserved helix-loop-helix ubiquitous kinase (Chuk) and Iκ kinase β. Other molecules of the complex have not been molecularly defined yet. As detailed in the text, transcription factors other than activating protein 1, as well as proteins not directly involved in transcriptional regulation, could also be activated by c-Jun *N*-terminal kinase/p38

Much effort is going into the investigation of the pharmacological applications of CpG-DNA. Once CpG-receptor-like structures are known, the question of the physiological role of CpG-DNA can be tackled.

## References

Abdel-Hafiz HA, Heasley LE, Kyriakis JM, Avruch J, Kroll DJ, Johnson GL, Hoeffler JP (1992) Activating transcription factor-2 DNA-binding activity is stimulated by phosphorylation catalyzed by p42 and p54 microtubule-associated protein kinases. Mol Endocrinol 6:2079–2089

Baeuerle PA, Henkel T (1994) Function and activation of NF-κB in the immune system. Annu Rev Immunol 12:141–179

Benimetskaya L, Loike JD, Khaled Z, Loike G, Silverstein SC, Cao L, el Khoury J, Cai TQ, Stein CA (1997) Mac-1 (CD11b/CD18) is an oligodeoxynucleotide-binding protein. Nat Med 3:414–420

Beyaert R, Cuenda A, Vanden Berghe W, Plaisance S, Lee JC, Haegeman G, Cohen P, Fiers W (1996) The p38/RK mitogen-activated protein kinase pathway regulates interleukin-6 synthesis response to tumor necrosis factor. EMBO J 15:1914–1923

Bird TA, Kyriakis JM, Tyshler L, Gayle M, Milne A, Virca GD (1994) Interleukin-1 activates p54 mitogen-activated protein (MAP) kinase/stress-activated protein kinase by a pathway that is independent of p21ras, Raf-1, and MAP kinase kinase. J Biol Chem 269:31836–31844

Bours V, Franzoso G, Azarenko V, Park S, Kanno T, Brown K, Siebenlist U (1993) The oncoprotein Bcl-3 directly transactivates through κB motifs via association with DNA-binding p50B homodimers. Cell 72:729–739

Chapman RE, Munro S (1994) Retrieval of TGN proteins from the cell surface requires endosomal acidification. EMBO J 13:2305–2312

Chen CY, Del Gatto-Konczak F, Wu Z, Karin M (1998) Stabilization of interleukin-2 mRNA by the c-Jun NH2-terminal kinase pathway. Science 280:1945–1949

Cuenda A, Rouse J, Doza YN, Meier R, Cohen P, Gallagher TF, Young PR, Lee JC (1995) SB 203580 is a specific inhibitor of a MAP-kinase homologue which is stimulated by cellular stresses and interleukin-1. FEBS Lett. 364:229–233

Derijard B, Raingeaud J, Barrett T, Wu IH, Han J, Ulevitch RJ, Davis RJ (1995) Independent human MAP-kinase signal transduction pathways defined by MEK and MKK isoforms. Science 267:682–685

DiDonato JA, Hayakawa M, Rothwarf DM, Zandi E, Karin M (1997) A cytokine-responsive IκB kinase that activates the transcription factor NF-κB. Nature 388:548–554

Drouet C, Shakhov AN, Jongeneel CV (1991) Enhancers and transcription factors controlling the inducibility of the tumor necrosis factor-α promoter in primary macrophages. J Immunol 147:1694–1700

Ganiatsas S, Kwee L, Fujiwara Y, Perkins A, Ikeda T, Labow MA, Zon LI (1998) SEK1 deficiency reveals mitogen-activated protein kinase cascade crossregulation and leads to abnormal hepatogenesis. Proc Natl Acad Sci USA 95:6881–6886

Ghosh S, May MJ, Kopp EB (1998) NF-κB and Rel proteins: evolutionarily conserved mediators of immune responses. Annu Rev Immunol 16:225–260

Grilli M, Chiu JJ, Lenardo MJ (1993) NF-κB and Rel: participants in a multiform transcriptional regulatory system. Int Rev Cytol 143:1–62

Gupta S, Campbell D, Derijard B, Davis RJ (1995) Transcription factor ATF2 regulation by the JNK signal transduction pathway. Science 267:389–393

Hacker H, Mischak H, Miethke T, Liptay S, Schmid R, Sparwasser T, Heeg K, Lipford GB, Wagner H (1998) CpG-DNA-specific activation of antigen-presenting cells requires stress kinase activity and is preceded by non-specific endocytosis and endosomal maturation. EMBO J 17:6230–6240

Hambleton J, Weinstein SL, Lem L, DeFranco AL (1996) Activation of c-Jun N-terminal kinase in bacterial lipopolysaccharide-stimulated macrophages. Proc Natl Acad Sci USA 93:2774–2778

Han J, Lee JD, Bibbs L, Ulevitch RJ (1994) A MAP kinase targeted by endotoxin and hyperosmolarity in mammalian cells. Science 265:808–811

Hibi M, Lin A, Smeal T, Minden A, Karin M (1993) Identification of an oncoprotein- and UV-responsive protein kinase that binds and potentiates the c-Jun activation domain. Genes Dev 7:2135–2148

Hipskind RA, Buscher D, Nordheim A, Baccarini M (1994) Ras/MAP kinase-dependent and -independent signaling pathways target distinct ternary complex factors. Genes Dev 8:1803–1816

Hiscott J, Marois J, Garoufalis J, D'Addario M, Roulston A, Kwan I, Pepin N, Lacoste J, Nguyen H, Bensi G (1993) Characterization of a functional NF-κ B site in the human interleukin 1β promoter: evidence for a positive autoregulatory loop. Mol Cell Biol 13:6231–6240

Ip YT, Davis RJ (1998) Signal transduction by the c-Jun N-terminal kinase (JNK) – from inflammation to development. Curr Opin Cell Biol 10:205–219

Karin M, Liu Zg, Zandi E (1997) AP-1 function and regulation. Curr Opin Cell Biol 9:240–246

Karin M (1998) The NF-κ B activation pathway: its regulation and role in inflammation and cell survival. Cancer J Sci Am 4 Suppl 1:S92–9

Kimura Y, Sonehara K, Kuramoto E, Makino T, Yamamoto S, Yamamoto T, Kataoka T, Tokunaga T (1994) Binding of oligoguanylate to scavenger receptors is required for oligonucleotides to augment NK cell activity and induce IFN. J Biochem (Tokyo) 116:991–994

Kirschning CJ, Wesche H, Merrill Ayres T, Rothe M (1998) Human toll-like receptor 2 confers responsiveness to bacterial lipopolysaccharide. J Exp Med 188:2091–2097

Kopp EB, Ghosh S (1995) NF-κ B and rel proteins in innate immunity. Adv Immunol 58:1–27

Krieg AM, Yi AK, Matson S, Waldschmidt TJ, Bishop GA, Teasdale R, Koretzky GA, Klinman DM (1995) CpG motifs in bacterial DNA trigger direct B-cell activation. Nature 374:546–549

Krieg AM, Wu T, Weeratna R, Efler SM, Love-Homan L, Yang L, Yi AK, Short D, Davis HL (1998) Sequence motifs in adenoviral DNA block immune activation by stimulatory CpG motifs. Proc Natl Acad Sci USA 95:12631–12636

Lee JC, Laydon JT, McDonnell PC, Gallagher TF, Kumar S, Green D, McNulty D, Blumenthal MJ, Heys JR, Landvatter SW (1994) A protein kinase involved in the regulation of inflammatory cytokine biosynthesis. Nature 372:739–746

Lewis TS, Shapiro PS, Ahn NG (1998) Signal transduction through MAP kinase cascades. Adv Cancer Res 74:49–139

Lin A, Minden A, Martinetto H, Claret FX, Lange-Carter C, Mercurio F, Johnson GL, Karin M (1995) Identification of a dual specificity kinase that activates the Jun kinases and p38-Mpk2. Science 268:286–290

Livingstone C, Patel G, Jones N (1995) ATF-2 contains a phosphorylation-dependent transcriptional activation domain. EMBO J 14:1785–1797

Macfarlane DE, Manzel L (1998) Antagonism of immunostimulatory CpG oligodeoxynucleotides by quinacrine, chloroquine, and structurally related compounds. J Immunol 160:1122–1131

Malinin NL, Boldin MP, Kovalenko AV, Wallach D (1997) MAP3K-related kinase involved in NF-κB induction by TNF, CD95 and IL-1. Nature 385:540–544

Mellman I, Fuchs R, Helenius A (1986) Acidification of the endocytic and exocytic pathways. Annu Rev Biochem 55:663–700

Messina JP, Gilkeson GS, Pisetsky DS (1991) Stimulation of in vitro murine lymphocyte proliferation by bacterial DNA. J Immunol 147:1759–1764

Muller JM, Rupec RA, Baeuerle PA (1997) Study of gene regulation by NF-κB and AP-1 in response to reactive oxygen intermediates. Methods 11:301–312

Muroi M, Muroi Y, Suzuki T (1994) The binding of immobilized IgG2a to Fcγ2a receptor activates NF-κB via reactive oxygen intermediates and tumor necrosis factor-α1. J Biol Chem 269:30561–30568

Nishina H, Bachmann M, Oliveira-dos-Santos AJ, Kozieradzki I, Fischer KD, Odermatt B, Wakeham A, Shahinian A, Takimoto H, Bernstein A, Mak TW, Woodgett JR, Ohashi PS, Penninger JM (1997) Impaired CD28-mediated interleukin 2 production and proliferation in stress kinase SAPK/ERK1 kinase (SEK1)/mitogen-activated protein kinase kinase 4 (MKK4)-deficient T lymphocytes. J Exp Med 186:941–953

Ohkuma S, Poole B (1981) Cytoplasmic vacuolation of mouse peritoneal macrophages and the uptake into lysosomes of weakly basic substances. J Cell Biol 90:656–664

Pisetsky DS (1996) Immune activation by bacterial DNA: a new genetic code. Immunity 5:303–310

Poltorak A, He X, Smirnova I, Liu MY, Huffel CV, Du X, Birdwell D, Alejos E, Silva M, Galanos C, Freudenberg M, Ricciardi-Castagnoli P, Layton B, Beutler B (1998) Defective LPS signaling in C3H/HeJ and C57BL/10ScCr mice: mutations in Tlr4 gene. Science 282:2085–2088

Pombo CM, Bonventre JV, Avruch J, Woodgett JR, Kyriakis JM, Force T (1994) The stress-activated protein kinases are major c-Jun amino-terminal kinases activated by ischemia and reperfusion. J Biol Chem 269:26546–26551

Price MA, Rogers AE, Treisman R (1995) Comparative analysis of the ternary complex factors Elk-1, SAP-1a and SAP-2 (ERP/NET) EMBO J 14:2589–2601

Prichett W, Hand A, Sheilds J, Dunnington D (1995) Mechanism of action of bicyclic imidazoles defines a translational regulatory pathway for tumor necrosis factor α. J Inflamm 45:97–105

Raingeaud J, Whitmarsh AJ, Barrett T, Derijard B, Davis RJ (1996) MKK3- and MKK6-regulated gene expression is mediated by the p38 mitogen-activated protein kinase signal transduction pathway. Mol Cell Biol 16:1247–1255

Regnier CH, Song HY, Gao X, Goeddel DV, Cao Z, Rothe M (1997) Identification and characterization of an IκB kinase. Cell 90:373–383

Rincon M, Enslen H, Raingeaud J, Recht M, Zapton T, Su MS, Penix LA, Davis RJ, Flavell RA (1998) Interferon-γ expression by Th1 effector T cells mediated by the p38 MAP-kinase signaling pathway. EMBO J 17:2817–2829

Roussel MF (1998) Key effectors of signal transduction and G1 progression. Adv Cancer Res 74:1–24

Sanchez I, Hughes RT, Mayer BJ, Yee K, Woodgett JR, Avruch J, Kyriakis JM, Zon LI (1994) Role of SAPK/ERK kinase-1 in the stress-activated pathway regulating transcription factor c-Jun. Nature 372:794–798

Schaeffer HJ, Catling AD, Eblen ST, Collier LS, Krauss A, Weber MJ (1998) MP1: a MEK binding partner that enhances enzymatic activation of the MAP kinase cascade. Science 281:1668–1671

Sluss HK, Barrett T, Derijard B, Davis RJ (1994) Signal transduction by tumor necrosis factor mediated by JNK protein kinases. Mol Cell Biol 14:8376–8384

Smeal T, Binetruy B, Mercola D, Grover-Bardwick A, Heidecker G, Rapp UR, Karin M (1992) Oncoprotein-mediated signalling cascade stimulates c-Jun activity by phosphorylation of serines 63 and 73. Mol Cell Biol 12:3507–3513

Sparwasser T, Miethke T, Lipford G, Erdmann A, Hacker H, Heeg K, Wagner H (1997) Macrophages sense pathogens via DNA motifs: induction of tumor necrosis factor-α-mediated shock. Eur J Immunol 27:1671–1679

Stacey KJ, Sweet MJ, Hume DA (1996) Macrophages ingest and are activated by bacterial DNA. J Immunol 157:2116–2122

Sweet MJ, Hume DA (1996) Endotoxin signal transduction in macrophages. J Leukoc Biol 60:8–26

Tonkinson JL, Stein CA (1994) Patterns of intracellular compartmentalization, trafficking and acidification of 5′-fluorescein labeled phosphodiester and phosphorothioate oligodeoxynucleotides in HL60 cells. Nucleic Acids Res 22:4268–4275

Ulevitch RJ (1999) Endotoxin opens the Tollgates to innate immunity [In Process Citation]. Nat Med 5:144–145

van Dam H, Duyndam M, Rottier R, Bosch A, de Vries-Smits L, Herrlich P, Zantema A, Angel P, van der Eb AJ (1993) Heterodimer formation of c-Jun and ATF-2 is responsible for induction of c-*jun* by the 243 amino acid adenovirus E1A protein. EMBO J 12:479–487

Vasquez KM, Wilson JH (1998) Triplex-directed modification of genes and gene activity. Trends Biochem Sci 23:4–9

Wagner H (1999) Bacterial CpG-DNA activates immune cells to signal 'infectious danger'. Advances in Immunology. 73, in press

Wesche H, Henzel WJ, Shillinglaw W, Li S, Cao Z (1997) MyD88: an adapter that recruits IRAK to the IL-1 receptor complex. Immunity 7:837–847

Westwick JK, Weitzel C, Minden A, Karin M, Brenner DA (1994) Tumor necrosis factor α stimulates AP-1 activity through prolonged activation of the c-Jun kinase. J Biol Chem 269:26396–26401

Whitmarsh AJ, Cavanagh J, Tournier C, Yasuda J, Davis RJ (1998) A mammalian scaffold complex that selectively mediates MAP kinase activation. Science 281:1671–1674

Whitmarsh AJ, Davis RJ (1998) Structural organization of MAP-kinase signaling modules by scaffold proteins in yeast and mammals. Trends Biochem Sci 23:481–485

Wloch MK, Pasquini S, Ertl HC, Pisetsky DS (1998) The influence of DNA sequence on the immunostimulatory properties of plasmid DNA vectors. Hum Gene Ther 9:1439–1447

Yang RB, Mark MR, Gray A, Huang A, Xie MH, Zhang M, Goddard A, Wood WI, Gurney AL, Godowski PJ (1998) Toll-like receptor-2 mediates lipopolysaccharide-induced cellular signalling. Nature 395:284–288

Yi AK, Klinman DM, Martin TL, Matson S, Krieg AM (1996) Rapid immune activation by CpG motifs in bacterial DNA. Systemic induction of IL-6 transcription through an antioxidant-sensitive pathway. J Immunol 157:5394–5402

Yi AK, Chang M, Peckham DW, Krieg AM, Ashman RF (1998a) CpG oligodeoxyribonucleotides rescue mature spleen B cells from spontaneous apoptosis and promote cell cycle entry. J Immunol 160: 5898–5906

Yi AK, Tuetken R, Redford T, Waldschmidt M, Kirsch J, Krieg AM (1998b) CpG motifs in bacterial DNA activate leukocytes through the pH-dependent generation of reactive oxygen species. J Immunol 160:4755–4761

Yi AK, Krieg AM (1998a) Rapid induction of mitogen-activated protein kinases by immune stimulatory CpG DNA. J Immunol 161:4493–4497

Yi AK, Krieg AM (1998b) CpG DNA rescue from anti-IgM-induced WEHI-231 B lymphoma apoptosis via modulation of IκBα and IκBβ and sustained activation of nuclear factor-κ B/c-Rel. J Immunol 160:1240–1245

Yoshimori T, Yamamoto A, Moriyama Y, Futai M, Tashiro Y (1991) Bafilomycin A1, a specific inhibitor of vacuolar-type H(+)-ATPase, inhibits acidification and protein degradation in lysosomes of cultured cells. J Biol Chem 266:17707–17712

Zandi E, Rothwarf DM, Delhase M, Hayakawa M, Karin M (1997) The IκB kinase complex (IKK) contains two kinase subunits, IKKα and IKKβ, necessary for IκB phosphorylation and NF-κB activation. Cell 91:243–252

# CpG DNA Co-Stimulates Antigen-Reactive T Cells

K. Heeg



## 1 Introduction

During the last several years, compelling evidence has accumulated that bacterial DNA is sensed as "infectious danger" by innate immune cells, such as macrophages and dendritic cells (DCs) (Krieg 1996b; Pisetsky 1996a,b, 1997; Lipford et al. 1998). Those cells are thought not only to represent central cellular restriction points during induction and execution of innate immune responses but also to control activation of cells of the acquired immune system, such as T lymphocytes (Matzinger 1994; Fearon and Locksley 1996; Medzhitov and Janeway 1997). Accordingly, processing and presentation of antigen (signal one) and the generation of permissive cytokine and co-stimulatory milieus (signal two) govern productive T-cell activation (Bretscher and Cohn 1970; Lafferty and Woolnough 1977;

Institute of Medical Microbiology and Hygiene, Philipps University, Pilgrimstein 2, D-35037 Marburg, Germany
E-mail: heeg@post.med.uni-marburg.de

Mueller et al. 1989). Evidently, sensing "infectious danger" and, thus, activation of innate immune cells represents a major prerequisite for successful adaptive immune responses of the B- and T-cell lineages.

The observation that synthetic, single-stranded oligodeoxynucletides (ODNs) mimic the effects of bacterial DNA in a DNA-sequence-specific manner (Yamamoto et al. 1992; Krieg et al. 1995) has led researchers to envisage its practical use (Lipford et al. 1997a; Klinman 1998; Krieg et al. 1998). First, immunostimulatory properties of bacterial DNA and ODNs could be attributed to unmethylated CpG-dinucleotide motifs, which occur at expected frequencies in bacterial DNA but at low frequencies in mammalian DNA (Bird 1980; Krieg 1996a; Pisetsky 1996a; Pesole et al. 1997). Second, ODNs displaying the classic canonical DNA motif 5′-purine-purine-CpG-pyrimidine-pyrimidine-3′ were shown to be often immunostimulatory (CpG ODN) (Krieg et al. 1995, 1996; Krieg 1996a). Finally, modification of the ODNs' backbones by phosphothioate linkages made the ODNs resistant to the action of nucleases (Agrawal et al. 1991; Gao et al. 1992) and, thus, allowed their application in animal models in vivo. The large body of evidence obtained so far suggests that CpG ODNs might be suitable for adjuvant use in prophylactic and therapeutic vaccination formulas for both B-cell and T-cell vaccinations. The following review summarizes our current knowledge of indirect (extrinsic) and direct (intrinsic) effects of CpG and non-CpG ODNs on the immune reactivity of T lymphocytes.

## 2 CpG ODNs Exert Extrinsic Effects on T Cells

### 2.1 CpG-Mediated Activation of APCs Augments T-Cell Responses in Vitro and in Vivo

Besides the mitogenic effects of CpG ODNs on B lymphocytes (Krieg et al. 1995), it was soon recognized that bacterial plasmid DNA and certain palindromic, synthetic ODNs (Yamamoto et al. 1992) are taken up by macrophages, which in turn are activated, a process associated with the translocation of nuclear factor κB (NFκB) (Stacey et al. 1996; Sparwasser et al. 1997). Further studies revealed that macrophages are directly activated by CpG ODNs to express co-stimulatory molecules (CD80, CD86, CD40) and produce pro-inflammatory [tumor necrosis factor (TNF), interleukin 6 (IL-6)] and effector cytokines (IL-12) (Sparwasser et al. 1997; Jakob et al. 1998; Sparwasser et al. 1998). Modifications of the ODN sequence defined CpG ODNs that preferentially induced IL-12 (but not TNF) in vivo and in vitro (Lipford et al. 1997b). Of note, such ODNs would be of value in devising ODNs for therapeutic use, since they lack potentially harmful effects mediated by the release of toxic amounts of TNF.

DCs disseminated in non-lymphatic tissue are immature yet phagocytically active (Cella et al. 1997; Vremec and Shortman 1997; Steinman et al. 1998).

Upon maturation of DCs, phagocytic activity and antigen uptake is lost and co-stimulatory activity is gained. During this step, DCs translocate in lymphatic organs, express high levels of major histocompatibility complex (MHC) and co-stimulatory molecules (Lutz et al. 1996; Cella et al. 1997) and, thus, are equipped for the activation of T lymphocytes (Schuler et al. 1997). In vitro, CpG ODNs were shown to induce both phenotypic maturation of bone-marrow-derived DCs (Sparwasser et al. 1998) and secretion of high levels of IL-12 (Jakob et al. 1998; Sparwasser et al. 1998). Thus, CpG ODNs are likely to induce a cellular and cytokine milieu that provides signals necessary for T-cell activation. These effects are primarily due to activation of antigen-presenting cells (APCs) by CpG ODNs (indirect or extrinsic effects). Examples of the extrinsic effects of CpG ODNs on T-cell activation in vitro are depicted in Fig. 1. When the number of APCs is limited in an allogeneic mixed-lymphocyte reaction, no T-cell proliferation can be induced. However, addition of CpG ODNs amplifies the stimulatory potency of the remaining APCs. As a consequence, T-cell responses to antigens are enhanced. In vivo, this effect might be responsible for the powerful activity of CpG ODNs as adjuvants in T-cell responses. Injection of "weak" proteinaceous antigens, together with CpG ODNs, allowed the generation of MHC-restricted peptide-specific $CD8^+$ T lymphocytes (Lipford et al. 1997a; Ehl et al. 1998).

## 2.2 CpG-Induced Cytokines Influence T-Cell Responses in Vivo

Unlike B cells, direct activation of T cells by bacterial CpG DNA or CpG ODNs has not been observed. Recently, however, it was found that lipopolysaccharides

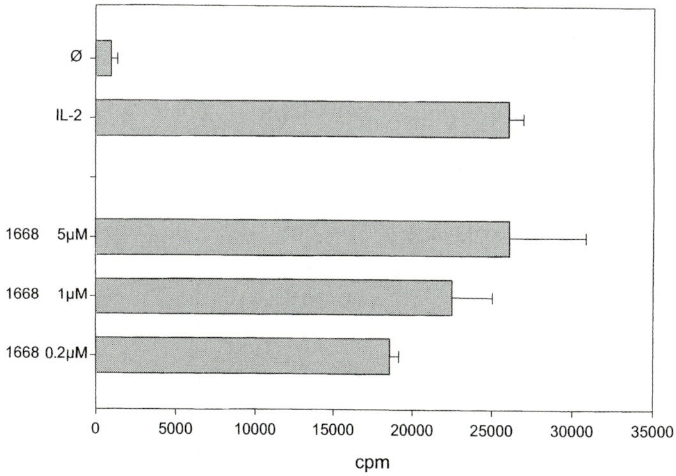


**Fig. 1.** Extrinsic effect of CpG oligodeoxynucleotides on mixed lymphocyte reactions in vitro. T cells from C57Bl/6 mice (20,000/well) were stimulated with 1000 irradiated BALB/c spleen cells in microtiter plates as indicated. After 4 days, the proliferative response was recorded

(LPS), a known mitogen for murine B cells, also activated T cells in vivo and in vitro (Mattern et al. 1994; Tough et al. 1997; Khoruts et al. 1998) in the apparent absence of T-cell-receptor (TCR) ligation. Several lines of evidence indicated that responsible cytokines were type-I interferons (IFN), known to be induced by LPS or poly(I:C) (Zhang et al. 1998). Moreover, these cytokines were found to induce the secretion of IL-15 in APC, which in turn caused bystander activation of memory (but not of naive T cells) in a TCR-independent fashion (Zhang et al. 1998). Since CpG ODNs also induce type-I IFNs, these TCR-independent extrinsic effects of CpG ODNs should also be operative in vivo. Indeed, within 20 h after subcutaneous injection of CpG ODNs, upregulation of CD69 (but not of IL-2R$\alpha$) on T cells of the draining lymph nodes can be observed (Sun et al. 1998b, S. Bendigs, personal communication). It follows that CpG-ODN-induced type-I IFNs may influence the signal threshold of antigen-reactive T cells in vivo. Again, this effect is due to the CpG-mediated activation of APCs.

Activation of T lymphocytes and subsequent differentiation into effector cells is guided by a distinctive milieu of cytokines. During T-helper cell development, IL-4 causes a bias for T-helper 2 (Th2) differentiation, while IL-12 instructs Th1 responses (Macatonia et al. 1995; Mosmann and Sad 1996). Immunizations performed with bacterial DNA, *Drosophila* DNA (Sun et al. 1998a) or CpG ODNs (Chu et al. 1997; Lipford et al. 1997a; Roman et al. 1997) revealed augmentation and a strong bias of the antibody response towards an immunoglobulin G2a (IgG2a) isotype characteristic of Th1 responses. These observations may be attributed to the strong induction of APC-derived IL-12 by CpG ODNs. The induction of antigen-specific $CD8^+$-cytolytic T-cell responses also suggests the preferential induction of Th1-dominated responses by CpG ODNs (Lipford et al. 1997a).

The same conclusions are based on experiments using murine models for infectious diseases. For example, infection of BALB/c mice with *Leishmania major* is followed by disease, because a strong Th2 response is triggered (Locksley and Louis 1992). In contrast, C57Bl/6 mice mount a Th1-biased immune response and control the infection via nitric oxide (NO) generation (Stenger et al. 1994). However, treatment of infected BALB/c mice with CpG ODNs resulted in a conversion of the disease, i.e., BALB/c mice mounted a strong Th1 response accompanied with a sustained expression of IL-12R$\beta$2 molecules and a T-helper cytokine pattern characteristic of a Th1 phenotype (Zimmermann et al. 1998). CpG ODNs were not only effective in preventing lethal infections (when injected together with *Leishmania*) but were also able to cure an established Th2-dominated lethal disease when the CpG ODNs were administered up to 3 weeks after infection (Zimmermann et al. 1998). Thus, CpG ODNs probably direct ongoing immune responses towards a Th1 phenotype due to the ODNs' ability to activate APCs and induce Th1-instructing cytokines. Similar conclusions can be drawn from models of murine experimental allergies (Kline et al. 1998). The adjuvant activity and Th1-promoting efficacy of CpG ODNs can be explained, at least in part, by their potent extrinsic activities on T cells.

# 3 Intrinsic Effects of CpG ODNs on T Cells

## 3.1 CpG-ODN-Mediated Co-Stimulation of T Cells in the Absence of APCs

To determine whether CpG ODNs can exert direct (intrinsic) effects on T cells, we prepared highly purified T lymphocytes (Bendigs et al. 1999). To achieve this, T lymphocytes were enriched via several rounds of negative selection by magnetic cell separation and then stimulated with plate-bound anti-CD3 monoclonal antibodies (mAbs). A typical experiment is shown in Fig. 2. The fact that T cells could be stimulated neither by anti-CD3 alone nor in the presence of anti-CD3 and LPS or concanavalin A (used as a control) indicated that our cultures were indeed devoid of APCs. Since addition of IL-2 caused proliferation in anti-CD3 mAb-stimulated cultures but not in the absence of anti-CD3, mAb signal 1 (TCR ligation) caused IL-2R expression and, thus, was operative. Of note, the requirement for IL-2 could be substituted by CpG ODNs (Fig. 2). This type of result allowed several conclusions. First, CpG ODNs did not directly activate T cells. In contrast to B cells and macrophages, purified T lymphocytes remained unresponsive to ODNs in the absence of TCR ligation. However, upon crosslinking of T-cell receptors (signal 1), addition of CpG ODNs induced proliferation of purified T cells in a dose-dependent manner (Fig. 2). In fact, upon TCR ligation, CpG ODNs induced expression of IL-2R, production of IL-2 and cytolytic activity in T cells (Bendigs et al. 1999). T-cell proliferation was dependent on IL-2, since anti-IL-2 antibodies completely abrogated proliferative and cytolytic responses. CpG-medi-

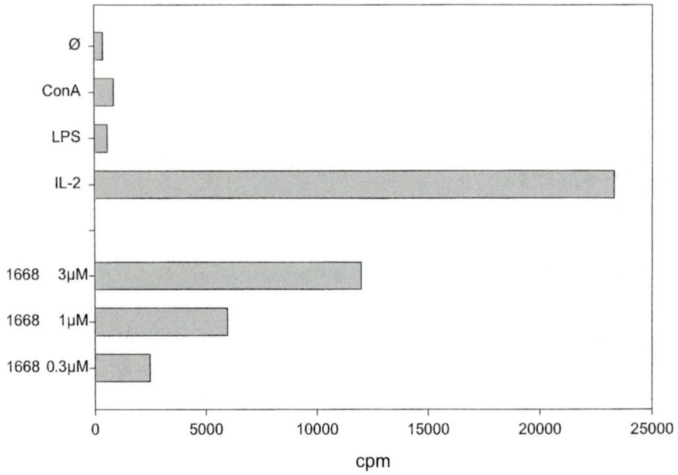


**Fig. 2.** CpG oligodeoxynucleotides co-stimulate T cells. Purified T cells (20,000/well) from C57Bl/6 mice were stimulated on anti-CD3-antibody-coated microtiter plates in the presence of the reagents indicated. After 4 days, the proliferative response was recorded

ated co-stimulation required a functional TCR signal (signal 1), since cyclosporin A (CsA), known to block $Ca^{++}$-dependent TCR signaling pathways (SIGAL and DUMONT 1992; MCCAFFREY et al. 1993), completely abolished T-cell proliferation co-stimulated by CpG ODNs (BENDIGS et al. 1999).

Collectively, these data show that CpG ODNs co-stimulated T cells to express IL-2R, produce IL-2, proliferate and finally develop into cytotoxic effector cells. In this respect, co-stimulation mediated by CpG ODNs phenotypically resembles co-stimulation of T cells mediated by CD28 (HARDING et al. 1992; SAGERSTRÖM et al. 1993). This notion was further supported by the observation that CpG ODNs substituted, at least in part, for the co-stimulatory defects of T lymphocytes from $CD28^{-/-}$ mice in vitro (BENDIGS et al. 1999). CpG ODNs also provided signal 2 to T lymphocytes from TCR-transgenic mice stimulated with antigenic peptides. To exclude possible extrinsic CpG-ODN effects, in these experiments APCs were fixed with glutaraldehyde prior to loading with peptides. If peptide-loaded fixed APCs were added, CpG ODN induced proliferation of T lymphocytes (BENDIGS et al. 1999). These results demonstrated that co-stimulation by CpG ODNs not only applies for experimental TCR stimulation with mAbs but also holds true for physiological TCR ligation induced by MHC plus peptide.

The restricted action of ODNs on T cells may also be prototypic for other cell types. Although it was initially reported that natural killer (NK) cells were responsive towards CpG ODNs (BALLAS et al. 1996), it was later found that IFN-γ release from NK cells was dependent on exogenous IL-12 (HALPERN et al. 1996; CHACE et al. 1997). Operationally, IL-12 could serve as signal 1 in NK cells to trigger sensitivity to CpG ODNs (signal 2). Similar principles might operate in B cells. B lymphocytes respond to ODNs in cultures of splenocytes with a short wave of cell divisions; nevertheless, proliferation, antibody production and cytokine secretion is several-fold enhanced after crosslinking of B-cell receptors (surface-bound Igs) (KRIEG et al. 1995). Thus, full sensitivity of B cells to CpG ODNs might develop only after receptor ligation (signal 1), as is the case for T cells. Overall, these results led us to conclude that APC-like macrophages or DCs are *a priori* sensitive to CpG ODNs, while cells of the lymphocytic lineage must develop sensitivity to CpG ODNs mediated by receptor ligation (signal 1). We anticipate that this functional divergence between APCs and lymphocytes might find its counterpart in different CpG-ODN receptors and/or distinct ODN-induced signal pathways.

## 3.2 ODN-Mediated Co-Stimulation Violates the Classical Rules of CpG-Mediated Stimulation of B Cells and APCs

### 3.2.1 The CpG Motif is Dispensable for Co-Stimulation of T Cells

When we analyzed the ODN-DNA-sequence requirements for T-cell co-stimulation, we were surprised by the finding that the rules of canonical CpG-DNA motifs, as defined for direct activation of B cells and APCs, did not apply for its co-stimulatory activity on T cells. First, in CpG motifs displaying ODNs like 1628, the central CpG could be methylated or replaced by a TpG without losing its co-stimulatory activity on T cells (Table 1). However, several ODNs (ODN 1720 or ODN ATT-Rep) were negative, indicating sequence dependency of the phosphothioate-modified ODNs towards T cells (BENDIGS et al. 1999). When we screened human DNA gene-sequence data for canonical CpG motifs, we also found CpG sequences in the human vertebrate genome (LIPFORD et al. 1997b, 1998). Some of these sequences were active to APCs [ODN IL-12p40, a CpG motif within the IL-12p40 gene (LIPFORD et al. 1997b)], yet some were inactive (ODN SP1, a CpG motif within the SP1-promoter site). However, ODN SP1 was active on T cells, i.e., it co-stimulated T cells (Table 1) (LIPFORD et al. 1999).

**Table 1.** Effects of oligodeoxynucleotides (ODN) on antigen-presenting cells (APCs) and T cells

| Designation | Sequence (5′–3′) | APC | | | T cells | |
|---|---|---|---|---|---|---|
| | | TNF | IL-12 | Inhibition of CpG effects | Co-stimulation | Adjuvant for in vivo CTL induction |
| 1668 | tccaGACGTTcct gatgct | + + + | + + + | | + | + + + |
| IL-12p40 | agctatGACGTT ccaagg | + | + + | | + | + + + |
| 1628 | GGGGtcAACGTT gaggggggg | Ø | Ø | n.t. | + + + | + |
| 1628M | GGGGtcAAMGTT gaggggggg | Ø | Ø | n.t. | + + + | n.t. |
| 1628GC | GGGGtcAAGCTT gaggggggg | Ø | Ø | n.t. | + + + | n.t. |
| SP1 | tCGatCGGGGCG GGGCGaGC | Ø | Ø | n.t. | + + + | + |
| SP1T | tTGatTGGGGTG GGGTGaGC | Ø | Ø | n.t. | + + + | n.t. |
| SP1M | tMGatMGGGGM GGGGMGaGC | Ø | Ø | n.t. | + + + | n.t. |
| 1720 | tccatgagcttcctgatgct | Ø | Ø | | Ø | Ø |
| ATT-Rep | attattattattattattat | Ø | Ø | | Ø | Ø |
| PZ31 | ctcctattGGGGG tttcctat | Ø | Ø | + + + | + + + | + |
| PZ33 | ctcctattGGGGT tttcctat | Ø | Ø | + + + | + + + | + |
| PZ34 | ctcctattGGTGG tttcctat | Ø | Ø | Ø | Ø | Ø |
| SB | tccatGACGTT cctgaatggctGGGG | Ø | Ø | + + | + + + | n.t. |
| SBDG | tccatGACGTT cctgaatggctGAGT | + + + | + + + | | + | n.t. |
| SBGC | tccatGAGCTT cctgaatggctGGGG | Ø | Ø | + + | + + + | n.t. |

*CTL*, cytotoxic T lymphocyte; *M*, methylated; *n.t.*, not tested.

### 3.2.2 G-Rich ODNs Co-Stimulate T Cells Yet Fail to Activate APCs

We soon realized that ODNs active on T cells either possessed a classical CpG motif, like ODN 1668, or displayed stretches of guanines, like ODN 1628 or ODN SP1 (G-rich ODN) (Table 1). In order to define a putative motif, we synthesized ODNs where a central G stretch was permutated with thymidine residues (PZ3, PZ31, PZ34; Table 1). This approach led to the definition of new ODN motifs active on T lymphocytes. One motif (four Gs or 5′-G-X-G-G-G-3′, where X is any base) characterizes ODNs that co-stimulate anti-CD3-triggered T cells. In retrospect, the active ODNs 1628 and SP1 also contain this G-rich motif (compare Table 1). Of note, ODNs embodying this G-rich motif failed entirely to activate APCs and did not trigger production of cytokines, such as IL-12 or TNF (Lipford et al. 1999). Further analyses revealed that G-rich ODNs actually antagonized the action of CpG ODNs on APCs (Häcker et al. 1998). Thus, G-rich ODNs effectively suppressed the activity of CpG ODNs on APCs. Detailed analyses with labeled ODNs revealed that G-rich ODNs are taken up into APCs with a much higher affinity and efficacy than were CpG ODNs. However, they competed for the uptake of CpG ODNs. Since G-rich ODNs fail to induce signaling cascades within APCs (Häcker et al. 1998) competition with CpG ODNs during uptake into APCs might explain the inhibitory effects of G-rich ODNs on CpG ODNs.

### 3.2.3 G-Rich Motifs in Hybrid ODNs are Dominant Negative on APCs Yet Co-Stimulate T Cells

When we synthesized hybrid ODNs consisting of a CpG motif (derived from ODN 1668) plus a G-rich motif at the 3′ end of the ODN, we found that these ODNs (ODN SB in Table 1) failed to activate APCs. Disrupting the G stretch (ODN SB-DG) fully restored reactivity on APCs. Thus, G motifs have dominant negative effects as hybrid ODNs when tested on APCs. The very same ODNs, however, effectively co-stimulated T lymphocytes. Thus, ODNs containing the G motif are agonists on T cells yet antagonists on APCs. In retrospect, the peculiar activities of ODN 1628 and ODN SP1 may now be explained. Although ODN 1628 and ODN SP1 contain a CpG motif, they exhibit only marginal activity on APCs. In contrast, these ODNs strongly co-stimulate T cells (Lipford et al. 1999; Table 1). We suggest that their G motifs antagonize the effects on APCs yet trigger T-cell co-stimulation.

### 3.2.4 ODN-Induced Signal Pathways on T Cells are Distinct from Those on APCs

ODNs differ in their affinity kinetics for uptake into APCs (Häcker et al. 1998; G. B. Lipford, personal communication). When we tested uptake of ODNs into T cells, we noticed that only a minute number of ODNs, if any, are taken up by naive resting T cells (unpublished data). In contrast, if T cells were stimulated with anti-CD3 antibodies, uptake of ODNs was enhanced. However, the efficacy of uptake was lower than for the uptake of ODNs in APCs (unpublished data). Some investigators have suggested that APCs take up ODNs via surface receptors, such as Mac-1 or the scavenger receptor (Benimetskaya et al. 1997). Since T cells express neither scavenger receptors nor Mac-1, different uptake mechanisms/receptors might be operative in partially activated T cells.

Recently, aspects of the CpG-ODN-induced signaling cascade within APCs have been unraveled (Häcker et al. 1998). ODNs are taken up into early endosomes and, within minutes, trigger the stress-activated protein kinase/c-Jun *N*-terminal kinase stress-kinase pathway followed by transcriptional activity of activator protein 1 and NFκB (Häcker et al. 1998). A critical restriction point during this signaling process represents the maturation of endosomes. Agents that block maturation of endosomes (chloroquine or bafilomycin) prevent ODN-mediated signaling and, thus, block all downstream effects on APCs (Häcker et al. 1998; Macfarlane and Manzel 1998). When we tested the outcome of endosomal blockade on T-cell co-stimulation, we found that neither chloroquine nor bafilomycin inhibited co-stimulation of T cells mediated by either CpG ODNs or G-rich ODNs (Lipford et al. 1999).

In conclusion, the mode of action of ODNs on T cells differs in several respects from their action on APCs. First, ODNs exert no direct stimulation on resting T cells. In order to develop sensitivity to ODNs, T cells require partial activation via TCR occupancy (signal 1). Second, the ODN-sequence requirements for T cells differ from those for ODNs active on APCs. Methylated and un-methylated CpG ODNs and G-rich ODNs co-stimulate T cells effectively. Third, the signal pathways operative in T cells appear to be distinct. While cellular uptake of ODNs is only minute in T cells, as compared with APCs, endosomal maturation is dispensable for co-stimulation of T cells and a condition for APC activation. It will, therefore, be rewarding to define in detail the receptors/signal cascades in T lymphocytes.

## 4 Effect of ODN on T-Cell Responses in Vivo

Proteinaceous antigens usually fail to elicit strong cytolytic T-cell responses in vivo unless CpG ODNs are used as adjuvants (Lipford et al. 1997a). As discussed above, the potent adjuvanticity of CpG ODNs may be attributed to their capacity to activate APCs (an extrinsic effect on T cells). However, a direct (intrinsic) effect of CpG ODNs and G-rich ODNs on T cells might contribute to their adjuvant activities on T cells in vivo. Since G-rich ODNs fail to activate APCs, these ODNs should unravel the relative importance of intrinsic ODN effects on T-cell activation in vivo. We therefore tested various CpG, G-rich and hybrid ODNs for their adjuvanticity during induction of $CD8^+$ cytotoxic T lymphocytes in vivo. We found that all three classes served as adjuvants in vivo (Lipford et al. 1999). However, the relative efficacy was highest with CpG ODNs. These results suggest that G-rich ODNs can probably be used as T-cell adjuvants, since co-stimulation of

T cells (in the absence of activated APCs) is sufficient to aid cellular immune responses in vivo. However, CpG ODNs activate (extrinsic effect) and co-stimulate (intrinsic effect) DCs. Therefore, CpG ODNs display the highest adjuvanticity.

An additonal discriminatory aspect of the action of CpG and G-rich ODNs emerged from these studies. In addition to their T-cell adjuvanticity, G-rich ODNs induced strong NK-like cytotoxic activity in vivo (unpublished data). Similar conclusions have previously been reported in the case of human NK cells (Ballas et al. 1996). Active ODNs showed at least one stretch of G either at the 5′ or 3′ end of the ODN. Thus, G-rich ODNs also provoke effector functions of the innate immune system. It is obvious that this property of G-rich ODNs might be of decisive importance during infection with NK-sensitive organisms. Preliminary studies in a murine infection model with *Toxoplasma gondii* indicate that this might be indeed the case (S. Zimmermann, personal communication).

The ability of CpG and G-rich ODNs to co-stimulate T cells might be of practical relevance in situations where APCs are not available, e.g. in tumor tissues. According to the two-signal model of T-cell activation and anergy induction (Mueller et al. 1989), tumor cells expressing antigen (signal 1) alone without co-stimulatory activity would be tolerogenic. Since CpG ODNs and G-rich ODNs co-stimulate T cells, these ODNs should bypass the tolerogenic effect and, thus, may activate anti-tumor T-cell responses. However, this beneficial outcome of T-cell co-stimulation by ODNs in vivo could include serious adverse consequences. ODNs might permit an APC-independent (re-)activation of autoreactive T cells.

## 5 Outlook

Analyses of the actions of ODNs on T cells enriched our knowledge in several respects. First, it was recognized that T cells respond to ODNs in an unperceived, restricted fashion. TCR ligation was necessary to render T cells permissive for the effects of ODNs. This mode of action could be prototypic for other cells of the lymphocytic lineage. Second, a new class of ODNs reactive on T cells and NK cells could be defined. The G-rich ODNs differ in many aspects from the classical CpG ODNs, yet they could be suitable for therapeutic use. In addition, the strong capacity of these ODNs to induce NK-like cytotoxicity could be advantageous in new therapeutic vaccines against NK-sensitive parasites. Third, beginning analyses of the signal cascade of ODNs indicate a signal pathway distinct from that of APCs. Collectively, the effects of ODNs on T-cell activation and immunization in vitro and in vivo could aid in the development of a new, safe and efficient adjuvant for therapeutic and prophylactic vaccines against infectious agents and cancer.

*Acknowledgements.* The work cited here was performed while K. Heeg served as staff member of the Institute of Medical Microbiology, Immunology and Hygiene, Trogerstr. 9, 81675 Munich/Germany. I am grateful to my former colleagues for sharing unpublished data and for help in preparing this review.

## References

Agrawal S, Temsamani J, Tang JY (1991) Pharmacokinetics, biodistribution, and stability of oligodeoxynucleotide phosphorothioates in mice. Proc Natl Acad Sci USA 88:7595–7599

Ballas ZK, Rasmussen WL, Krieg AM (1996) Induction of NK activity in murine and human cells by CpG motifs in oligodeoxynucleotides and bacterial DNA. J Immunol 157:1840–1845

Bendigs S, Salzer U, Lipford GB, Wagner H, Heeg K (1999) CpG-oligodeoxynucleotides costimulate primary T cells in the absence of APC. Eur J Immunol 29:1209–1218

Benimetskaya L, Loike JD, Khaled Z, Loike G, Silverstein SC, Cao L, El Khoury J, Cai TQ, Stein CA (1997) Mac-1 (CD11b/CD18) is an oligodeoxynucleotide-binding protein. Nature Med 3:414–420

Bird AP (1980) DNA methylation and the frequency of CpG in animal DNA. Nucleic Acids Res 8: 1499–1504

Bretscher P, Cohn M (1970) A theory of self-nonself discrimination. Science 169:1042–1049

Cella M, Sallusto F, Lanzavecchia A (1997) Origin, maturation and antigen presenting function of dendritic cells. Curr Opin Immunol 9:10–16

Chace JH, Hooker NA, Mildenstein KL, Krieg AM, Cowdery JS (1997) Bacterial DNA-induced NK cell IFN-γ production is dependent on macrophage secretion of IL-12. Clinical Immunology 84:185–193

Chu RS, Targoni OS, Krieg AM, Lehmann PV, Harding CV (1997) CpG oligodeoxynucleotides act as adjuvants that switch on T helper 1 (Th1) immunity. J Exp Med 186:1623–1631

Ehl S, Hombach J, Aichele P, Rülicke T, Odermatt B, Hengartner H, Zinkernagel RM, Pircher H (1998) Viral and bacterial infections interfere with peripheral tolerance induction and activate CD8+ T cells to cause immunopathology. J Exp Med 187:763–774

Fearon DT, Locksley RM (1996) The instructive role of innate immunity in the acquired immune response. Science 272:50–54

Gao WY, Han FS, Storm C, Egan W, Cheng YC (1992) Phosphorothioate oligonucleotides are inhibitors of human DNA polymerases and RNase H: implications for antisense technology. Mol Pharmacol 41:223–229

Häcker H, Mischak H, Miethke T, Liptay S, Schmid R, Sparwasser T, Heeg K, Lipford GB, Wagner H (1998) CpG-DNA specific activation of antigen presenting cells requires stress kinase activity and is proceeded by non-specific endocytosis and endosomal maturation. EMBO J 17:6230–6240

Halpern MD, Kurlander RJ, Pisetsky DS (1996) Bacterial DNA induces murine interferon-gamma production by stimulation of interleukin 12 and tumor necrosis factor-alpha. Cell Immunol 167:72–78

Harding FA, McArthur JG, Gross JA, Raulet DH, Allison JP (1992) CD28-mediated signalling co-stimulates murine T cells and prevents induction of anergy in T cells clones. Nature 356:607–609

Jakob T, Walker PS, Krieg AM, Udey MC, Vogel JC (1998) Activation of cutaneous dendritic cells by CpG-containing oligodeoxynucleotides: a role for dendritic cells in the augmentation of Th1 responses by immunostimulatory DNA. J Immunol 161:3042–3049

Khoruts A, Mondino A, Pape KA, Reiner SL, Jenkins MK (1998) A natural immunological adjuvant enhances T cell clonal expansion through a CD28-dependent, Interleukin (IL)-2-independent mechanism. J Exp Med 187:225–236

Kline JN, Waldschmidt TJ, Businga TR, Lemish JE, Weinstock JV, Thorne PS, Krieg AM (1998) Modulation of airway inflammation by CpG oligodesoxynucleotides in a murine model of asthma. J Immunol 160:2555–2559

Klinman DM (1998) Therapeutic applications of CpG-containing oligodeoxynucleotides. Antisense Nucleic Acid Drug Dev 8:181–184

Krieg AM, Yi AK, Matson S, Waldschmidt TJ, Bishop GA, Teasdale R, Koretzky GA, Klinman DM (1995) CpG motifs in bacterial DNA trigger direct B-cell activation. Nature 374:546–549

Krieg AM (1996a) Lymphocyte activation by CpG dinucleotide motifs in prokaryotic DNA. Trends Microbiol 4:73–76

Krieg AM (1996b) An innate immune defense mechanism based on the recognition of CpG motifs in microbial DNA. J Lab Clin Med 128:128–133

Krieg AM, Matson S, Fisher E (1996) Oligodeoxynucleotide modifications determine the magnitude of B cell stimulation by CpG motifs. Antisense Nucleic Acid Drug Dev 6:133–139

Krieg AM, Yi AK, Schorr J, Davis H (1998) The role of CpG oligonucleotides in DNA vaccines. Trends Microbiol 6:23–27

Lafferty KJ, Woolnough J (1977) The origin and mechanism of the allograft reaction. Immunol Rev 35:231–262

Lipford GB, Bauer M, Blank C, Reiter R, Wagner H, Heeg K (1997a) CpG-containing synthetic oligonucleotides promote B and cytotoxic T cell responses to protein antigen: a new class of vaccine adjuvants. Eur J Immunol 27:2340–2344

Lipford GB, Sparwasser T, Bauer M, Zimmermann S, Koch ES, Heeg K, Wagner H (1997b) Immunostimulatory DNA: sequence dependent production of potentially harmful or useful cytokines. Eur J Immunol 27:3420–3426

Lipford GB, Heeg K, Wagner H (1998) Bacterial DNA as immune cell activator. Trends Microbiol 6:496–500

Lipford GB, Bendigs S, Wagner H, Heeg K (1999) Bacterial CpG DNA mediated costimulation: Sequence dependence but independence of APC. submitted

Locksley RM, Louis JA (1992) Immunology of leishmaniasis. Curr Opin Immunol 4:413–418

Lutz MB, Girolomoni G, Ricciardi-Castagnoli P (1996) The role of cytokines in functional regulation and differentiation of dendritic cells. Immunobiology 195:431–455

Macatonia SE, Hosken NA, Litton M, Vieira P, Hsieh CS, Culpepper JA, Wysocka M, Trinchieri G, Murphy KM, O'Garra A (1995) Dendritic cells produce IL-12 and direct the development of Th1 cells from naive CD4+ T cells. J Immunol 154:5071–5079

Macfarlane DE, Manzel L (1998) Antagonism of immunostimulatory CpG-Oligodeoxynucleotides by quinacrine, chloroquine, and structurally related compounds. J Immunol 160:1122–1131

Mattern T, Thanhäuser A, Reiling N, Toellner KM, Duchrow M, Kusumoto S, Rietschel ET, Ernst M, Brade H, Flad HD, Ulmer AJ (1994) Endotoxin and lipid A stimulate proliferation of human T cells in the presence of autologous monocytes. J Immunol 153:2996–3004

Matzinger P (1994) Tolerance, danger, and the extended family. Annu Rev Immunol 12:991–1045

McCaffrey PG, Luo C, Kerppola TK, Jain J, Badalian TM, Ho AM, Burgeon E, Lane WS, Lambert JN, Curran T, Verdine GL, Rao A, Hogan PG (1993) Isolation of the cyclosporin-sensitive T cell transcription factor NFATp. Science 262:750–754

Medzhitov R, Janeway CA (1997) Innate immunity: the virtues of a nonclonal system of recognition. Cell 91:295–298

Mosmann TR, Sad S (1996) The expanding universe of T cell subsets: Th1, Th2 and more. Immunol Today 17:138–146

Mueller DL, Jenkins MK, Schwartz RH (1989) Clonal expansion versus functional clonal inactivation: a costimulatory signalling pathway determines the outcome of T cell antigen receptor occupancy. Annu Rev Immunol 7:445–480

Pesole G, Liuni S, Grillo G, Saccone C (1997) Structural and compositional features of untranslated regions of eukaryotic mRNAs. Gene 205:95–102

Pisetsky DS (1996a) Immune activation by bacterial DNA: a new genetic code. Immunity 5:303–310

Pisetsky DS (1996b) The immunologic properties of DNA. J Immunol 156:421–423

Pisetsky DS (1997) Immunostimulatory DNA: a clear and present danger? Nature Med 3:829–831

Roman M, Martin-Orozco E, Goodman JS, Nguyen M-D, Sato Y, Ronaghy A, Kornbluth RS, Richman DD, Carson DA, Raz E (1997) Immunostimulatory DNA sequences function as T helper-1-promoting adjuvants. Nature Med 3:849–854

Sagerström CG, Kerr EM, Allison JP, Davis MM (1993) Activation and differentiation requirements of primary T cells in vitro. Proc Natl Acad Sci USA 80:8987–8991

Schuler G, Thurner B, Romani N (1997) Dendritic cells: from ignored cells to major players in T cell mediated immunity. Int Arch Allergy Immunol 112:317–322

Sigal NH, Dumont FJ (1992) Cyclosporin A, FK-506, and Rapamycin: pharmacologic probes of lymphocyte signal transduction. Annu Rev Immunol 10:519–560

Sparwasser T, Miethke T, Lipford GB, Erdmann A, Häcker H, Heeg K, Wagner H (1997) Macrophages sense pathogens via DNA motifs: Induction of TNF-alpha mediated shock. Eur J Immunol 27: 1671–1679

Sparwasser T, Koch ES, Vabulas RM, Heeg K, Lipford GB, Ellwart J, Wagner H (1998) Bacterial DNA and immunostimulating CpG oligonucleotides trigger maturation and activation of murine dendritic cells. Eur J Immunol 28:2045–2054

Stacey KJ, Sweet MJ, Hume DA (1996) Macrophages ingest and are activated by bacterial DNA. J Immunol 157:2116–2122

Steinman RM, Pack M, Inaba K (1998) Dendritic cells in the T cell areas of lymphoid organs. Immunol Rev 156:25–37

Stenger S, Thüring H, Röllinghoff M, Bogdan C (1994) Tissue expression of inducible nitric oxide synthase is closely associated with resistance to *Leishmania major*. J Exp Med 180:783–793

Sun S, Kishimoto H, Sprent J (1998a) DNA as an adjuvant: Capacity of insect DNA and synthetic oligodeoxynucleotides to augment T-cell responses to specific antigen. J Exp Med 187:1145–1150

Sun S, Zhang X, Tough DF, Sprent J (1998b) Type I interferon-mediated stimulation of T cells by CpG DNA. J Exp Med 188:2335–2342

Tough DF, Sun S, Sprent J (1997) T cell stimulation in vivo by lipopolysaccharide (LPS). J Exp Med 185:2089–2094

Vremec D, Shortman K (1997) Dendritic cell subtypes in mouse lymphoid organs; Cross-correlation of surface markers, changes with incubation, and differences among thymus, spleen, and lymph nodes. J Immunol 159:565–573

Yamamoto S, Yamamoto T, Kataoka T, Kuramoto E, Yano O, Tokunaga T (1992) Unique palindromic sequences in synthetic oligonucleotides A required to induce IFN and augment IFN-mediated natural killer activity. J Immunol 148:4072–4076

Zhang X, Sun S, Hwang I, Tough DF, Sprent J (1998) Potent and selective stimulation of memory-phenotype CD8+ T cells in vivo by IL-15. Immunity 8:591–599

Zimmermann S, Egeter O, Hausmann S, Lipford GB, Röcken M, Wagner H, Heeg K (1998) CpG oligodeoxynucleotides trigger protective and curative Th1 responses in lethal murine leishmaniasis. J Immunol 160:3627–3630

# Role of Type I Interferons in T Cell Activation Induced by CpG DNA

S. Sun[1] and J. Sprent[2]



## 1 Introduction

As a result of early studies on mycobacteria (Tokunaga et al. 1992; Yamamoto et al. 1992), it is now well established that DNA from nonvertebrates, as well as certain synthetic oligodeoxynucleotides (ODNs), can cause strong activation of lymphoid cells, especially B cells, natural killer (NK) cells and antigen-presenting cells (APCs) (Messina et al. 1991; Krieg et al. 1995; Ballas et al. 1996; Stacey et al. 1996; Sparwasser et al. 1998); such activation leads to B-cell proliferation and enhanced production of cytokines by APCs and NK cells. The immunostimulatory properties of DNA/ODNs are controlled by unmethylated CpG dinucleotide motifs with optimal flanking sequences (Krieg et al. 1995). Reflecting the paucity of these motifs in vertebrate DNA – which is generally heavily methylated – purified DNA from humans, mice, frogs and fish (and also plants) are nonstimulatory for lymphoid cells, at least in terms of eliciting B-cell proliferation (Sun et al. 1997). By contrast, DNA from nonvertebrates, which is largely unmethylated, is strongly

[1] R.W. Johnson Pharmaceutical Research Institute, 3210 Merryfield Row, San Diego, CA 92121, USA
[2] Department of Immunology, IMM4, The Scripps Research Institute, 10550 N. Torrey Pines Road, La Jolla, California 92037, USA
E-mail: jsprent@scripps.edu

stimulatory for B cells (SUN et al. 1997); this applies to DNA from bacteria, insects, nematodes, yeasts and mollusks. For bacteria, insect and yeast DNA, selective methylation of CpG motifs abolished immunostimulatory activity (KRIEG et al. 1995; SUN et al. 1996, 1997).

The discovery that the immunostimulatory properties of DNA are restricted to unmethylated CpG motifs (plus appropriate flanking sequences) could explain the well-known phenomenon of "CpG suppression" found in vertebrate DNA (BIRD 1980). Thus, the low frequency of CpG motifs found in vertebrate DNA may reflect an evolutionary pressure to avoid DNA-induced stimulation of the immune system (SUN et al. 1997).

In the case of B cells and APCs, stimulation by CpG motifs in nonvertebrate DNA or ODNs (henceforth referred to as CpG DNA/ODNs) appears to involve endocytosis of the ligands, followed by initiation of various intracellular signalling pathways (HACKER et al. 1998; YI and KRIEG 1998; YI et al. 1998). Such signalling causes B cells to proliferate and differentiate into antibody-secreting cells, whereas APCs are induced to synthesize various cytokines, such as interleukin 1 (IL-1), IL-12, tumor necrosis factor α, interferon α (IFNα) and IFNβ, and also to upregulate the expression of major histocompatibility complex (MHC) class II and co-stimulatory molecules (SUN et al. 1996; SPARWASSER et al. 1998). Information on the capacity of CpG DNA/ODNs to stimulate other types of lymphoid cells is still sparse.

For T cells, recent evidence suggests that CpG DNA/ODNs act as efficient adjuvants for stimulating antigen-specific proliferative responses of $CD4^+$ and $CD8^+$ cells (CHU et al. 1997; LIPFORD et al. 1997; ROMAN et al. 1997; WEINER et al. 1997; DAVIS et al. 1998; SUN et al. 1998a); for $CD4^+$ cells, the adjuvant effect promotes differentiation into T-helper 1 cells and preferentially stimulates production of immunoglobulin $G2_a$ ($IgG2_a$) and $IgG2_b$ antibodies. Whether the adjuvant effect of CpG DNA/ODNs reflects a direct action on T cells, APCs or both is still unclear. Evidence favoring a direct effect on T cells has come from the finding that CpG ODNs can augment the proliferative response of purified T cells to anti-T-cell-receptor (TCR) monoclonal antibodies (mAbs) in vitro (LIPFORD et al. 1997), implying that CpG ODNs can provide a co-stimulatory signal to T cells. Alternatively, CpG DNA/ODNs may act largely by improving the function of APCs, e.g. by upregulating the expression of co-stimulatory molecules and/or inducing synthesis of stimulatory cytokines. On this second possibility, we have recently found that CpG DNA/ODNs can activate T cells via APC-production of type-I (α,β) interferons (IFN-I) (SUN et al. 1998b). The data are summarized below.

## 2 T-Cell Activation in Vivo Following Injection of Insect DNA

In prior studies, we observed that proliferative responses of naïve $CD4^+$ and $CD8^+$ cells to specific antigens in vivo were augmented when the antigen was co-injected with CpG DNA (insect DNA) or CpG ODNs (SUN et al. 1998a); control ODNs containing methylated rather than unmethylated CpG motifs (ZpG ODNs) were ineffective. This adjuvant effect on proliferation was not seen when CpG DNA/ODNs were injected without antigens. In fact, CpG DNA/ODNs proved incapable of inducing proliferation of naïve T cells. To our surprise, however, injection of CpG DNA/ODNs caused moderate to strong upregulation of a variety of cell-surface molecules, including CD69, B7–2, class I, Ly6C and intercellular adhesion molecule 1 (ICAM-1). This finding applied to $CD4^+$ and $CD8^+$ T cells, B cells and APCs (macrophages and dendritic cells) and was most prominent within the first 24h of injection; certain other molecules, e.g. CD25 (IL-2Rα), were not upregulated. Data illustrating upregulation of surface markers on $CD4^+$ and $CD8^+$ T cells and $B220^+$ B cells in the draining lymph nodes (LNs) of normal C57BL/6 (B6) mice injected with insect DNA 18h before are shown in Fig. 1.

For normal T cells, the expression of certain cell-surface markers, notably CD62L, CD45R and CD44, splits these cells into discrete subsets of naïve- and memory-phenotype cells (Tough and Sprent 1994). Thus, for CD44 expression, typical naïve T cells express low to intermediate levels of CD44 ($CD44^{lo/int}$), whereas memory cells express high levels of CD44 ($CD44^{hi}$). As shown in Fig. 2, the upregulation of surface markers on T cells following injection of either insect DNA or CpG ODNs was not obviously skewed to particular subsets of T cells. Thus, both for $CD4^+$ and $CD8^+$ cells, upregulation of CD69 applies to naïve ($CD44^{lo/int}$) T cells as well as to memory ($CD44^{hi}$) cells.

## 3 The Role of APC

As in vivo, the capacity of CpG DNA/ODNs to induce upregulation of surface markers on T cells also occurs in vitro (Sun et al. 1998b). Significantly, however, this finding only applied when the cells cultured contained APCs, e.g. when CpG DNA/ODNs were added to unseparated spleen cells. Thus, when highly purified T cells were cultured with CpG DNA/ODNs, upregulation of surface markers on T cells was virtually undetectable. However, strong expression of these markers occurred when purified T cells were supplemented with APCs, e.g. spleen cells from

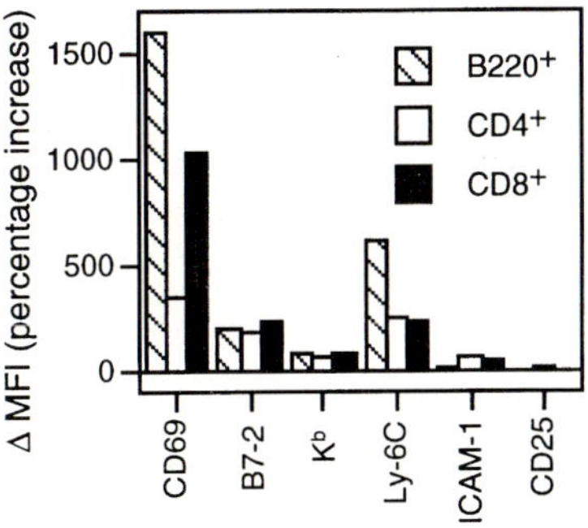


**Fig. 1.** DNA-induced upregulation of various cell-surface molecules on lymphoid subsets from B6 mice. B6 mice were injected with *Drosophila melanogaster* DNA (100μg/mouse); 18h later, draining lymph-node cells were double stained for expression of the markers shown and also for B220, CD4 or CD8 expression. The data are expressed as the percentage increase in the mean fluorescence intensity (ΔMFI) of staining for cells from DNA-injected mice versus phosphate-buffered-saline-injected mice. Adapted from Sun et al. (1998b)

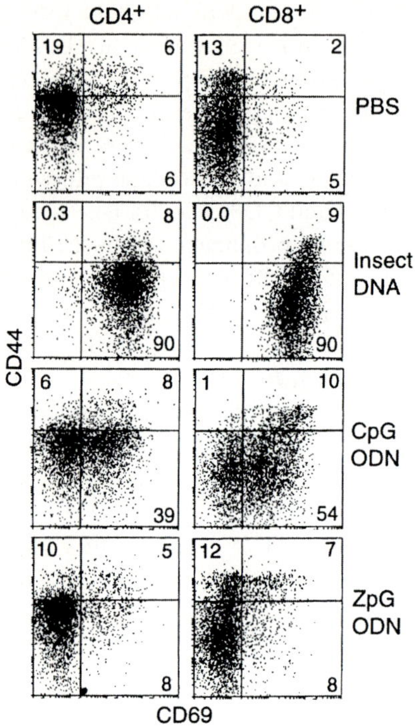


**Fig. 2.** CD69 upregulation of T cells in response to *Drosophila melanogaster* DNA or synthetic oligodeoxynucleotides (ODNs) in vivo. B6 mice were injected with *D. melanogaster* DNA (100μg/mouse) or ODNs (50μg/mouse); 18h later, draining lymph-node cells were triple stained for CD4 or CD8 and also for CD44 and CD69 expression. Adapted from Sun et al. (1998b)

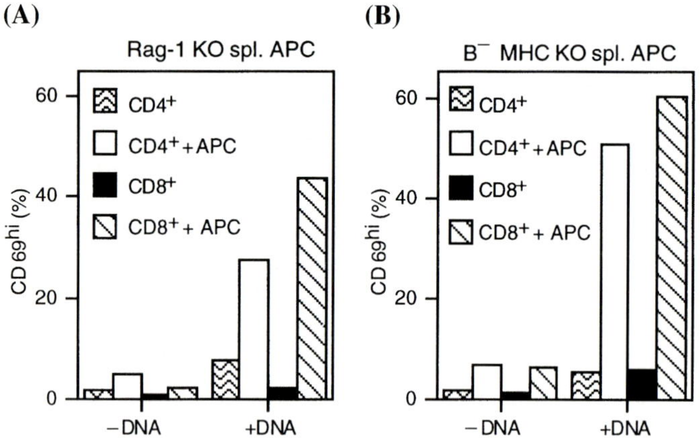


**Fig. 3A,B.** DNA-induced upregulation of CD69 on T cells requires antigen-presenting cells. Purified $CD4^+$ or $CD8^+$ cells ($10^6$ cells/well) were cultured overnight in the presence or absence of *Drosophila melanogaster* DNA (100μg/ml), either alone or in the presence of Rag-$1^{-/-}$ spleen cells (**A**) or B-cell-depleted spleen cells from mahor histocompatibility complex $I^{-/-}$ $II^{-/-}$ mice (**B**); spleen cells were added at $2\times10^6$ cells/well. Cells were then stained for CD8 and CD69. The data show the percentage of gated $CD8^+$ cells that were positive for CD69 expression relative to unstimulated $CD8^+$ cells. Adapted from Sun et al. (1998b)

RAG-$1^{-/-}$ mice (which lack T and B cells) (Fig. 3A) or B-depleted spleen cells from mice lacking MHC class-I and -II molecules (Fig. 3B).

These findings led to three conclusions. First, CpG DNA/ODNs appeared to have no effect on purified T cells. Second, CpG DNA/ODNs acted on T cells when accompanied by APCs. Third, to activate T cells, APCs did not have to express MHC molecules, implying that the function of APCs did not depend upon TCR/MHC interaction.

## 4 The Role of IFN-I

Since CpG DNA/ODNs are known to induce APC to synthesize various cytokines, it seemed likely that APC-dependent activation of T cells by CpG DNA/ODNs was mediated by cytokines. In considering which particular cytokines might be involved, it is notable that the surface markers upregulated by CpG DNA/ODNs included Ly6C. This finding is of interest, because only one cytokine, namely IFN-I, is known to induce Ly6C upregulation (Dumont and Coker 1986).

To seek direct evidence for the role of IFN-I, we examined the effects of injecting insect DNA into mice lacking receptors for IFN-I, i.e. IFN-IR$^{-/-}$ mice; since these mice were on a 129 background, normal 129 mice were used as a control. The results were clear-cut (Fig. 4). Thus, when insect DNA was injected in vivo (Fig. 4A) or added to a mixture of T cells and APCs in vitro (Fig. 4B), surface-marker upregulation on T cells was prominent with normal 129 cells but very low or undetectable with 129 IFN-IR$^{-/-}$ cells; this finding applied to B cells as well as to CD4$^{+}$ and CD8$^{+}$ T cells.

The above findings indicated that APC-dependent stimulation of T cells in response to CpG DNA was mediated largely or solely by IFN-I, presumably by a direct action of IFN-I on T cells. In favor of this idea, the failure of IFN-IR$^{-/-}$ T cells to respond to insect DNA in vitro could not be overcome by addition of normal IFN-IR$^{+}$ APCs, i.e. RAG-1$^{-/-}$ spleen cells (Fig. 4B).

A key prediction of the above findings was that adding IFN-I to purified normal T cells in vitro would induce upregulation of surface markers in the absence of APCs. This was indeed the case. Moreover, the range of surface markers upregulated when purified T cells were cultured with IFN-I (IFN-β) was virtually the same as when APC-containing T-cell populations were exposed to CpG DNA/ODNs. The capacity of IFN-I to induce upregulation of CD69 and B7-2 on purified CD8$^{+}$ cells in vitro is shown in Fig. 5. In control studies, culturing purified T cells with IFN-γ induced modest upregulation of ICAM-1 but no upregulation of CD69, B7-2, Ly6C or MHC class I.

## 5 The Adjuvant Effect of IFN-I

In prior studies, we observed that poly-I:C, a powerful inducer of IFN-I, acted as an adjuvant for the primary proliferative response of 2C TCR transgenic CD8$^{+}$

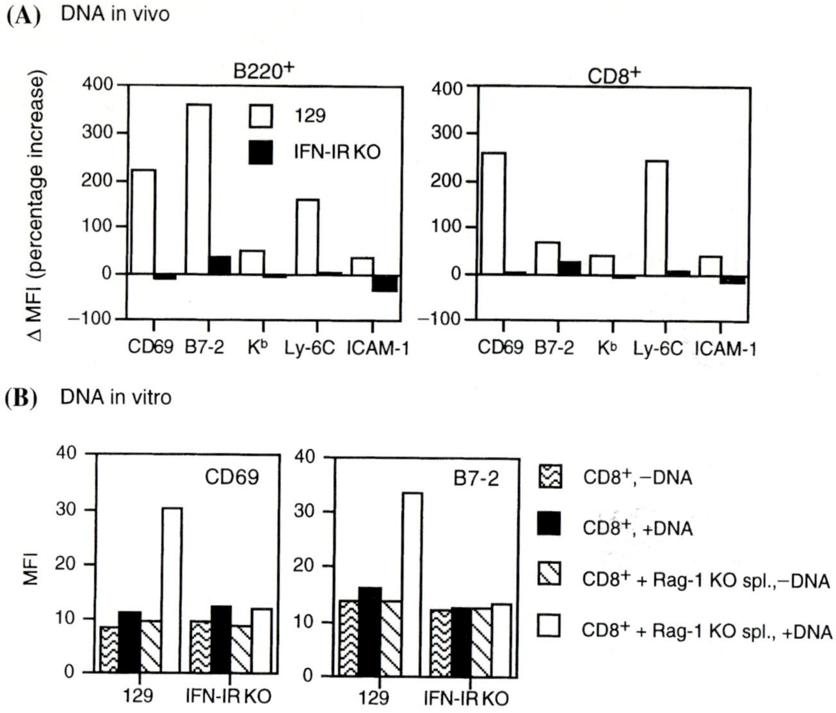


**Fig. 4A,B.** DNA-induced upregulation of various cell surface molecules in interferon-I receptor (IFN-IR) knockout (KO) mice versus 129 wild-type control mice. **A** IFN-IR KO mice and 129 wild-type mice were injected with *Drosophila melanogaster* DNA (100μg/mouse); 18h later, draining lymph-node cells were double stained for expression of the markers shown and for B220, CD4 or CD8 expression. The data are expressed as the percentage increase in the mean fluorescence intensity (ΔMFI) of staining for cells from DNA-injected mice versus phosphate-buffered-saline-injected mice. **B** Purified $CD8^+$ cells from 129 or IFN-IR KO mice ($10^6$ cells/well) were cultured with or without *D. melanogaster* DNA (100μg/ml), either alone or with RAG-$1^{-/-}$ spleen antigen-presenting cells ($2 \times 10^6$ cells/well), then stained for CD8 and CD69 or B7-2 18h later. The data are expressed as the MFI of staining for the markers shown, gating on $CD8^+$ cells. Adapted from Sun et al. (1998b)

cells responding to specific antigens (Tough et al. 1996). Similar results were found with CpG DNA/ODNs (Sun et al. 1998a). In both situations, co-injection of poly-I:C or CpG DNA/ODNs considerably augmented the T proliferative responses to specific antigens. In light of these findings, we have been assessing the adjuvant effects of purified IFN-I (IFN-β), using IFN-γ as a control. A summary of our unpublished data on this topic is given below.

In initial experiments, we examined whether IFN-I could amplify the in vitro proliferative response of 2C TCR transgenic $CD8^+$ cells to specific peptides. This was indeed the case. Thus, when 2C $CD8^+$ cells were cultured with peptides presented by normal spleen APCs, addition of IFN-I considerably augmented the T proliferative response; IFN-γ, by contrast, had no effect. The adjuvant effect of IFN-I also applied in vivo. Thus, as found previously with poly-I:C (Tough et al.

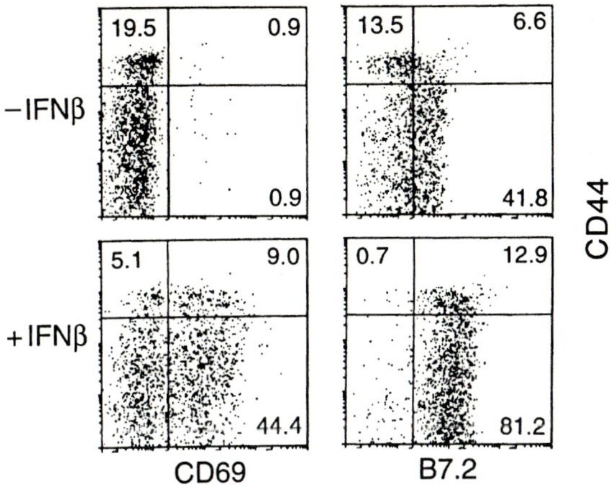


**Fig. 5.** Upregulation of CD69 and B7-2 on purified $CD8^+$ cells in response to interferon β (IFNβ) in vitro. Purified $CD8^+$ cells ($2 \times 10^6$ cells/well) were cultured with or without IFNβ (10,000 units/ml) overnight; cells were then triple stained for CD8, CD44 and CD69 or B7-2. The data shown are for gated $CD8^+$ cells

1996) and CpG DNA/ODNs (Sun et al. 1998a), injection of IFN-I enhanced the proliferative response of 2C $CD8^+$ cells to specific peptides in vivo. To our surprise, however, under certain experimental conditions, IFN-I did not act as an adjuvant but instead inhibited the T proliferative response.

## 6 Anti-Proliferative Effects of IFN-I

It should be noted that the adjuvant effect of IFN-I was only apparent during the later stages of the T proliferative response, i.e. on day 3 and thereafter; this applied both in vivo and in vitro. At earlier stages, i.e. on day 2, T proliferative responses were markedly inhibited by IFN-I (but not by IFN-γ). Under certain conditions, IFN-I also inhibited late (day 3, 4) proliferative responses. Here, the presence or absence of viable APCs seemed to be the key. Thus, with T proliferative responses driven by peptides presented by normal viable spleen APCs, IFN-I inhibited early responses but augmented late responses. By contrast, when responses occurred in the absence of viable APCs, IFN-I inhibited both early and late proliferative responses. This latter finding applied to naïve T-cell proliferative responses elicited by cross-linked anti-TCR plus anti-CD28 mAbs; it also applies to proliferation of memory-phenotype ($CD44^{hi}$) $CD8^+$ cells in response to a cytokine, IL-15. Similar findings applied to proliferation of 2C $CD8^+$ cells to peptides presented by transfected *Drosophila* cells as APCs; these cells die rapidly at 37°C (Cai et al. 1996) and are presumably unresponsive to IFN-I.

Our interpretation of these findings is that IFN-I has two opposing effects on T-cell function. In the absence of viable APCs, IFN-I acts directly on T cells, presumably via IFN-IR, and downregulates the capacity of the cells to mount a proliferative response. When viable APCs are present, however, stimulation of APCs by IFN-I somehow counters the inhibitory effects of IFN-I on T cells and leads to enhanced proliferative responses.

## 7 Concluding Comments

As with other nonspecific stimuli, such as lipopolysaccharides (LPS) and poly-I:C, the effects of exposing lymphoid cells to CpG DNA/ODNs are highly complex. For some cell types, especially B cells and dendritic cells, CpG DNA/ODNs appear to act directly and cause upregulation of certain cell-surface markers (including co-stimulatory molecules, such as B7) and synthesis of various cytokines.

Direct activation of APCs would seem the most likely explanation for the strong adjuvant function of CpG DNA/ODNs. This notion rests on the assumption that resting APCs function poorly and need to be activated in order to present antigens effectively to naïve T cells. However, the precise difference between "efficient" and "inefficient" APCs is still unclear. The simplest idea is that the efficiency of APCs is a direct reflection of the range and density of the co-stimulatory/adhesion molecules on the cell surface. The data on the APC function of transfected *Drosophila* cells are in accordance with this idea. Thus, in terms of presenting specific peptides to purified naïve 2C transgenic $CD8^+$ cells, class-I-($L^d$)-transfected *Drosophila* cells are highly efficient APCs when these cells co-express B7 (B7-1 or B7-2) and ICAM-1 but are totally nonfunctional without these co-stimulatory molecules (Cai et al. 1996). Interestingly, these data only apply when purified T cells are used. With unseparated 2C spleen cells as responders, the results are quite different (Sun et al. 1996). Here, significant T proliferative responses to peptides occur, with *Drosophila* cells expressing only class-I molecules alone. The explanation for this paradox is that CpG DNA released from the *Drosophila* cells acts on the B cells and APCs in spleen and induces upregulation of B7 and ICAM-1. Provided that high concentrations of peptides are used, these activated cells then provide bystander co-stimulation for the responding $CD8^+$ cells. It is worth noting that, even with high concentrations of peptides, *Drosophila* APCs expressing class-I molecules alone are totally non-stimulatory for purified $CD8^+$ cells. Hence, in this system, we have not found evidence that CpG DNA can provide a direct co-stimulatory signal for T cells (Sun et al. 1996; Lipford et al. 1997).

In considering the above data, the point to emphasize is that co-expression of just two co-stimulatory molecules, B7 and ICAM-1, enables class-I-transfected *Drosophila* cells to display strong APC function for purified T cells in the absence of added cytokines. Since the endogenous "cytokines" made by *Drosophila* cells are unlikely to affect mammalian cells (T cells), it would seem to follow that, for normal APCs, efficient presentation of peptides to T cells simply requires that the APCs express high levels of appropriate co-stimulatory molecules. Hence, if adjuvants are able to induce direct upregulation of costimulatory molecules on APCs – which seems to be the case – concomitant production of cytokines by APC may be redundant. But is this the case in a normal physiological setting?

With normal APCs in intact animals, assessing whether the adjuvant effect of CpG DNA/ODNs is due in part to cytokine production is clearly difficult, because the range of cytokines released by APCs after activation by adjuvants is considerable. One approach to this problem is to investigate whether individual APC-derived cytokines have adjuvant activity. Based on the effects of IFN-I considered earlier, this is indeed the case. Thus, both in vitro and in vivo, purified IFN-I acts as a strong adjuvant for APC-induced T proliferative responses (unpublished data of the authors). In light of this finding, it would seem quite likely that the adjuvant function of compounds such as CpG DNA/ODNs is not due solely to direct upregulation of co-stimulatory molecules on APCs but also involves the production of stimulatory cytokines, such as IFN-I.

How IFN-I exerts its adjuvant activity is still unclear. It seems that the most likely possibility is that, like CpG DNA/ODNs, IFN-I functions by augmenting upregulation of co-stimulatory molecules on APCs. This idea may be an oversimplification, however, because in our experiments, IFN-$\gamma$ induced strong upregulation of co-stimulatory molecules on APCs but, at least in vitro, failed to display adjuvant activity. Hence, one has to consider other possibilities. For example, IFN-I may induce APCs to produce cytokines with direct co-stimulatory functions for T cells, e.g. IL-15 and IL-6. IFN-I could also act at the T-cell level, e.g. by causing T cells to express receptors for stimulatory cytokines released by APCs.

Although direct evidence for these and other possibilities is still sparse, it is of interest that IFN-I does act directly on T cells, including naïve cells, and causes these cells to enter a state of partial activation. Such activation is associated with upregulation of various cell-surface markers but does not lead to cell division. Indeed, in the absence of viable APCs, it is striking that the partial activation of T cells induced by IFN-I is associated with a reduced capacity to mount T proliferative responses. The cells are not simply "anergic", however, because we have found no evidence that IFN-I inhibits T-cell production of IL-2. Based on studies with cell lines (Grander et al. 1997), the anti-proliferative function of IFN-I on naïve T cells may reflect intracellular production of cell-cycle inhibitors. If so, the paradox remains that the presence of viable APCs reverses the anti-proliferative effect of IFN-I and causes enhanced T proliferative responses. How APCs negate the anti-proliferative function of IFN-I has yet to be resolved.

As a whole, the data on the adjuvant effects of CpG DNA/ODNs suggest that these compounds act primarily at the APC level and lead both to direct upregulation of co-stimulatory molecules and release of various cytokines. For naïve T cells responding to specific antigens, the combination of APC activation and cytokine production elicited by CpG DNA/ODNs and other adjuvants enhances the T proliferative response and promotes differentiation into long-lived memory T cells. It is important to emphasize that T-cell activation and proliferation in this situation is heavily antigen dependent; without antigens, T cells undergo partial activation, e.g. in response to IFN-I, but do not enter the cell cycle. However, this scenario applies only to naïve T cells. Thus, when memory-phenotype T cells are exposed to adjuvants, the cytokines released from activated APCs can act directly on T cells and induce proliferation in the apparent absence of antigen (Tough et al. 1996). Such "bystander" proliferation of memory-phenotype T cells is especially prominent for $CD8^+$ cells ($CD44^{hi}$ $CD8^+$ cells) and can be evoked not only by CpG DNA/ODNs but also by other compounds with adjuvant activity, notably poly-I:C and LPS (Tough et al. 1996, 1997). How these agents – and also certain cytokines, such as IFN-I – induce proliferation of $CD44^{hi}$ $CD8^+$ cells in vivo is still unclear, but secondary production of stimulatory cytokines, such as IL-15, is a likely possibility.

*Acknowledgements.* We thank Ms. Barbara Marchand for typing the manuscript. This work was supported by grants CA38355, CA25803, AI21487, AI32068 and AG01743 from the United States Public Health Service. This is publication no. 12245-IMM from the Scripps Research Institute.

## References

Ballas ZK, Rasmussen WL, Krieg AM (1996) Induction of NK activity in murine and human cells by CpG motifs in oligodeoxynucleotides and bacterial DNA. J Immunol 157:1840–1845

Bird AP (1980) DNA methylation and the frequency of CpG in animal DNA. Nucleic Acids Res 8: 1499–1504

Cai Z, Brunmark A, Jackson MR, Loh D, Peterson PA, Sprent J (1996) Transfected Drosophila cells as a probe for defining the minimal requirements for stimulating unprimed $CD8^+$ T cells. Proc Natl Acad Sci USA 93:14736–14741

Chu RS, Targoni OS, Krieg AM, Lehman PV, Harding CV (1997) CpG oligodeoxynucleotides act as adjuvants that switch on T helper 1 (Th1) immunity. J Exp Med 186:1623–1631

Davis HL, Weeranta R, Walsschmidt TJ, Tygrett L, Schorr J, Krieg AM (1998) CpG DNA is a potent enhancer of specific immunity in mice immunized with recombinant Hepatitis B surface antigen. J Immunol 160:870–876

Dumont FJ, Coker LZ (1986) Interferon-$\alpha/\beta$ enhances the expression of Ly-6 antigens on T cells in vivo and in vitro. Eur J Immunol 16:735–740

Grander D, Sangfelt O, Erickson S (1997) How does interferon exert its cell growth inhibitory effect? Eur J Haematol 59:129–135

Hacker H, Mischak H, Miethke T, Liptay S, Schmid R, Sparwasser T, Heeg K, Lipford GB, Wagner H (1998) CpG-DNA-specific activation of antigen-presenting cells requires stress kinase activity and is preceded by non-specific endocytosis and endosomal maturation. Embo J 17:6230–6240

Krieg AM, Yi AK, Matson S, Waldschmidt TJ, Bishop GA, Teasdale R, Koretzky GA, Klinman DM (1995) CpG motifs in bacterial DNA trigger direct B-cell activation. Nature 374:546–549

Lipford GB, Bauer M, Blank C, Reiter R, Wagner H, Heeg K (1997) CpG-containing synthetic oligonucleotides promote B and cytotoxic T cell responses to protein antigen: a new class of vaccine adjuvants. Eur J Immunol 27:2340–2344

Messina JP, Gilkeson GS, Pisetsky DS (1991) Stimulation of in vitro murine lymphocyte proliferation by bacterial DNA. J Immunol 147:1759–1764

Roman M, Martin-Orozco E, Goodman JS, Nguyen M-D, Sato Y, Ronaghy A, Kornbluth RS, Richman DD, Carson DA, Raz E (1997) Immunostimulatory DNA sequences function as T helper-1-promoting adjuvants. Nature Med 3:849–854

Sparwasser T, Koch ES, Vabulas RM, Heeg K, Lipford GB, Ellwart JW, Wagner H (1998) Bacterial DNA and immunostimulatory CpG oligonucleotides trigger maturation and activation of murine dendritic cells. Eur J Immunol 28:2045–2054

Stacey KJ, Sweet MJ, Hume DA (1996) Macrophages ingest and are activated by bacterial DNA. J Immunol 157:2116–2122

Sun S, Beard C, Jaenisch R, Jones P, Sprent J (1997) Mitogenicity of DNA from different organisms for murine B cells. J Immunol 159:3119–3125

Sun S, Cai Z, Langlade-Demoyen P, Kosaka H, Brunmark A, Jackson MR, Peterson PA, Sprent J (1996) Dual function of *Drosophila* cells as APC for naive $CD8^+$ T cells: implications for tumor immunotherapy. Immunity 4:555–564

Sun S, Kishimoto H, Sprent J (1998a) DNA as an adjuvant: capacity of insect DNA and synthetic oligodeoxynucleotides to augment T cell responses to specific antigen. J Exp Med 187:1145–1150

Sun S, Zhang X, Tough DF, Sprent J (1998b) Type I interferon-mediated stimulation of T cells by CpG DNA. J Exp Med 188:2335–2342

Tokunaga T, Yano O, Kuramoto E, Kimura Y, Yamamoto T, Kataoka T, Yamamoto S (1992) Synthetic oligonucleotides with particular base sequences from the cDNA encoding proteins of mycobacterium bovis BCG induce interferons and activate natural killer cells. Microbiol Immunol 36:55–66

Tough DF, Borrow P, Sprent J (1996) Induction of bystander T cell proliferation by viruses and type I interferon in vivo. Science 272:1947–1950

Tough DF, Sprent J (1994) Turnover of naive- and memory-phenotype T cells. J Exp Med 179:1127–1135

Tough DF, Sun S, Sprent J (1997) T cell stimulation in vivo by lipopolysaccharide (LPS). J Exp Med 185:2089–2094

Weiner GJ, Liu HM, Wooldridge JE, Dahle CE, Krieg AM (1997) Immunostimulatory oligodeoxynucleotides containing the CpG motif are effective as immune adjuvants in tumor antigen immunization. Proc Natl Acad Sci USA 94:10833–10837

Yamamoto S, Yamamoto T, Shimada S, Kuramoto E, Yano O, Kataoka T, Tokunaga T (1992) DNA from bacteria, but not from vertebrates, induces interferons, activates natural killer cells and inhibits tumor growth. Microbiol Immunol 36:983–997

Yi AK, Krieg AM (1998) CpG DNA rescue from anti-IgM-induced WEHI-231 B lymphoma apoptosis via modulation of IκBα and IκBβ and sustained activation of nuclear factor-κ B/c-Rel. J Immunol 160:1240–1245

Yi AK, Tuetken R, Redford T, Waldschmidt M, Kirsch J, Krieg AM (1998) CpG motifs in bacterial DNA activate leukocytes through the pH-dependent generation of reactive oxygen species. J Immunol 160:4755–4761

# Hematopoietic Remodeling Triggered by CpG DNA

G.B. LIPFORD and T. SPARWASSER



## 1 Introduction

During acute bacterial infection, significant hematological alterations occur, most of which are simply monitored as blood leukocytosis (SELIG and NOTHDURFT 1995). Hematopoietic growth factors acting as regulators maintain the steady state between production and consumption of mature blood cells. Particularly under stress conditions, such as infection, these factors play a major role in cellular adaptation processes (CANNISTRA and GRIFFIN 1988). Many of these growth factors not only influence the proliferation and commitment at the stem cell and progenitor cell level, but also signal for a rapid mobilization of hematopoietic progenitor cells from the bone marrow (BM) into the blood and distant hematopoietic organs, thus remodeling cellular compartmentalization (MORRISON et al. 1995). Additionally, growth factors and other cytokines play a major role in functional adaptation at the level of the mature cell compartment, stimulating metabolic, cytotoxic, or phagocytic activities. Because, at any given time, progenitors in different hematopoietic organs are exposed to different cytokines and growth factors, the adaptive demand induced by infection on cellular composition and status can be met.

Institute of Medical Microbiology, Immunology and Hygiene, Technical University Munich, Trogerstr. 9, D-81675 Munich, Germany

The mobilization of stem cells is evolutionarily conserved in all species tested thus far, including mouse, dog, monkey, and human. However, the selective forces for the conservation of this phenomenon are not immediately apparent (Morrison et al. 1995). It has long been recognized, in experimental animals, that splenomegaly, mobilization of hematopoietic precursors into the blood, and increased numbers of colony-forming cells (CFCs) are induced not only by gram-positive and gram-negative bacteria but also by bacterial subcellular fractions, such as the mycobacterial component of complete Freund's adjuvant (CFA), lipopolysaccharides (LPSs), and muramyl dipeptide. Additionally, synthetic double-stranded polyribonucleotides, such as Poly I–Poly C, show enhancement in CFCs. We add to this list bacterial CpG DNA, recently shown to trigger hematopoietic events after injection in mice. Apparently, recognition of bacterial CpG DNA as a foreign bacterial product signals infectious danger and induces the release of a variety of colony-stimulating factors (CSFs), cytokines, and chemokines that facilitate hematopoietic events.

## 2 Infection Danger Induces Hematopoietic Mobilization

Macrophages and dendritic cells (DCs) are equipped with pattern-recognition receptors (PRRs), which recognize conserved molecular structures shared by large groups of pathogens. The main difference between PRRs and the clonally distributed antigen receptors of T and B cells is that the specificities of PRRs are germline encoded. Thus, one parameter imposed on PRRs is the recognition of non-self through pathogen-associated structural patterns. The innate immune system controls the initiation of the adaptive immune response by regulating the expression of co-stimulatory activity on antigen-presenting cells (APCs) and the release of effector cytokines, which in turn instruct the adaptive immune system to develop a particular T-helper 1 (Th1)- or Th2-effector response (T. Sparwasser and G.B. Lipford, this issue). Here, we focus on an additional responsibility of the innate immune system: elaboration of hematopoietically active cytokines and growth factors. Upon stimulation with bacterial products, macrophages and DCs are capable of releasing granulocyte CSF, monocyte CSF, granulocyte–monocyte CSF (GM-CSF), interferon α (IFN-α), tumor necrosis factor α (TNF-α), transforming growth factor β, interleukin 1 (IL-1), IL-6, IL-8, IL-10, IL-12, macrophage inflammatory protein 1 (MIP-1), MIP-2, and others. Many of these substances have dramatic hematopoietic consequences (Broxmeyer 1995).

It has long been known that certain adjuvants induce hematopoietic events. McNeill demonstrated that intraperitoneal (i.p.) injection of CFA stimulated the multiplication and affected the distribution of cells in vivo that form granulocytic and monocytic colonies in vitro (McNeill 1970). The results showed that a 100-fold rise in spleen CFCs occurs at the same time as a CFC rise in the blood, but the level of CFCs in the BM is only two- to threefold greater than control levels. The rise in CFCs in the spleen was associated with leukocytosis in the peripheral blood. It was hypothesized that, following injection of foreign materials granulocyte–monocyte progenitor cells in the marrow are stimulated to divide, and many of these cells leave the marrow and are trapped in the spleen. These cells then undergo further differentiation into more mature cells, which appear in the peripheral blood and cause leukocytosis. A likely explanation was given by the observation that CFA induced increased serum CSF levels within 6–8h after injection. Similar observations and explanations were subsequently provided when either LPS or muramyl dipeptide was used as a stimulant (MOATAMED et al. 1975; APTE and PLUZNIK 1976a,b; STABER and METCALF 1980; WUEST and WACHSMUTH 1982; GALELLI and CHEDID 1983). In an early study utilizing LPS, it was shown that LPS-non-responder mice (C3H/HeJ) had strongly diminished CFC levels due to their reduced ability to produce CSFs in response to LPS (APTE and PLUZNIK 1976b). These observations imply that several bacterial cell products stimulate the release of hematopoietically active factors that expand and mobilize progenitor cells.

Several bacterial infections have been shown to lead to very similar phenomena. These micro-organisms include both gram-positive and gram-negative bacteria. Salmonella infection induces a marked increase in splenic and BM CFCs 2–3 days after infection (WILSON et al. 1982). Others have claimed, however, that BM CFCs were depressed after Salmonella infection (MIYANOMAE et al. 1983; KIRIKAE et al. 1986). Of additional interest, the hematopoietic response to Salmonella infection could be monitored in LPS-unresponsive C3H/HeJ mice (MIYANOMAE et al. 1983). Similar observations have been made after *Listeria monocytogenes*, *Brucella abortus*, and *Escherichia coli* infections (WING et al. 1985; CHEERS and YOUNG, 1987; ROTHSTEIN et al. 1987). Bacillus Calmette-Guerin (BCG), injected intravenously (i.v.), was noted to quickly push the quiescent stem cells of BM into the S phase of the cell cycle (POUILLART et al. 1975). In later studies, it was noted that i.v. injection of BCG into C57Bl/6 mice resulted in a rapid development of transient anemia associated with an increased number of granulocytes and monocytes (MARCHAL and MILON 1986). In IFN-$\gamma$-deficient mice infected with mycobacteria, a dramatic remodeling of the hematopoietic system was noted (MURRAY et al. 1998). Myeloid cell proliferation proceeded unchecked throughout the course of infection, resulting in a transition to extramedullary hematopoiesis. The splenic architecture was altered by expansion of macrophages, granulocytes, and extramedullary hematopoietic tissue. These features coincided with splenomegaly, an increase in splenic myeloid CFC, and granulocytosis in the peripheral blood. Systemic levels of cytokines were elevated, particularly IL-6 and GM-CSF.

## 3 CpG DNA Signals Infectious Danger

Yamamoto, Tokunaga and colleagues discovered immunostimulatory DNA by a series of studies originally aimed at analyzing BCG-mediated tumor resistance in mice. A fraction extracted from BCG (designated MY-1) was shown to exhibit anti-tumor activity in vivo, augment natural killer (NK) cell activity, and trigger type-I and type-II IFN release from murine spleen cells or human peripheral blood lymphocytes (PBL) in vitro (TOKUNAGA et al. 1984, 1988; MASHIBA et al. 1988; YAMAMOTO et al. 1988). These activities can be destroyed by DNase pre-treatment of MY-1, but not by RNase treatment. Pisetsky and co-workers independently observed that normal mice, as well as humans, respond to bacterial DNA (but not vertebrate DNA) by producing anti-DNA antibodies (MESSINA et al. 1991). They realized that bacterial DNA was mitogenic for murine B cells and postulated that this activity resulted from "non-conserved structural determinants". The differential stimulative capacity of bacterial DNA versus vertebrate DNA was also demonstrated for induction of NK cell activity by Yamamoto et al. (YAMAMOTO et al. 1992b).

High-performance liquid chromatography analysis of BCG extracts showed that the MY-1 fraction was composed of a broad size range of DNA fragments, with a peak at 45 bases. Synthetic 45-mer oligodeoxynucleotides (ODNs) derived from BCG cDNA sequences were positive for IFN-inducing capacity and augmentation of NK cytotoxicity (KATAOKA et al. 1992; YAMAMOTO et al. 1992a, 1992b). Subsequently, Krieg et al. formulated a hypothetical framework for understanding the pattern recognition of bacterial or synthetic DNA (KRIEG et al. 1995). Using sequence-specific CpG-containing ODNs mediated mitogenicity to B cells as an assay, they discovered that CpG dinucleotides with selective flanking bases were important and, specifically, that DNA motifs displaying a 5′-Pu-Pu-CpG-Pyr-Pyr-3′ base sequence were biologically active. Thus, it was speculated that immune cells sense unique base sequences of pathogen-associated DNA. The realization that these sequences are under-represented in vertebrate DNA offers an explanation for several biological observations in the context of non-self pattern recognition by the immune system (WAGNER 1999).

CpG ODNs directly activate immature DCs and macrophages (STACEY et al. 1996; SPARWASSER et al. 1997b, 1998; JAKOB et al. 1998). In utilizing PRRs, macrophages recognize non-self DNA through CpG motifs and initiate inflammatory responses. In early work by Yamamoto et al. it was concluded that IFN-$\alpha/\beta$ produced by immunostimulatory-ODN-stimulated spleen cells might have originated in an adherent cell population (TOKUNAGA et al. 1988; YAMAMOTO et al. 1988). We and others have discovered that DNA from gram-negative and gram-positive bacteria, plasmid DNA, or synthetic CpG ODNs triggered macrophages to activate the transcription factor nuclear factor (NF$\kappa$B), to transcribe cytokine mRNAs, and to secrete pro-inflammatory cytokines, such as GM-CSF, TNF-$\alpha$, IL-1, IL- 6, Il-12, and IL-18 (HALPERN et al. 1996; SATO et al. 1996; STACEY et al. 1996; CHACE et al. 1997; LIPFORD et al. 1997b; ROMAN et al. 1997; SPARWASSER et al. 1997a,b).

## 4 Induction of Splenomegaly by CpG ODNs

Splenomegaly is a well-recognized phenomenon accompanying some oligonucleotide injections. Branda et al. observed that mice developed massive splenomegaly and polyclonal hypergammaglobulinemia within 2 days after intravenous injection of a phosphorothioate oligomer that was antisense to a portion of the rev region of the human immunodeficiency virus 1 genome (Branda et al. 1993). Histological examination of spleens from injected animals showed marked expansion of a uniform-appearing population of small lymphocytes. Flow-cytometry analysis indicated that the responding cells were predominantly B lymphocytes. Mojcik et al. observed that injection of mice with antisense to the initiation region of the env gene resulted in (i) increased spleen cell numbers, primarily due to an increase in splenic B cells, (ii) increased class-II major histocompatibility complex expression on B cells, (iii) increased RNA and DNA synthesis, and (iv) increased numbers of immunoglobulin (Ig)-producing cells (Mojcik et al. 1993). They concluded that products of certain endogenous retroviral sequences regulate lymphocyte activation in vivo. McIntyre et al. in efforts to test the efficacy of NFκB p65 oligonucleotides in vivo, unexpectedly observed that the control p65-sense (but not the p65-antisense) oligonucleotides caused massive splenomegaly in mice (McIntyre et al. 1993). In this study, they demonstrated a sequence-specific stimulation of splenic cell proliferation, both in vivo and in vitro, by treatment with p65-sense oligonucleotides. Cells expanded by this treatment were primarily B-220+, sIg+ B cells. The secretion of Ig by the p65-sense oligonucleotide-treated splenocytes was also enhanced. In addition, the p65-sense-treated splenocytes (but not several other cell lines) showed an upregulation of NFκB-like activity in the nuclear extracts, an effect not dependent on new protein or RNA synthesis. Zhao et al. concluded that phosphorothioated-ODNs induce splenomegaly due to B-cell proliferation (Zhao et al. 1996b). In a follow-up, they observed that administration of the 27-mer-phosphorothioate oligonucleotide in mice resulted in splenomegaly and an increase in IgM production 48h after administration (Zhao et al. 1996a).

Bacterial DNA and synthetic ODNs containing unmethylated CpG dinucleotides induce murine B cells to proliferate and secrete Ig in vitro and in vivo (Krieg et al. 1995). This activation is very much enhanced by the antigen-receptor ligation. Optimal B-cell activation requires a DNA motif in which an unmethylated CpG dinucleotide is flanked by two 5′ purines and two 3′ pyrimidines. ODNs containing this CpG motif induce more than 95% of all spleen B cells to enter the cell cycle. In a study by Monteith et al., treatment of rodents with phosphorothioate ODNs induced a form of immune stimulation characterized by splenomegaly, lymphoid hyperplasia, hypergammaglobulinemia, and mixed mononuclear cellular infiltrates in numerous tissues (Monteith et al. 1997). Immune stimulation was evaluated in mice with in vivo and in vitro studies, including a review of historical data. All phosphorothioate ODNs evaluated induced splenomegaly and B-lymphocyte proliferation. Splenomegaly and B-lymphocyte proliferation increased with the dose or concentration of ODN. The overriding evidence provided by the literature thus concludes that the phenomena of splenomegaly induced by ODNs is probably sequence dependent and explained by B-cell mitogenicity.

Zhao et al. administered to mice a 27-mer phosphorothioate oligonucleotide (sequence 5′-TCG TCG CTG TCT CCG CTT CTT CTT GCC-3′), which had previously been shown to cause splenomegaly and hypergammaglobulinemia on in vivo administration in mice, and studied the pattern and kinetics of cytokine production at both the splenic mRNA and serum protein levels (ZHAO et al. 1997). Following i.p. administration of high doses (50mg/kg) of oligonucleotide, significant increases in the splenic mRNA levels of IL-6, IL-12p40, IL-1β, and IL-1Ra and serum levels of IL-6, IL-12, MIP-1β, and MCP-1 were observed. In contrast, no significant differences in splenic mRNA levels of IL-2, IL- 4, IL-5, IL-9, IL-13, IL-15, IFN-γ, or migration inhibition factor or serum levels of IL-2, IL-4, IL-5, IL-10, IFN-γ, or GM-CSF were detected. These studies show a distinct pattern and kinetics of cytokine production following oligonucleotide administration and further demonstrate that cytokine induction is not a general property of phosphorothioate oligonucleotides, but is dependent on a given sequence and dose of the oligonucleotides. Serum release of IL-1, IL-6, IL-12, and TNF-α has been confirmed (LIPFORD et al. 1997b and unpublished data). Contrary to Zhao et al. Klinman et al. have demonstrated GM-CSF production by macrophages in vitro, and the production of IFN-γ has been demonstrated by several authors (KLINMAN et al. 1996; YI et al. 1996).

## 5 Splenomegaly is Associated with Extramedullary Hematopoiesis

The cytokine repertoire induced by low doses of CpG ODNs is Th1 in nature, and ample evidence suggests that this strongly biases subsequent immune-response development to Th1 (T. Sparwasser and G.B. Lipford, this issue). This type of cellular-oriented immune response becomes particularly important in infections caused by intracellular pathogens. CpG ODNs have been demonstrated to protect mice efficiently in different infection models, such as listeriosis, leishmaniasis, and pathology caused by *Francisella tularensis* (KRIEG et al. 1998; ZIMMERMANN et al. 1998; ELKINS et al. 1999). Interestingly, in addition to acute protective effects primarily mediated by activation of APCs, lasting effects up to 2 weeks after injection of ODNs have been observed (KRIEG et al. 1998; our own unpublished results). These findings suggest a propensity for CpG DNA to induce longer-term changes in the immune system.

Pro-inflammatory cytokines secreted by innate immune cells upon stimulation with bacterial CpG DNA, such as TNF-α and IL-1, can trigger CSF production by accessory cells locally as well as in the bone-marrow microenvironment. Compared to hematopoietic growth factors induced in the BM, CSFs generated at peripheral sites of infection have different functions. Here, CSFs are most likely to influence activation and maturation of myeloid cells necessary for the acute defense against the invading pathogen, whereas the proliferation and differentiation of myeloid progenitor cells and hematopoietic remodeling provide for an adequate second line of defense (CANNISTRA and GRIFFIN 1988).

When we examined the protective effects of CpG ODNs, we observed a dramatic but transient splenomegaly, with a maximum increase of spleen weight at day 6 (SPARWASSER et al. 1999). This increase of spleen weight by more than three times the normal weight was dose- and sequence-dependent and normalized after 12–14 days. In contrast to previous studies by others analyzing the splenic cellular composition within the first 4 days after ODN challenge, we could not explain our observations by B-cell proliferation. Histologically, splenic architecture was altered, showing large immature blast within the red pulp. Facs staining of day-6 spleen cells revealed a more than tenfold expansion of a non-B-/non-T-cell fraction. This population of spleen cells was highly enriched for hematopoietic progenitors. Colony assays measuring granulocyte–macrophage colony-forming units (GM-CFUs) demonstrated an increase in splenic myeloid progenitors paralleling the increase in splenic cell count. Kinetically, a discrete increase of GM-CFUs in BM preceding the splenic changes could be detected, as if a mobilization of BM-derived progenitor cells to the spleen may have taken place. A single application of CpG ODN (60μg/injection) far exceeded the documented hematopoietic stimulus LPS. Furthermore, the number of early erythrocyte progenitors (burst-forming units, BFU-Es) in the spleen was also increased after i.p. injection of CpG ODNs. Transfer of spleen cells from CpG-ODN pre-treated animals into lethally irradiated syngenic mice yielded increased spleen colony forming units (CFU-Ss). CFU-Ss are an indicator for the existence of primitive stem cells, and it is controversial whether these cells respond to cytokines (MORRISON et al. 1995). These data suggested the possibility of reconstituting lethally irradiated mice by an adoptive transfer of CFU-Ss contained in the spleens of CpG-DNA-treated animals.

## 6 CpG-DNA-Mediated Protection from Immunosuppression

Proliferation, mobilization and differentiation of hematopoietic progenitor cells are of major therapeutic importance in clinical situations of immunosuppression. Recently, we described that the hematopoietic effects of CpG ODNs could be used to mitigate irradiation-induced damage to the hematopoietic system (SPARWASSER et al. 1999). CpG-DNA challenge of sublethally irradiated mice caused radioprotective effects, in that recovery of GM-CFU and cytotoxic T-cell (CTL) function was enhanced. Interestingly, a single i.p. injection after sublethal irradiation caused sufficient hematopoietic remodeling to compensate for radiation-induced damage to the lympho-hematopoietic system. CpG-ODN challenge within 30min after sublethal irradiation led to a fourfold increase of splenic GM-CFUs after 2 weeks. Irradiated mice immunized with soluble ovalbumin (OVA) exhibited significantly enhanced OVA-specific CTL activity if CpG DNA was given therapeutically. The increase in GM-CFUs and CTL function correlated with enhanced resistance to *Listeria* infection. Mice were infected at day 14 after irradiation, and survival was recorded for 30 days. If CpG ODNs were applied, these mice were protected against a normally lethal challenge with the pathogen, implying an increased resistance due to CpG-DNA-triggered cellular-adaption processes. Preliminary data suggest that these results can also be extended to immunosuppression caused by chemotherapeutic agents, such as 5-fluorouracil (our own unpublished data).

## 7 Conclusion

CpG DNA signals infectious danger not only to immune cells but also to hematopoietic cells and thereby communicates the need for increased hematopoiesis necessary to combat infections. APC-derived cytokines and CSFs appear to be responsible for a transient phase of hematopoietic remodeling comprising mobilization and differentiation of BM progenitor cells. To date, it is not clear which of these factors, which can also act synergistically, are relevant for the hematopoietic effects observed in mice. It is also feasible that CpG ODNs directly target BM stroma cells, thereby inducing the production and release of hematopoietically active growth factors, cytokines, and chemokines.

The strong hematopoietic potency of bacterial CpG-DNA motifs, which far exceeds that of high doses of LPS in promoting hematopoietic changes in the murine system, could explain why e.g. Salmonella infections have similar effects in LPS-susceptible and LPS-resistant mice (Miyanomae et al. 1983). Furthermore, the well-known hematopoietic effect of CFA containing dead mycobacteria may depend strongly on the presence of bacterial DNA (McNeill 1970). Most importantly, CpG ODNs appear to display a dual function. When used as a "new class of adjuvants" with proteinaceous antigen (Lipford et al. 1997a), CpG-ODNs cause an acute and strong Th1-biasing effect on emanating immune responses via DC activation and local production of cytokines, such as IL-12. However, long-term effects include crucial changes in the cellular compartmentalization of innate and adaptive immune cells; these changes are antigen independent. As a consequence, the system is "primed" for protective Th1-polarized immune responses, as shown in CTL responses to OVA and resistance of BALB/c mice against *Leishmania major* (our own unpublished data). Protective immunity against various pathogens has been demonstrated to depend on acute APC activation caused by the pattern recognition of pathogen DNA (Krieg et al. 1998; Zimmermann et al. 1998; Elkins et al. 1999). Conversely, its ability to cause the release of hematopoietic factors that control the steady state between production and consumption of immune cells may be a critical factor contributing to the observed lasting priming effects (Th1) of CpG DNA. The cellular and molecular basis of the priming effect is poorly understood. However, because of its possible importance in immunosuppressive situations, the therapeutic use of CpG ODNs to mitigate damage to the hematopoietic system by e.g. chemotherapy, radiotherapy, or accidental radiation exposure might be an interesting alternative to combinations of recombinant growth factors.

## References


Akhtar S, Agrawal S (1997) In vivo studies with antisense oligonucleotides. Trends Pharmacol Sci 18: 12–18

Apte RN, Pluznik DH (1976a) Control mechanisms of endotoxin and particulate material stimulation of hemopoietic colony forming cell differentiation. Exp Hematol 4:10–18

Apte RN, Pluznik DH (1976b) Genetic control of lipopolysaccharide induced generation of serum colony stimulating factor and proliferation of splenic granulocyte/macrophage precursor cells. J Cell Physiol 89:313–324

Branda RF, Moore AL, Mathews L, McCormack JJ, Zon G (1993) Immune stimulation by an antisense oligomer complementary to the rev gene of HIV-1. Biochem Pharmacol 45:2037–2043

Broxmeyer HE (1995) Role of cytokines in hematopoiesis. In: Aggarwal B, Puri R (eds). Human cytokines: their role in disease and therapy (Cambridge: Blackwell Science), pp 27–35

Cannistra SA, Griffin JD (1988) Regulation of the production and function of granulocytes and monocytes. Semin Hematol 25:173–188

Chace JH, Hooker NA, Mildenstein KL, Krieg AM, Cowdery JS (1997) Bacterial DNA-induced NK cell IFN-gamma production is dependent on macrophage secretion of IL-12. Clin Immunol Immunopathol 84:185–193

Cheers C, Young AM (1987) Serum colony stimulating activity and colony forming cells in murine brucellosis: relationship to immunopathology. Microb Pathog 3:185–194

Elkins KL, Rhinehart-Jones TR, Stibitz S, Conover JS, Klinman DM (1999) Bacterial DNA containing CpG motifs stimulates lymphocyte-dependent protection of mice against lethal infection with intracellular bacteria. J Immunol 162:2291–2298

Galelli A, Chedid L (1983) Modulation of myelopoiesis in vivo by synthetic adjuvant-active muramyl peptides: induction of colony-stimulating activity and stimulation of stem cell proliferation. Infect Immun 42:1081–1085

Halpern MD, Kurlander RJ, Pisetsky DS (1996) Bacterial DNA induces murine interferon-gamma production by stimulation of interleukin-12 and tumor necrosis factor-alpha. Cell Immunol 167:72–78

Jakob T, Walker PS, Krieg AM, Udey MC, Vogel JC (1998) Activation of cutaneous dendritic cells by CpG-containing oligodeoxynucleotides: a role for dendritic cells in the augmentation of Th1 responses by immunostimulatory DNA. J Immunol 161:3042–3049

Kataoka T, Yamamoto S, Yamamoto T, Kuramoto E, Kimura Y, Yano O, Tokunaga T (1992) Antitumor activity of synthetic oligonucleotides with sequences from cDNA encoding proteins of *Mycobacterium bovis* BCG. Jpn J Cancer Res 83:244–247

Kirikae T, Yoshida M, Sawada H, Tezuka H, Fujita J, Mori KJ (1986) Effects of splenectomy on the retention of *Salmonella enteritidis* and on the hemopoietic response to *Salmonella* infection. Biomed Pharmacother 40:6–10

Klinman DM, Yi AK, Beaucage SL, Conover J, Krieg AM (1996) CpG motifs present in bacteria DNA rapidly induce lymphocytes to secrete interleukin 6, interleukin 12, and interferon gamma. Proc Natl Acad Sci USA 93:2879–2883

Krieg AM, Love-Homan L, Yi AK, Harty JT (1998) CpG DNA induces sustained IL-12 expression in vivo and resistance to *Listeria monocytogenes* challenge. J Immunol 161:2428–2434

Krieg AM, Yi AK, Matson S, Waldschmidt TJ, Bishop GA, Teasdale R, Koretzky GA, Klinman DM (1995) CpG motifs in bacterial DNA trigger direct B-cell activation. Nature 374:546–549

Lipford GB, Bauer M, Blank C, Reiter R, Wagner H, Heeg K (1997a) CpG-containing synthetic oligonucleotides promote B and cytotoxic T cell responses to protein antigen: a new class of vaccine adjuvants. Eur J Immunol 27:2340–2344

Lipford GB, Sparwasser T, Bauer M, Zimmermann S, Koch ES, Heeg K, Wagner H (1997b) Immunostimulatory DNA: Sequence dependent production of potentially harmful or useful cytokines. Eur J Immunol 27:3420–3426

Marchal G, Milon G (1986) Control of hemopoiesis in mice by sensitized L3T4+ Lyt2-lymphocytes during infection with bacillus Calmette-Guerin. Proc Natl Acad Sci USA 83:3977–3981

Mashiba H, Matsunaga K, Tomoda H, Furusawa M, Jimi S, Tokunaga T (1988) In vitro augmentation of natural killer activity of peripheral blood cells from cancer patients by a DNA fraction from Mycobacterium bovis BCG. Jpn J Med Sci Biol 41:197–202

McIntyre KW, Lombard-Gillooly K, Perez JR, Karsch C, Sarmiento UM, Larigan JD, Landreth KT, Narayanan R (1993) A sense phosphorothioate oligonucleotide directed to the initiation codon of

transcription factor NF-κB p65 causes sequence-specific immune stimulation. Antisense Res Dev 3:309–322
McNeill TA (1970) Antigenic stimulation of bone marrow colony forming cells. Immunology 18:61–72
Messina JP, Gilkeson GS, Pisetsky DS (1991) Stimulation of in vitro murine lymphocyte proliferation by bacterial DNA. J Immunol 147:1759–1764
Miyanomae T, Mori KJ, Seto A (1983) Hemopoietic responses of LPS-unresponsive C3H/HeJ mice to salmonella infection. Biomed Pharmacother 37:65–68
Moatamed F, Karnovsky MJ, Unanue ER (1975) Early cellular responses to mitogens and adjuvants in the mouse spleen. Lab Invest 32:303–312
Mojcik CF, Gourley MF, Klinman DM, Krieg AM, Gmelig Meyling F, Steinberg AD (1993) Administration of a phosphorothioate oligonucleotide antisense to murine endogenous retroviral MCF *env* causes immune effects in vivo in a sequence-specific manner. Clin Immunol Immunopathol 67:130–136
Monteith DK, Henry SP, Howard RB, Flournoy S, Levin AA, Bennett CF, Crooke ST (1997) Immune stimulation – a class effect of phosphorothioate oligodeoxynucleotides in rodents. Anticancer Drug Des 12:421–432
Morrison SJ, Uchida N, Weissman IL (1995) The biology of hematopoietic stem cells. Annu Rev Cell Dev Biol 11:35–71
Murray PJ, Young RA, Daley GQ (1998) Hematopoietic remodeling in interferon-gamma-deficient mice infected with mycobacteria. Blood 91:2914–2924
Pouillart P, Palangie T, Schwarzenberg L, Brugerie H, Lheritier J, Mathe G (1975) Letter: Effect of BCG on haemopoietic stem cells. Biomedicine 23:469–471
Roman M, Martin-Orozco E, Goodman JS, Nguyen MD, Sato Y, Ronaghy A, Kornbluth RS, Richman DD, Carson DA, Raz E (1997) Immunostimulatory DNA sequences function as Th1-promoting adjuvants. Nat Med 3:849–854
Rothstein G, Christensen RD, Nielsen BR (1987) Kinetic evaluation of the pool sizes and proliferative response of neutrophils in bacterially challenged aging mice. Blood 70:1836–1841
Sato Y, Roman M, Tighe H, Lee D, Corr M, Nguyen MD, Silverman GJ, Lotz M, Carson DA, Raz E (1996) Immunostimulatory DNA sequences necessary for effective intradermal gene immunization. Science 273:352–354
Selig C, Nothdurft W (1995) Cytokines and progenitor cells of granulocytopoiesis in peripheral blood of patients with bacterial infections. Infect Immun 63:104–109
Sparwasser T, Hültner L, Koch ES, Luz A, Lipford GB, Wagner H (1999) Immunostimulatory CpG-oligodeoxynucleotides cause extramedullary murine hemopoiesis. J Immunol 162:2368–2374
Sparwasser T, Koch ES, Vabulas RM, Lipford GB, Heeg K, Ellwart JW, Wagner H (1998) Bacterial DNA and immunostimulatory CpG oligonucleotides trigger maturation and activation of murine dendritic cells. Eur J Immunol 28:2045–2054
Sparwasser T, Miethke T, Lipford G, Borschert K, Häcker H, Heeg K, Wagner H (1997a) Bacterial DNA causes septic shock. Nature 386:336–337
Sparwasser T, Miethke T, Lipford G, Erdmann A, Häcker H, Heeg K, Wagner H (1997b) Macrophages sense pathogens via DNA motifs: induction of tumor necrosis factor-α-mediated shock. Eur J Immunol 27:1671–1679
Staber FG, Metcalf D (1980) Cellular and molecular basis of the increased splenic hemopoiesis in mice treated with bacterial cell wall components. Proc Natl Acad Sci USA 77:4322–4325
Stacey KJ, Sweet MJ, Hume DA (1996) Macrophages ingest and are activated by bacterial DNA. J Immunol 157:2116–2122
Tokunaga T, Yamamoto H, Shimada S, Abe H, Fukada T, Fujisawa Y, Furutani Y, Yano O, Kataoka T, Sudo T, Makiguchi N, Suganuma T (1984) Antitumor activity of deoxyribonucleic acid fraction from *Mycobacterium bovis* BCG. I. Isolation, physicochemical characterization, and antitumor activity. J Natl Cancer Inst 72:955–962
Tokunaga T, Yamamoto S, Namba K (1988). A synthetic single-stranded DNA, poly(dG,dC), induces interferon-α/β and -γ, augments natural killer activity, and suppresses tumor growth. Jpn J Cancer Res 79:682–686
Wagner H (1999) Bacterial CpG-DNA activates immune cells to signal 'infectious danger'. Adv Immunol 73:329–368
Wilson BM, Rosendaal M, Plant JE (1982) Early haemopoietic responses to *Salmonella typhimurium* infection in resistant and susceptible mice. Immunology 45:395–399
Wing EJ, Barczynski LC, Waheed A, Shadduck RK (1985) Effect of *Listeria monocytogenes* infection on serum levels of colony-stimulating factor and number of progenitor cells in immune and nonimmune mice. Infect Immun 49:325–328

Wuest B, Wachsmuth ED (1982) Stimulatory effect of *N*-acetyl muramyl dipeptide in vivo: proliferation of bone marrow progenitor cells in mice. Infect Immun 37:452–462

Yamamoto S, Kuramoto E, Shimada S, Tokunaga T (1988) In vitro augmentation of natural killer cell activity and production of interferon-alpha/beta and -gamma with deoxyribonucleic acid fraction from *Mycobacterium bovis* BCG. Jpn J Cancer Res 79:866–873

Yamamoto S, Yamamoto T, Kataoka T, Kuramoto E, Yano O, Tokunaga T (1992a) Unique palindromic sequences in synthetic oligonucleotides are required to induce IFN and augment IFN-mediated natural killer activity. J Immunol 148:4072–4076

Yamamoto S, Yamamoto T, Shimada S, Kuramoto E, Yano O, Kataoka T, Tokunaga T (1992b) DNA from bacteria, but not from vertebrates, induces interferons, activates natural killer cells and inhibits tumor growth. Microbiol Immunol 36:983–997

Yi AK, Klinman DM, Martin TL, Matson S, Krieg AM (1996) Rapid immune activation by CpG motifs in bacterial DNA: Systemic induction of IL-6 transcription through an antioxidant-sensitive pathway. J Immunol 157:5394–5402

Zhao Q, Temsamani J, Iadarola PL, Agrawal S (1996a) Modulation of oligonucleotide-induced immune stimulation by cyclodextrin analogs. Biochem Pharmacol 52:1537–1544

Zhao Q, Temsamani J, Iadarola PL, Jiang Z, and Agrawal S (1996b) Effect of different chemically modified oligodeoxynucleotides on immune stimulation. Biochem Pharmacol 51:173–182

Zhao Q, Temsamani J, Zhou RZ, Agrawal S (1997) Pattern and kinetics of cytokine production following administration of phosphorothioate oligonucleotides in mice. Antisense Nucleic Acid Drug Dev 7:495–502

Zimmermann S, Egeter O, Hausmann S, Lipford GB, Röcken M, Wagner H, Heeg K (1998) CpG oligonucleotides trigger curative Th1 responses in lethal murine *Leishmaniasis*. J Immunol 160: 3627–3630

# CpG DNA Augments the Immunogenicity of Plasmid DNA Vaccines

D.M. Klinman, K.J. Ishii, and D. Verthelyi



## 1 Introduction: Immunostimulatory Properties of CpG Oligodeoxynucleotides

Vaccine development has been revolutionized by the use of antigen-encoding DNA plasmids to induce cellular and humoral immune responses against pathogenic viruses, parasites, bacteria and tumors (Cox et al. 1993; Ulmer et al. 1993; Wang et al. 1993; Sedegah et al. 1994). DNA vaccines are composed of an antigen-encoding gene whose expression is regulated by a strong mammalian promoter incorporated into the plasmid backbone of bacterial DNA (Wolff et al. 1990; Manthorpe et al. 1993; Ulmer et al. 1993). When injected intramuscularly or intradermally, DNA vaccines are transcribed and translated, and the protein they encode is presented to the immune system in the context of self major histocompatibility complex (MHC) (Wolff et al. 1990; Cox et al. 1993; Ulmer et al. 1993).

Although the nature, magnitude and duration of the immune response elicited by DNA vaccines is influenced by multiple factors, it has been repeat-

Section of Retroviral Immunology, Building 29A, Room 3D10, Center for Biologics Evaluation and Research, Food and Drug Administration, Bethesda, MD 20892, USA edly shown that intramuscular delivery stimulates a T helper 1 (Th1)-driven response characterized by cytotoxic T lymphocyte (CTL) induction and the release of interferon γ (IFNγ) and antigen-specific immunoglobulin G2a (IgG2a) antibodies (Ulmer et al. 1993; Mor et al. 1995; Klinman et al. 1998). Numerous animal studies (particularly in mice) demonstrate that DNA-vaccine-induced immunity can confer protection against pathogen challenge (Ulmer et al. 1993; Sedegah et al. 1994). Nevertheless, it has proven difficult to elicit equally strong and protective immune responses in primates (Calarota et al. 1998; Wang et al. 1998). This has prompted widespread efforts to improve the immunogenicity of DNA vaccines, including (i) altering membrane expression of the encoded gene, (ii) substituting stronger promoters, (iii) targeting the plasmid to transfect specific cell types, and/or (iv) co-administering DNA vaccines with adjuvants (including plasmids encoding cytokines and/or co-stimulatory molecules).

Investigators in the field of autoimmunity were the first to demonstrate that bacterial DNA has immunostimulatory properties (Gilkeson et al. 1989; Messina et al. 1991). Yamamoto et al. were the first to report that synthetic oligodeoxynucleotides (ODNs) with sequences patterned after those found in bacterial DNA could activate natural killer (NK) cells to secrete IFNγ (Yamamoto et al. 1992). They hypothesized that palindromic sequences present in the synthetic ODNs were responsible for this stimulation. In collaboration with Dr. Krieg, my lab demonstrated that specific sequence motifs present in bacterial DNA consisting of an unmethylated CpG dinucleotide flanked by two 5′ purines (optimally, GpA) and two 3′ pyrimidines elicited an "innate" immune response (Krieg et al. 1995; Halpern et al. 1996; Klinman et al. 1996; Sato et al. 1996). These motifs are 20 times less common in mammalian than in microbial DNA, due to differences in frequency of utilization and methylation pattern of CpG dinucleotides in eukaryotes versus prokaryotes (Razin and Friedman 1981; Cardon et al. 1994). We showed that these CpG motifs activate a variety of immunologically relevant cells to proliferate and/or secrete, including macrophages and B, T and NK cells (Krieg et al. 1995; Klinman et al. 1996; Sato et al. 1996). ODNs containing immunostimulatory CpG motifs induced a significant rise in the number of cells secreting interleukin 6 (IL-6), IL-12 and IFNγ within 10h and a rise in the number of IgM-secreting cells within 36h. In contrast, mammalian DNA and ODNs in which the critical CpG dinucleotide was eliminated by inversion or methylation did not stimulate cytokine or Ig secretion (Table 1; Klinman et al. 1996). Other investigators subsequently showed that IL-18 and tumor-necrosis-factor production were also induced by this motif (Halpern et al. 1996). The immunostimulatory capacity of CpG motifs may be of use in a variety of therapeutic applications. This review will examine the contribution of CpG motifs as immune adjuvants when used with DNA vaccines, emphasizing the extensive studies conducted in mice.

**Table 1.** Immunostimulatory effect of CpG DNA. BALB/c spleen cells incubated with 50μg/ml of heat-denatured *Escherichia coli* DNA or calf thymus DNA, or with 1μM of stimulatory or control phosphorothioate oligodeoxynucleotides (ODNs)

| | Fold increase in cytokine-secreting cell number | | | |
|---|---|---|---|---|
| | IL-6 | IL-12 | IFNγ | IgM |
| *E. coli* DNA | 3.2 ± 0.2 | 3.8 ± 0.4 | 4.7 ± 2.3 | 3.9 ± 1.1 |
| Calf-thymus DNA | 0.8 ± 0.2 | 1.1 ± 0.2 | 0.8 ± 0.3 | 0.7 ± 0.2 |
| CpG ODN | 5.5 ± 1.1 | 8.3 ± 1.7 | 4.7 ± 1.1 | 4.2 ± 1.6 |
| CpG ODN (methylated) | 0.9 ± 0.2 | 1.2 ± 0.3 | 0.8 ± 0.2 | 1.1 ± 0.2 |
| CpG ODN (DNase) | 1.3 ± 0.2 | 0.8 ± 0.2 | 1.1 ± 0.2 | 0.9 ± 0.2 |
| GpC ODN | 1.2 ± 0.3 | 1.3 ± 0.3 | 1.2 ± 0.3 | 1.3 ± 0.3 |

The effect on cytokine production was determined after 10h by ELIspot assay (Mor et al. 1995; Klinman et al. 1996). Data represent the fold increase in the number of cytokine-secreting cells over background. Results represent the mean ± SD of at least three independent experiments.
IFN, interferon; Ig, immunoglobulin; IL, interleukin.

# 2 Immunostimulatory Properties of DNA Plasmids

## 2.1 Cytokine Production

A considerable body of evidence indicates that plasmid DNA vaccines primarily elicit IFNγ- and IgG2a-dominated immune responses when injected intramuscularly (Mor et al. 1995; Sato et al. 1996). We examined whether this reflected the activity of CpG motifs in the bacterial plasmid backbones of these vaccines.

Initially, the ability of DNA plasmids (with and without protein-encoding inserts) to induce cytokine production was studied in vitro. Two different plasmid constructs were examined: a vaccine encoding the circumsporozoite protein (CSP) of *P. yoelli* malaria known as 1012/PyCSP and manufactured by VICAL, Inc. (Mor et al. 1995) and a vaccine encoding the gp160 envelope protein of the human immunodeficiency virus 1 virus, known as pCMV160. When incubated in vitro with spleen cells from normal mice, both DNA vaccines induced a 5–8-fold increase in the number of cells secreting IgM, IL-6, IL-12 and IFNγ (Table 2 and data not shown). Similar levels of immune stimulation were observed when the experiment was repeated using plasmid vectors from these vaccines (i.e., devoid of antigen-encoding insert DNA). Of interest, each vector contained over a dozen immunostimulatory CpG motifs.

Further experiments were undertaken to determine whether the CpG motifs in the backbones of these DNA vaccines contributed to the observed cytokine production. To reduce the possibility that some form of bacterial contamination was responsible for the observed immunostimulation, we tested three different plasmids prepared in three different laboratories. Equivalent amounts of each plasmid elicited similar levels of cytokine production in vitro (data not shown). Treating these plasmids with DNAse or Sss I methylase (the latter selectively methylates the cytosine of CpG dinucleotides) uniformly eliminated cytokine production, indi-

**Table 2.** Immunostimulatory effect of CpG motifs in DNA plasmids

| | Fold increase in cytokine secreting cell number | | | |
|---|---|---|---|---|
| | IL-6 | IL-12 | IFNγ | IgM |
| 1012 vector | 5.3 ± 1.1 | 7.6 ± 1.4 | 5.0 ± 1.2 | 4.6 ± 0.6 |
| 1012/PyCSP DNA vaccine | 5.9 ± 1.3 | 8.1 ± 1.9 | 4.6 ± 1.5 | 5.2 ± 1.6 |
| DNAse treated 1012/PyCSP | 0.9 ± 0.2 | 1.0 ± 0.1 | 1.1 ± 0.2 | 1.2 ± 0.2 |
| 20% methylated 1012/PyCSP | 4.7 ± 1.3 | 7.7 ± 1.6 | 4.0 ± 0.9 | ND |
| 80% methylated 1012/PyCSP | 1.9 ± 0.3 | 2.2 ± 0.4 | 2.6 ± 0.4 | ND |
| 97% methylated 1012/PyCSP | 1.2 ± 0.2 | 1.3 ± 0.1 | 0.9 ± 0.1 | 0.8 ± 0.2 |

BALB/c spleen cells were incubated with 50μg/ml of plasmid DNA. The effect on cytokine production was determined as described in the legend to Table 1.
CSP, cirsumsporozoite protein; IFN, interferon; Ig, immunoglobulin; IL, interleukin; ND, not determined.

**Table 3.** Addition of CpG motifs increases DNA-vaccine immunogenicity

| | Fold increase | | |
|---|---|---|---|
| | Ab titer | G2a:G1 ratio | IFNγ |
| 50μg of 1012/PyCSP | 680 | 3.6 | 4.6 |
| 4μg of 1012/PyCSP | 64 | 1.4 | 1.2 |
| 4μg of 1012/PyCSP + 50μg of CpG ODN | 410 | 2.7 | 3.9 |
| 4μg of 1012/PyCSP + 50μg of GpC ODN | 58 | 1.3 | 1.7 |
| 4μg of 1012/PyCSP + 100μg of 1012 vector | 440 | 2.9 | 5.2 |
| 2μg of 2534/PyCSP | 260 | 3.4 | 3.5 |

BALB/c mice were primed and boosted with each plasmid. Serum immunoglobulin G (IgG) anti-CS.1 levels and frequency of P16-responsive interferon-γ (IFNγ)-secreting T cells was monitored 3 weeks post boost.
Ab, antibody; CSP, cirsumsporozoite protein; ODN, oligodeoxynucleotides.

cating that DNA, rather than some unknown contaminant, was responsible for the cytokine release (Table 2).

These findings led us to examine the nature of the cytokines produced by antigen-specific T cells derived from DNA-plasmid-vaccinated mice. We found that mice primed and boosted with DNA vaccines mount an IFNγ-dominated Th1 immune response (Mor et al. 1995), an observation confirmed in other systems (Sato et al. 1996). We then took spleen cells from animals immunized with 1012/PyCSP and re-stimulated them in vitro with P16, an immunodominant T-cell epitope present on the CSP. As seen in Table 3, spleen cells from vaccinated mice responded to antigen re-stimulation by a significant increase in IFNγ (but not IL-4) production, consistent with a CpG-mediated effect.

## 2.2 Antibody Production

The isotype of antigen-specific antibodies induced by plasmid DNA vaccination was also examined. Animals immunized with CS1 protein emulsified in complete

Freund's adjuvant primarily produce IgG1 anti-CSP antibodies (IgG1:IgG2a ratio of 5.6). In contrast, intramuscular injection of CSP-encoding DNA vaccines preferentially stimulated IgG2a-antibody production [IgG1:IgG2a ratios less than 0.7 for three different CSP-encoding plasmids (data not shown)]. Since IFNγ promotes IgM-to-IgG2a isotype switching (SNAPPER and PAUL 1987; FINKLEMAN et al. 1988), these findings are consistent with CpG-motif-induced IFNγ production contributing to the preferential production of IgG2a antibodies in immunized mice.

## 3 Co-Administration of CpG ODNs Improves the Immunogenicity of DNA Vaccines

If CpG motifs present in DNA plasmids contribute to vaccine immunogenicity, then co-administering additional motifs (in the form of CpG ODNs) would be expected to boost the immune response elicited by DNA vaccines. To examine this hypothesis, we primed and boosted mice with only 4μg of 1012/PyCSP (a suboptimal vaccine dose was utilized to maximize the likelihood of detecting a CpG effect) (KLINMAN et al. 1997). As seen in Table 3, this sub-optimal dose stimulated a detectable but reduced anti-CSP response. Co-administering 50μg of CpG ODNs significantly improved both IgG anti-CSP serum levels in vivo and P16-dependent IFNγ production in vitro (Table 3). This observation has since been confirmed using CpG ODNs with other DNA vaccines (DAVIS et al. 1998). Control ODNs lacking immunostimulatory CpG motifs had no effect. Of interest, co-administering 100μg of vector alone (without CSP-encoding inserts) also improved the immune response elicited by 1012/PyCSP (KLINMAN et al. 1997). Presumably, CpG motifs present in the vector backbone acted as adjuvants in a fashion similar to the CpG ODNs. This observation raises the interesting possibility that higher doses of plasmid vaccine or co-administration of multiple DNA vaccines encoding different antigens may boost the immune response to each element due to the synergistic immunostimulatory effect of the additional CpG motifs.

In addition to their direct effects on cytokine and Ig-secreting lymphocytes, CpG ODNs also contribute to the development of an immune response by up-regulating cell-surface expression of MHC class-II molecules (KRIEG et al. 1995; JAKOB et al. 1998; SPARWASSER et al. 1998). Originally demonstrated in B cells, this effect has also been observed in professional antigen-presenting cells. Indeed, CpG ODNs up-regulate the expression of a variety of co-stimulatory molecules in dendritic cells, including CD40 and CD86 (JAKOB et al. 1998; SPARWASSER et al. 1998), with the fraction of stimulated antigen-presenting cells APCs rising as a function of CpG-ODN concentration. Of particular importance, CpG-ODN-mediated activation of these APCs increased their functional capacity, as reflected by an improved ability to stimulate alloreactive T cells (JAKOB et al. 1998).

## 4 Engineering CpG Motifs into the Plasmid Vector Improves DNA-Vaccine Immunogenicity

A more direct approach to assessing whether CpG motifs contribute to the immunogenicity of DNA vaccines involved engineering more CpG motifs into the DNA-vaccine backbone. This approach was pioneered by Sato et al., who substituted a CpG-containing *ampR* gene for a *kanR* selectable marker in a β-galactosidase-encoding plasmid. They found that the re-engineered plasmid elicited a higher IgG-antibody response, more CTLs and greater IFNγ production than the original vector (Sato et al. 1996). Our lab utilized a series of vectors engineered by VICAL, Inc., who inserted multiple AACGTT motifs into the 1012 plasmid vector. Mice were primed and boosted with 50, 10 or 2μg of the 1012/PyCSP- or CpG-enriched 2534/PyCSP plasmid. As seen in Table 3, both vaccines stimulated strong antibody responses at optimal doses of 50μg/mouse. However, low doses of 2534/PyCSP elicited an IgG anti-CSP response significantly higher than that caused by a similar dose of 1012/PyCSP. These findings indicate that additional CpG motifs decreased the amount of vaccine required to induce antigen-specific antibody production.

In the same way, spleen cells from mice immunized with each vaccine were tested for reactivity to P16 stimulation in vitro. Optimal doses of both 1012/PyCSP and 2534/PyCSP generated P16-responsive IFNγ-secreting cells. At a low dose, only cells from 2534/PyCSP-immunized mice were responsive to P16. As expected, no cytokine production resulted when cells from non-immunized mice (or mice treated with vector alone) were exposed to P16, nor did cells from immunized mice respond to stimulation by an unrelated peptide (data not shown).

It is noteworthy that CpG motifs were limited in their ability to augment antibody and cytokine production in vivo. For example, the immune response induced by the 2534/PyCSP plasmid was no greater than that of the 1020/PyCSP plasmid when both were administered at an optimal dose of 50μg. Similarly, increasing the concentration of CpG ODN to above 1μg/ml had little additional effect on the number of spleen cells activated to secrete cytokine in vitro (data not shown). Work by Krieg et al. indicates that adding too many CpG motifs to the plasmid backbone may actually reduce immunogenicity. In that work, introducing 16 additional CpG motifs into a plasmid improved the humoral immune response of the DNA vaccine, while introducing 50 such motifs reduced the response (Krieg et al. 1998). These findings suggest that the maximal stimulatory effect of CpG motifs may require relatively low doses of DNA.

## 5 Deleting CpG Motifs from the Plasmid Vector Reduces DNA-Vaccine Immunogenicity

When the CpG motifs present in bacterial DNA or synthetic ODNs are methylated, their ability to induce cytokine production is significantly reduced (Tables 1, 2;

KRIEG et al. 1995; KLINMAN et al. 1996). To further examine the contribution of CpG motifs to the immunogenicity of DNA vaccines, we treated the 1012/PyCSP plasmid with Sss I CpG methylase. This reagent selectively methylates the cytosines of CpG dinucleotides (NUR et al. 1985). The duration of Sss-I treatment could be adjusted to achieve different levels of CpG methylation. Increasing the percent methylation of CpG dinucleotides reduced the ability of the 1012/PyCSP vaccine to activate cytokine-secreting cells in vitro. For example, 80% methylated vaccine induced 59–83% less IL-6, IL-12 and IFNγ than did native 1012/PyCSP, whereas 97% methylated material induced virtually no cytokine production (Table 4 and data not shown). Similarly, highly methylated DNA vaccines are less immunogenic in vivo. As seen in Table 4, the IgG anti-PyCSP antibody response induced by 97% methylated material in vivo was only 3% of the control values.

The effect of CpG methylation on vaccine immunogenicity could have been due to either (i) the elimination of immunostimulatory CpG motifs or (ii) reduced activity of the plasmid's promoter region, resulting in decreased antigen expression (HUG et al. 1996). To evaluate the latter possibility, HeLa cells were transfected in vitro with native or methylated 1012/PyCSP. As seen in Table 5, cells transfected with 20, 80 and 97% methylated vaccine produced 21, 42 and 75% less CSP protein, respectively, than cells treated with the same amount of native plasmid. Thus, methylation reduced but did not eliminate promoter-region activity. Based on these in vitro results, we calculated that 40μg of 80% methylated 1012/PyCSP would induce the same level of in vivo antigen expression as 23μg of native vaccine. However, dose-titration experiments showed that BALB/c mice immunized and boosted with 23μg of 1012/PyCSP mounted an IgG anti-CSP antibody response nearly fivefold greater than that induced by 40μg of 80% methylated plasmid (Table 5). These findings support the conclusion that immunostimulatory CpG motifs play a role in vaccine immunogenicity.

To further establish that decreased promoter region function did not account for the full effect of methylation on plasmid immunogenicity, we co-administered the methylated 1012/PyCSP vaccine with 1012 vector alone or with CpG-containing ODNs. Although neither the vector nor the ODN encoded CSP protein, both were able to boost the CPS-specific immune response induced by native and

**Table 4.** Elimination of CpG motifs reduces DNA-vaccine immunogenicity

| | Fold increase | | |
|---|---|---|---|
| | Ab titer | G2a:G1 ratio | IFNγ |
| Plasmid DNA vaccine | 39 | 3.1 | 5.1 |
| DNAse treated DNA vaccine | 2 | ND | 1.1 |
| 80% methylated DNA vaccine | 6 | ND | 1.8 |
| 97% methylated DNA vaccine | 1 | ND | 0.9 |

BALB/c mice were immunized with 50μg of plasmid DNA. The effect on serum immunoglobulin G anti-CS.1 antibody titer and on the number of T cells activated by in vitro stimulation by P16 is shown 3 weeks post immunization.
Ab, antibody; IFN, interferon; ND, not determined.

**Table 5.** Effect of methylation on promoter-region function and vaccine immunogenicity

| % Methylation of 1012/PyCSP | In vitro protein production (units) | Equivalent dose of DNA vaccine | Predicted IgG anti-CSP titer | Observed IgG anti-CSP titer | Difference |
|---|---|---|---|---|---|
| 0% (Native) | 100 | 40 | 3200 | 3200 ± 280 | 0 |
| 20% Methylated | 79 | 32 | 3000 | 3000 ± 340 | 0 |
| 80% Methylated | 58 | 23 | 2800 | 610 ± 110 | 4.6 |
| 97% Methylated | 25 | 10 | 2100 | 160 ± 60 | 13.1 |

An 80% confluent monolayer of HeLa cells was transfected overnight with 1μg of plasmid DNA. Production of the encoded circumsporozoite protein (CSP) was monitored by incorporation of sulfur-35-methionine followed by immunoprecipitation with anti-CSP antibodies (Abs). Relative protein content was determined by Phospho-Imager analysis of sodium dodecyl sulfate-polyacrylamide gels. The dose of native vaccine that was equivalent to 40μg of methylated plasmid (in terms of protein production) was calculated. The titer of immunoglobulin G (IgG) anti-CSP Ab produced by BALB/c mice immunized and boosted with that dose of native vaccine was compared to the titer obtained when mice were immunized and boosted with 40μg of methylated vaccine (3 mice/group). The difference between the calculated and observed anti-CSP Ab response presumably reflects the contribution of CpG motifs to vaccine immunogenicity.

methylated 1012/PyCSP (Klinman et al. 1997). This finding demonstrates that partial methylation of the plasmid's promoter region does not prevent expression or recognition of the encoded CSP antigen and confirms that CpG motifs contribute to vaccine immunogenicity.

## 6 CpG ODNs Improve the Immunogenicity of Protein Antigens

We examined whether CpG ODNs could boost the immune response elicited by protein antigens. CpG- and non-CpG-containing ODNs were mixed with ovalbumin (OVA) and injected intraperitoneally into BALB/c mice. As seen in Table 6, soluble OVA alone stimulated a modest IgG anti-OVA response. Antibody production was increased approximately threefold when the OVA was co-administered with CpG (but not control) ODNs.

In these experiments, both the OVA and ODNs could freely diffuse from the site of injection. We reasoned that the adjuvant-like effect of CpG ODNs might be improved if their proximity to antigen were maintained. Towards that end, ODNs and OVA were emulsified in incomplete Freund's adjuvant (IFA). As expected, inclusion of CpG ODNs with OVA in IFA boosted IgG anti-OVA production by tenfold (Table 6). This effect was not observed with OVA in IFA alone (Klinman et al. 1998). Other laboratories have made similar observations, showing that co-delivery of CpG ODNs with antigens in a lipid/liposome formulation induced a substantial increase in the immune response (Chu et al. 1997; Lipford et al. 1997; Weiner et al. 1997).

To eliminate the requirement for IFA in this system, we sought to conjugate synthetic ODNs to these antigens. Unfortunately, efficiently linking DNA to pro-

**Table 6.** CpG-oligodeoxynucleotides (ODN) increase the immune response to a protein antigen

| | Fold increase | | |
|---|---|---|---|
| | Ab titer | G2a:G1 ratio | IFNγ |
| OVA alone | 18 | 0.1 | 2.3 |
| OVA + CpG ODN | 46 | 0.4 | 4.7 |
| OVA + GpC ODN | 16 | 0.1 | 2.1 |
| OVA conjugated to GpC ODN | 180 | 0.9 | 9.4 |
| (OVA + CpG ODN) emulsified in IFA | 140 | 0.8 | 8.8 |

BALB/c mice were immunized and boosted with 20μg of ovalbumin (OVA) plus 50μg of ODN. The ODN were either mixed in the same syringe, conjugated to the OVA via biotin–avidin bridges, or emulsified in incomplete Freund's adjuvant (IFA). Three weeks after treatment, antigen-specific serum antibody (Ab) titers were determined by enzyme-linked immunosorbent assay. At the same time, the number of cells stimulated to secrete interferon γ(IFNγ) was monitored by ELIspot assay.

tein proved difficult. As an alternative, we biotinylated both CpG ODNs and OVA and used multivalent avidin cross-linkers to create stable complexes between these molecules. As seen in Table 6, OVA conjugated to CpG ODN was extremely immunogenic, with IgG anti-OVA antibody production being increased tenfold over OVA–avidin or OVA alone (Klinman et al. 1998). This effect was eliminated when the complexes were treated with DNAse, demonstrating that the effect was due to the CpG-containing DNA (data not shown).

Consistent with results involving DNA vaccines, CpG ODNs altered the isotypes of antibodies elicited by OVA immunization. OVA alone elicited a primarily IgG1 antibody response (IgG2a:IgG1 ratio of 0.1, Table 6). Addition of CpG ODNs significantly increased the production of IgG2a antibodies and increased the IgG2a:IgG1 ratio from 0.1 to 0.9. A similar shift in isotype profile has been observed using a variety of other antigens (including polysaccharide antigens) administered in the context of CpG ODNs by our laboratory and others (Chu et al. 1997; Lipford et al. 1997; Weiner et al. 1997).

Finally, we examined the effects of CpG ODNs on the activation of antigen-specific IFNγ-producing cells. Co-administration of CpG ODNs with OVA (emulsified in IFA or complexed through biotin–avidin bridges) increased the number of spleen cells actively secreting IFNγ in vivo by twofold when compared to mice immunized with OVA alone (Klinman et al. 1998). To establish that this effect was antigen specific, cells from immunized mice were stimulated in vitro with 20μg/ml of OVA. There was a significant dose-related increase in IFNγ production accompanying the co-administration of CpG ODNs with OVA (Table 6). Co-administration of CpG ODNs with OVA emulsified in IFA had a similar effect on IFNγ production (data not shown). In contrast, non-CpG ODNs did not elicit the development of cytokine-producing spleen cells.

Several other laboratories have also observed that CpG motifs have adjuvant-like properties. Consistent with our results, those labs found that adjuvant activity was improved by linking ODNs directly to antigens (either by covalently linking ODNs to proteins or by adding motifs to the backbones of DNA vaccines) (Roman et al. 1997). These results confirm the intuitive expectation that optimal stimulation occurs when antigen and adjuvant are presented to the immune system in close spatial and temporal proximity.

## 7 Immunosuppressive Motifs in DNA Vaccines

Whereas CpG-containing bacterial DNA causes immune stimulation in vivo and in vitro, when mammalian DNA is co-administered with CpG DNA, it can block such activation in a dose-dependent manner. This suppression may account for the inability of mammalian DNA, which contains some CpG motifs (albeit at much lower frequency than bacterial DNA), to induce immune stimulation. Several labs have shown that a subset of non-stimulatory ODNs actually suppress the activation induced by CpG ODNs. Recent work by Krieg et al. indicates that eliminating suppressive motifs from the backbone of a DNA vaccine can improve its immunogenicity by threefold (Krieg et al. 1998). These observations suggest that DNA can both stimulate and suppress the immune system.

## 8 Conclusion

In the studies reviewed above, several different immunostimulatory CpG ODNs were analyzed (AACGTTGAACGTTCGC, GCTAGACGTTAGCGT and TCAACGTT). All of these exhibited adjuvant-like properties, whether co-administered with DNA vaccines or proteins. It is unlikely that the effects observed were due to non-DNA contaminants, since (i) CpG-containing and control ODNs were prepared, processed and administered under identical conditions, yet only the former were active, (ii) all of the reagents used in these studies were carefully purified (endotoxin contamination was under 0.4IU/μg) and (iii) the activity of these CpG-containing ODNs was eliminated by methylation and by DNAse treatment.

Our findings, and those of several other investigators, indicate that the CpG motifs present in DNA vaccines serve an immunostimulatory function, triggering an innate immune response that promotes humoral and/or cell-mediated responses against environmental and plasmid-encoded antigens. This model is consistent with that of Fearon and Locksley, who postulated that an innate immune response could create an immune milieu conducive to the development of antigen-specific immunity (Fearon and Locksley 1996).

DNA vaccines have shown enormous promise in animals studies, yet their activity in humans has been less impressive. Thus, vaccine researchers continue to search for additional methods of improving DNA-vaccine efficacy. Evidence suggests that the motifs that induce optimal immune stimulation in mice differ from those that are most active in humans. As our understanding of the processes underlying CpG-induced immune stimulation mature, we will be better able to harness their activity for the improvement of DNA vaccines.

*Acknowledgements.* The assertions herein are the private ones of the authors and are not to be construed as official or as reflecting the views of the Food and Drug Administration at large. The experiments reported herein were conducted according to the principles set forth in the Guide for the Care and Use of Laboratory Animals, 1985. This work was supported by a grant from the National Vaccine Program.

## References

Calarota S, Bratt G, Nordlund S, Hinkula J, Leandersson AD, Sandstrom E, Wahren B (1998) Cellular cytotoxic response induced by DNA vaccination in HIV-1 infected patients. Lancet 351:1320–1325

Cardon LR, Burge C, Clayton DA, Karlin S (1994) Pervasive CpG suppression in animal mitochondrial genomes. Proc Natl Acad Sci 91:3799–3803

Chu RS, Targoni OS, Krieg AM, Lehmann PV, Harding CV (1997) CpG oligonucleotides act as adjuvants that switch on T helper 1 (Th1) immunity. J Exp Med 186:1623–1631

Cox GJ, Zamb TJ, Babiuk LA (1993) Bovine herpesvirus 1: immune responses in mice and cattle injected with plasmid DNA. J Virol 67:5664–5667

Davis HL, Weeranta R, Waldschmidt TJ, Tygrett L, Schorr J, Krieg AM (1998) CpG DNA is a potent enhancer of specific immunity in mice immunized with recombinant hepatitis B surface antigen. J Immunol 160:870–876

Fearon DT, Locksley RM (1996) The instructive role of innate immunity in the acquired immune response. Science 272:50–53

Finkleman FD, Katona IM, Mosmann TR, Coffman RC (1988) Interferon gamma regulates the isotype of Ig secreting during in vivo humoral responses. J Immunol 140:1022–1030

Gilkeson GS, Grudier JP, Karounos DG, Pisetsky DS (1989) Induction of anti-double stranded DNA antibodies in normal mice by immunization with bacterial DNA. J Immunol 142:1482–1486

Halpern MD, Kurlander RJ, Pisetsky DS (1996) Bacterial DNA induces murine interferon-γ production by stimulation of IL-12 and tumor necrosis factor-α. Cell. Immunol 167:72–78

Hug M, Silke J, Georgiev O, Rusconi S, Schaffner W, Matsuo K (1996) Transcriptional repression by methylation: cooperativity between a CpG cluster in the promoter and remote CpG-rich regions. FEBS Letters 379:251–254

Jakob T, Walker PS, Krieg AM, Udey MC, Vogel JC (1998) Activation of cutaneous dendritic cells by CpG containing oligodeoxynucleotides: A role for dendritic cells in the augmentation of Th1 responses by immunostimulatory DNA. J Immunol 161:3042–3049

Klinman DM, Barnhart KM, Conover J (1998) CpG motifs as immune adjuvants. Vacc

Klinman DM, Conover J, Bloom ET, Weiss W (1998) Immunogenicity and efficacy of DNA vaccination in aged mice. J Gerontol (In press)

Klinman DM, Yamshchikov G, Ishigatsubo Y (1997) Contribution of CpG motifs to the immunogenicity of DNA vaccines. J Immunol 158:3635–3642

Klinman DM, Yi A, Beaucage SL, Conover J, Krieg AM (1996) CpG motifs expressed by bacterial DNA rapidly induce lymphocytes to secrete IL-6, IL-12 and IFNγ. Proc Natl Acad Sci USA 93:2879–2883

Krieg AM, Wu T, Weeratna R, Efler SM, Love L, Yang L, Yi A, Short D, Davis HL (1998) Sequence motifs in adenoviral DNA block immune activation by stimulatory CpG motifs. Proc Natl Acad Sci 95:12631–12636

Krieg AM, Yi A, Matson S, Waldschmidt TJ, Bishop GA, Teasdale R, Koretzky GA, Klinman DM (1995) CpG motifs in bacterial DNA trigger direct B-cell activation. Nature 374:546–548

Lipford GB, Bauer M, Blank C, Reiter R, Wagner H, Heeg K (1997) CpG-containing synthetic oligonucleotides promote B and cytotoxic T cell responses to protein antigen: a new class of vaccine adjuvants. Eur J Immunol 27:2340–2344

Manthorpe M, Cornefert JF, Hartikka J, Felgner J, Rundell A, Margalith M, Dwarki V (1993) Gene therapy by intramuscular injection of plasmid DNA: studies on firefly luciferase gene expression in mice. Hum Gene Ther 4:419–431

Messina JP, Gilkeson GS, Pisetsky DS (1991) Stimulation of in vitro murine lymphocyte proliferation by bacterial DNA. J Immunol 147:1759–1764

Mor G, Klinman DM, Shapiro S, Hagiwara E, Sedegah M, Norman JA, Hoffman SL, Steinberg AD (1995) Complexity of the cytokine and antibody response elicited by immunizing mice with Plasmodium yoelii circumsporozoite protein plasmid DNA. J Immunol 155:2039–2046

Nur I, Szyf M, Razin A, Glaser G, Rottem S, Razin S (1985) Procaryotic and eucaryotic traits of DNA methylation in spiroplasmas (mycoplasmas). Journal of Bacteriology 164:19–24

Razin A, Friedman J (1981) DNA methylation and its possible biological roles. Prog Nucleic Acid Res Mol Biol 25:33–52

Roman M, Martin-Orozco E, Goodman JS, Mguyen M, Sato Y, Ronaghy A, Kornbluth RS, Richman DD, Carson DA, Raz E (1997) Immunostimulatory DNA sequences function as T helper-1 promoting adjuvants. Nature Medicine 3:849–854

Sambrook J, Fritsch EF, Maniatis T Analysis and cloning of eukaryotic genomic DNA. In: Sambrook J, Fritsch EF, Maniatis T, eds. Molecular Cloning: A Laboratory Manuel. Cold Spring Harbor, NY. Cold Spring Harbor Laboratory Press, 1989:1–62

Sato Y, Roman M, Tighe H, Lee D, Corr M, Nguyen M, Carson DA, Raz E (1996) Immunostimulatory DNA sequences necessary for effective intradermal gene immunization. Science 273:352–354

Sedegah M, Hedstrom R, Hobart P, Hoffman SL (1994) Protection against malaria by immunization with plasmid DNA encoding circumsporozoite protein. Proc Natl Acad Sci USA 91:9866–9870

Snapper CM, Paul WE (1987) Interferon gamma and B cell stimulatory factor-1 reciprocally regulate Ig isotype production. Science 236:944–947

Sparwasser T, Koch, Vabulas RM, Heeg K, Lipford GB, Ellwart JW, Wager H (1998) Bacterial DNA and immunostimulatory CpG oligonucleotides trigger maturation and activation of murine dendritic cells. Eur J Immunol 28:2045–2054

Ulmer JB, Donnelly JJ, Parker SE, Rhodes GH, Felgner PL, Dwarki VJ, Gromkoski SH, Deck RR, DeWitt CM, Friedman A (1993) Heterologous protection against influenza by injection of DNA encoding a viral protein. Science 259:1745–1749

Wang B, Ugen KE, Srikantan V, Agadjanyan MG, Dang K, Refaeli Y, Sato A, Boyer J, Williams WV, Weiner DB (1993) Gene inoculation generates immune responses against human immunodeficiency virus type 1. Proc Natl Acad Sci USA 90:4156–4160

Wang R, Doolan DL, Le TP, Hedstrom RC, Coonan KM, Charoenvit Y, Hoffman SL (1998) Induction of antigen specific cytotoxic T lymphocytes in humans by a malaria DNA vaccine. Science 282: 476–480

Weiner GJ, Liu HM, Wooldridge JE, Dahle CE, Krieg AM (1997) Immunostimulatory oligodeoxynucleotides containing the CpG motif are effective as immune adjuvants in tumor antigen immunization. Proc Natl Acad Sci USA 94:10833–10837

Wolff JA, Malone RW, Williams P, Chong W, Ascadi G, Jani A, Felgner PL (1990) Direct gene transfer into mouse muscle in vivo. Science 247:1465–1468

Yamamoto S, Yamamoto T, Shimada S, Kuramoto E, Yano O, Kataoka T, Tokunaga T (1992) DNA from bacteria, but not vertebrates, induces interferons, activated NK cells and inhibits tumor growth. Microbiol Immunol 36:983–997

# The Role of Bacterial DNA in Autoantibody Induction

D.S. Pisetsky



## 1 Introduction

Antibodies to DNA (anti-DNA) are prototypic autoantibodies that occur prominently in systemic lupus erythematosus (SLE). These antibodies target sites on single- as well as double-stranded DNA and serve as markers of diagnostic and prognostic significance. In addition, anti-DNA play a direct role in disease pathogenesis, provoking renal injury through immune-complex deposition (Pisetsky 1992). Because of their close association with SLE, anti-DNA have been considered virtually synonymous with autoimmunity and key to understanding disease mechanisms (Eilat and Anderson 1994; Radic and Weigert 1994).

While anti-DNA are a cardinal feature of SLE, immune reactivity to DNA extends beyond this disease. As shown by provocative studies from many laboratories, DNA, depending on base sequence, can induce responses in normal as well as aberrant immunity, with DNA from bacteria displaying potent immunostimulatory activities. These activities include induction of specific antibodies as well as the stimulation of cytokine production and B-cell activation (Pisetsky 1996a,

VA Medical Center, Box 151G, Room E-1008, 508 Fulton Street, Durham, NC 27705, USA
E-mail: dpiset@acpub.duke.edu

1996b). In its effects on innate immunity, bacterial DNA resembles endotoxin, suggesting a role as a danger signal in eliciting host defense (Pisetsky 1997).

Since the immunostimulatory properties of bacterial DNA are discussed in detail elsewhere in this volume, this chapter will focus on their relevance to the induction of anti-DNA autoantibodies. As this account will indicate, bacterial DNA displays unique immunological properties that could allow it to drive both anti-self as well as anti-foreign responses to DNA antigens. These properties reflect a variety of structural elements and suggest novel mechanisms accounting for the serological features of SLE.

## 2 DNA as an Autoantigen

Until recently, analysis of the immunological properties of DNA was guided by two main ideas. The first idea is that the relevant immunological determinants of DNA are conserved determinants present on all DNA. Indeed, most studies on lupus serology have dichotomized DNA antigens into either single stranded (ss) or double stranded (ds) forms, with antibodies to dsDNA the most diagnostic of disease. In the case of dsDNA, the epitopes recognized by autoantibodies have been considered charge arrays along the phosphodiester backbone. While this binding is dependent on charge–charge interactions, it requires the spatial orientation of the classic double helical structure, or B-DNA (Stollar 1992).

In contrast to dsDNA, ssDNA is structurally flexible, because the rigidity conferred by base pairing of the two strands is lost. Nevertheless, autoantibody binding to ssDNA also appears to be a consequence of charge, albeit in a less ordered format. Many antibodies to DNA crossreact with ss- and dsDNA, suggesting a common determinant exposed on both DNA forms. While interactions with phosphate groups in the backbone provide the preponderance of bond energy for antibody binding, interactions with bases themselves may also contribute.

From the molecular analysis of SLE antibody binding, DNA emerges as a relatively uniform antigen, with a charged backbone providing the major site of immune recognition. As such, anti-DNA autoantibodies can bind foreign as well as self DNA as long as the backbone structure is present. This notion is implicit in the technology of anti-DNA assays. Assays for anti-DNA have used many different DNA preparations, essentially interchangeably, with assays based on mammalian and bacterial DNA (including supercoiled plasmids). These assays all perform well clinically, suggesting that the major autoantibody specificities are directed to conserved determinants rather than non-conserved sites exclusive to self DNA (Pisetsky 1998).

The second idea underlying much research in SLE concerns the immunogenic potential of DNA. Following the recognition of DNA's role as a major autoantigen, investigators attempted to replicate lupus in normal animals by immunizing them with DNA. These studies showed that naturally occurring DNA is a poor immunogen. When administered alone, DNA fails to stimulate antibody production although, when complexed with a protein carrier and presented in adjuvant, it can elicit a limited response to ssDNA. Immunization experiments, with some exceptions (see below), failed to induce antibodies to dsDNA, the serological hallmark of lupus (MADAIO et al. 1984).

In contrast to natural DNA, certain synthetic DNA and chemically modified DNA are immunogenic. For example, normal mice can generate a robust antibody response to Z-DNA. Z-DNA is a helical structure formed by tracts of alternating purine and pyrimidine bases; the presence of certain divalent cations promotes transition to this structure. In Z-DNA, the helix displays a zig-zag structure with a left-handed orientation, in contrast to the right-handed orientation of conventional B-DNA. When used as an immunogen, Z-DNA induces antibodies that are specific for this helical form and do not crossreact with B-DNA (LAFER et al. 1981; STOLLAR 1994). Other unusual DNA structures (DNA cruciforms) can also elicit responses in normal animals, although the induced antibodies are specific for the immunizing DNA and do not crossreact with natural DNA (FRAPPIER et al. 1989).

The immunization experiments have suggested that natural dsDNA is immunologically inert and unable to induce a significant antibody response. This paltry activity could reflect two phenomena. The first concerns tolerance and suggests that, because of anergy or clonal deletion, B cells that recognize dsDNA are absent from the normal B-cell repertoire. As such, the normal animal would be unable to produce anti-DNA even when stimulated with DNA in the context of a protein carrier providing T-cell help. The failure to produce autoantibodies to DNA by immunization contrasts with the generation, in normal animals, of autoantibody responses to protein antigens, such as collagen or myelin basic protein. With protein autoantigens, immunization can induce both autoimmune responses and clinical manifestations of disease. This result suggests that tolerance for nucleic-acid antigens may be more complete than for proteins, perhaps reflecting their structure or in vivo metabolism.

As noted above, while natural DNA is a weak immunogen, unusual DNA structures can elicit appreciable responses under conditions in which B-DNA is inactive. The difference in response patterns among nucleic acids may reflect the concentration of unusual DNA structures and their ability to induce tolerance. Z-DNA exists at most transiently in cells and is present in only low concentrations because of sequence requirements. As a result, the contact of Z-DNA with immune cells may be ephemeral or below a threshold necessary to establish tolerance. When presented in a stabilized form in high concentration, Z-DNA may, therefore, induce responses.

While tolerance mechanisms could explain the poor immunogenicity of natural DNA, other properties of DNA could influence the induction of responses. DNA is readily digestible in vivo because of nucleases in serum and cells. Furthermore, DNA in the circulation is rapidly cleared by cells of the reticulo-endothelial system, suggesting that DNA used for immunization may be short lived (EMLEN and MANNIK 1984; YOSHIDA et al. 1996). Indeed, there is evidence that the immunogenicity of DNA is related to its stability, with immunogenicity enhanced by backbone modifications that protect against nuclease digestion. Among these modifications, compounds with phosphorothioate backbones are more effective immunogens than phosphodiesters; phosphorothioates are nuclease resistant (BRAUN and LEE 1988).

In evaluating the immunogenicity of DNA, the protein carrier has received little attention as a factor influencing responsiveness. Compounds such as methylated bovine serum albumin (mBSA) and methylated gammaglobulin have been commonly used in these experiments, because they can complex DNA by charge. Studies of a synthetic protein called Fus-1 indicate that the carrier may affect the immunogenicity of natural DNA. Fus-1 is 27 amino acids long and bears a sequence corresponding to a protein from *Trypanosoma cruzi*. Complexes of Fus-1 with sheared mammalian DNA can induce antibodies to dsDNA in normal mice. Since these induced antibodies are similar to lupus anti-DNA in their binding properties, these results suggest that normal mice may produce anti-DNA autoantibodies under certain circumstances (DESAI et al. 1993; KRISHNAN and MARION 1993).

The impact of Fus-1 on responsiveness to DNA is not well understood, though it will be important to later considerations on the response to bacterial DNA (see below). Assuming that normal mice have at least some capacity for anti-DNA production, the ability of Fus-1 complexes to induce anti-dsDNA could reflect at least three factors: formation of complexes that protect dsDNA from degradation; induction of more vigorous T-cell help; and more effective presentation of a dsDNA epitope for stimulation of B cells. Even with a carrier like Fus-1, however, the immunogenicity of mammalian DNA appears limited and dependent on physical–chemical form.

## 3 Antibody Response to Bacterial DNA

The categorization of DNA as ss or ds is an obvious simplification that ignores the structural richness imparted by the base sequence. Theoretically, such sequences could serve as epitopes for antibody binding just as amino acid sequences serve as epitopes for antibody binding to protein antigens. In our laboratory, we have been interested in the possibility that anti-DNA antibodies can bind non-conserved sequences and conserved backbone conformations. Our efforts to find antibodies to DNA sequences provided some of the initial evidence for the immunological heterogeneity of DNA and the unique properties of bacterial DNA antigens.

To identify antibodies directed to base sequences as opposed to backbone, we tested as antigens a series of mammalian and bacterial DNA. These DNA differed in base composition and presumably presented a range of sequences to allow detection of antibodies to sequence motifs. While many different DNA preparations have been used as antigens in clinical assays, there have been few studies directly comparing the antigenicities of natural DNA varying in species origin

(Stollar et al. 1962). In these experiments, with ssDNA as an antigen, sera from SLE patients showed similar levels of binding to the various DNA preparations in an enzyme-limited immunosorbent assay (ELISA) (Karounos et al. 1988). These findings indicate predominant reactivity to a conserved determinant widely expressed among natural DNA. This determinant could involve both base sequences and backbone; the particular sequences, however, would have to be ubiquitous among the DNA.

Although anti-DNA responses have been widely viewed as exclusive to SLE, we nevertheless tested sera of normal human subjects (NHS) using the same antigen preparations. Results of these studies were entirely unexpected and refute the notion that anti-DNA are exclusive to SLE (Karounos et al., 1988). As these data indicated, the reactivity of NHS sera with DNA is not uniform (Fig. 1). Thus, NHS sera, while non-reactive to mammalian DNA, bind significantly to DNA from certain bacterial DNA, including *Micrococcus lysodeikticus* (MC) and *Staphylococcus epidermidis* (SE). The antigenicity of these preparations is fully sensitive to digestion with DNase, indicating that DNA is the relevant antigen. In NHS sera, as in sera from SLE patients, the anti-DNA antibodies display the immunoglobulin G (IgG) isotype and differ from natural autoantibodies, which are IgM. These results demonstrate clearly that anti-DNA responses are a feature of both normal and aberrant immunity.

As shown by subsequent experiments, anti-DNA antibodies in NHS differ from SLE antibodies in both specificity and immunochemical properties. Whereas SLE anti-DNA bind DNA independent of species origin, NHS anti-DNA bind

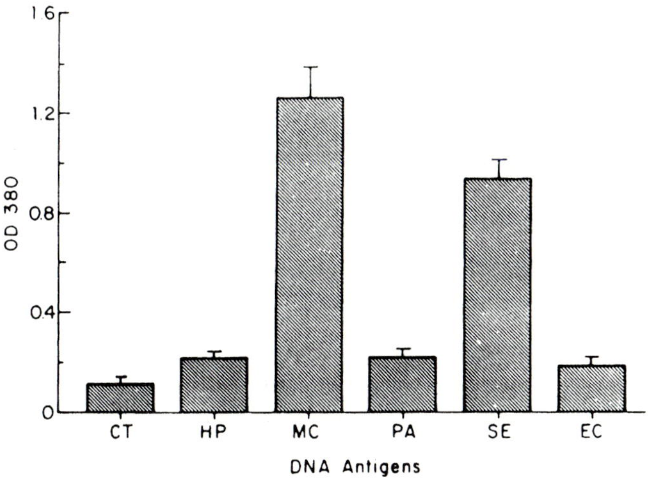


**Fig. 1.** Anti-DNA antibodies in sera of normal human subjects. Sera of 12 normal human subjects were screened by enzyme-linked immunosorbent assay for binding to single-stranded DNA from calf thymus (*CT*), human placenta (*HP*), *Micrococcus lysodeikticus* (*MC*), *Pseudomonas aeruginosa* (*PA*), *Staphylococcus epidermidis* (*SE*), and *Escherichia coli* (*EC*). Results are presented as mean ± SD. Reproduced with permission from Journal of Immunology

selectively to bacterial DNA and recognize epitopes with limited distribution among bacterial species. This selectivity suggests predominant reactivity with bases as opposed to the phosphodiester backbone. Indeed, binding to NHS antibodies to bacterial DNA is much less sensitive to salt concentration than binding to SLE anti-DNA, consistent with less dependence on ionic interactions (PISETSKY and ROBERTSON 1992; WU et al. 1997).

In addition to epitope specificity, antibodies to DNA in NHS and SLE sera differ in isotype. Whereas SLE anti-DNA are primarily IgG1 and IgG3, NHS anti-DNA display a preponderance of the IgG2 isotype (ROBERTSON et al. 1992; WU et al. 1997). In normal immunity, the IgG2 isotype characterizes the response to bacterial carbohydrate antigens. These findings suggest that bacterial DNA may resemble polysaccharides in immune activity, displaying a T-independent antibody induction in contrast to the T-dependent induction of anti-DNA autoantibodies.

While not predicted by prevailing ideas on SLE, immune responses to bacterial DNA can be readily explained using concepts derived from the studies of protein antigens and unusual DNA structures. Thus, bacterial DNA, like foreign proteins, displays sequences that are species-specific and absent from the host. These sequences cannot establish tolerance and can, therefore, induce responses when presented in an immunogenic form during either colonization or infection. Since NHS sera contain antibodies directed to DNA from only certain bacteria, other factors must determine the response; at present, the rules for antigenicity are not clear. Interestingly, among bacterial DNA, *Escherichia coli* (EC) DNA does not show appreciable antibody levels in NHS sera, although this organism is ubiquitous in the gut (WU et al. 1997).

In addition to DNA from bacteria, DNA from certain viruses show reactivity with normal sera (REKVIG et al. 1992). For both viral and bacterial DNA, the antibodies in NHS sera appear to be highly specific for each DNA. Together, these findings indicate that antibody responses to foreign DNA are a common feature of normal immunity, with previous failure to recognize these responses reflecting the range of DNA antigens tested. While certain bacterial or viral DNA show similar levels of antibodies in normal individuals and lupus patients, responses to B-DNA are nevertheless specific for lupus and retain their diagnostic utility.

## 4 Induction of Anti-DNA Antibodies by Immunization with Bacterial DNA

Although studies on normal humans strongly suggest that bacterial DNA is immunogenic, more direct proof for this idea was sought from animal experiments. Normal mice were, therefore, immunized with preparations of EC dsDNA complexed with mBSA in Freund's adjuvant, and their sera was assayed for antibodies both to the immunizing bacterial DNA and mammalian calf thymus (CT) DNA. This format was similar in design to previous studies indicating that mammalian DNA is immunologically inactive.

As these studies indicated, EC dsDNA can induce responses in normal mice under conditions in which mammalian DNA is without effect (GILKESON et al. 1989a). The induced antibodies, however, are highly specific for bacterial DNA and do not crossreact with mammalian DNA. In this respect, the induced antibodies resemble antibodies in NHS sera that target non-conserved sites on bacterial DNA; they differ from SLE anti-DNA that target conserved sites present on both bacterial DNA and mammalian DNA. The epitopes recognized by the induced antibodies are most likely sequences, although charges from the backbone may contribute to binding energy.

Immunization with bacterial ssDNA produces a somewhat different response pattern than dsDNA (GILKESON et al. 1989b). Whereas, in normal mice, bacterial dsDNA elicits antibodies that bind exclusively to the immunizing antigen, bacterial ssDNA stimulates antibodies that crossreact with mammalian ssDNA. The induction of these crossreactive antibodies may suggest that-B cell tolerance for ss- and dsDNA differ, with normal individuals retaining a greater capacity for producing autoantibodies to the ssDNA form. It is important to note, however, that the levels of antibodies induced by bacterial ssDNA are far greater than those induced by mammalian ssDNA, indicating that bacterial DNA enhances responsiveness even to conserved determinants.

In view of the immunological activities of bacterial DNA, the immunogenicity of this nucleic acid appears related to its intrinsic immunostimulatory activity and its content of antigenic sequences that are exclusive to bacterial DNA. As discussed elsewhere in this volume, bacterial DNA is mitogenic for B cells and can stimulate a variety of cytokines, including interleukin 12 (IL-12), tumor necrosis factor $\alpha$, IL-6, interferon $\alpha/\beta$ (IFN-$\alpha/\beta$) and IFN-$\gamma$. These activities result from short six base motifs called CpG motifs or ISS (immunostimulatory sequences) that have the general structure of two $5'$ purines, an unmethylated CpG motif and two $3'$ pyrimidines. These sequences occur much more commonly in bacterial than in mammalian DNA because of CpG suppression in mammalian DNA and the frequent methylation of cytosine in this position in mammalian DNA (YAMAMOTO et al. 1992a; KRIEG et al. 1995). The content of ISS differs among bacterial DNA, suggesting that bacterial DNA vary in immunogenicity (YAMAMOTO et al. 1992b).

Taken together with the serological studies, these considerations suggest that DNA displays three immunologically relevant determinants: conformational determinants on ds- or ssDNA that function as B-cell epitopes in SLE; sequential determinants on bacterial ss- or dsDNA that function as B-cell epitopes in normal immunity; and ISS on bacterial DNA that serve as internal adjuvants. While an ISS could theoretically be a target of antibody reactivity, the reactivity of NHS suggests that these sequences are not commonly recognized. Instead NHS antibodies appear to target sites that are species-specific rather than targeting an ISS, which would be expected to have a broader distribution.

## 5 The Role of Bacterial DNA in Driving Autoantibody Production

The immune properties of bacterial DNA, while unanticipated by previous research, can nevertheless explain a conundrum that has long influenced SLE research: the mechanism by which a putatively non-immunogenic molecule can drive autoantibody production. As shown clearly in studies on both mouse and man, DNA is neither uniform nor inert. Rather, bacterial DNA, unlike mammalian DNA, is immunologically potent and can drive production of antibodies to sequential determinants. The intriguing question is whether it can also drive the production of autoantibodies to conformational determinants. If so, this finding would suggest that bacterial DNA could either initiate an autoantibody response to DNA or boost its levels during active SLE.

To explore the capacity of bacterial DNA to drive autoantibody production, we used as a model immunization of NZB/NZW mice. These animals develop a lupus-like illness characterized by high levels of anti-DNA production in association with glomerulonephritis. Using protocols developed from normal mice, we immunized female NZB/NZW mice with bacterial or mammalian dsDNA prior to the onset of spontaneous disease. Sera were assayed by ELISA for antibodies to both bacterial and mammalian dsDNA (Gilkeson et al. 1995).

The results of these studies provide important insight into the potential of bacterial DNA to promote SLE antibody production. In the preautoimmune mice, EC dsDNA elicited significant production of anti-dsDNA antibodies. Unlike antibodies in normal mice, however, the induced antibodies had autoantibody activity and bound both mammalian and bacterial dsDNA. This crossreactivity is reminiscent of the SLE response and signifies binding to a conserved determinant. In contrast, immunization of NZB/NZW mice with mammalian dsDNA was without effect, suggesting that unique structural features of bacterial DNA are required for autoantibody production (Gilkeson et al. 1995). The failure of mammalian DNA to induce anti-DNA in NZB/NZW mice is surprising, because these mice are destined to produce such antibodies and would be expected to have appropriate B-cell precursors.

Two conclusions from these studies are relevant for models for the mechanisms of anti-DNA production in SLE: (1) the induction of crossreactive antibodies in autoimmune mice occurs under conditions in which normal mice generate a response exclusive for foreign DNA; and (2) bacterial DNA can stimulate anti-DNA autoantibody responses under conditions in which mammalian DNA is inactive. As suggested above, the production of autoantibodies to DNA may reflect abnormalities in the B-cell repertoire in lupus mice. Because of tolerance defects in autoimmunity, this repertoire may retain precursors that can bind dsDNA and be stimulated to autoantibody production under appropriate immunization conditions. In this view, the difference between normal and autoimmune mice would reside in the B-cell-precursor population with the autoimmune repertoire skewed to reactivity to conserved conformational determinants (as opposed to non-conserved sequential determinants) (Gilkeson et al. 1993).

The failure of mammalian DNA to induce these precursors in NZB/NZW mice is more difficult to explain. Since the immunizations involve a protein carrier and adjuvant, induction of adequate T cells would be expected. In this situation, either mammalian or bacterial DNA should be able to stimulate the B-cell response, because both DNA sources would bear the crossreactive backbone determinant. If, however, the generation of T-cell help in NZB/NZW mice was limiting, the adjuvant properties of the bacterial DNA could be a critical ingredient allowing induction of adequate levels of help. It is also possible that bacterial DNA show less degradation in vivo than mammalian DNA, providing a higher antigen content.

The studies using Fus-1 as a carrier may be relevant to these considerations. As described previously, Fus-1 as a carrier may allow induction of responses under conditions in which mBSA fails. Fus-1, because of its sequence, may induce a more robust T-cell response and allow emergence of an autoantibody response from rare precursors. Similarly, bacterial DNA may facilitate these responses, because its adjuvant activity can enhance T-cell responses to mBSA and thereby allow adequate help for anti-DNA with that carrier. Whether the induction of T-cell help for anti-DNA results from properties of the carrier or adjuvant properties of the DNA, autoantibody production appears to occur much more readily in mice from an autoimmune background than in those of a normal strain.

## 6 Antibody Responses to Bacterial DNA in SLE

Theoretically, an individual with lupus can produce two types of anti-DNA: anti-DNA autoantibodies that bind both bacterial and mammalian DNA and anti-DNA antibodies that bind exclusively to bacterial DNA. The balance between these responses would be a function of the composition of both the B-cell precursor population and the role of foreign vs self DNA in driving these responses. Determination of the nature of this balance should, therefore, help distinguish the role of different DNA antigens in disease induction.

To assess the expression of different types of anti-DNA in SLE, we performed absorption experiments using DNA affinity columns to deplete anti-DNA populations (Pisetsky and Drayton 1997). SLE sera with high titers of anti-DNA autoantibodies were absorbed with CT DNA and then tested for binding to both CT and MC DNA. Control NHS sera were similarly absorbed. The results of these experiments were clear-cut. Absorptions that removed antibodies to CT DNA from SLE sera essentially completely removed antibodies to MC DNA; this absorption had no effect on the binding of NHS sera to MC DNA. Furthermore, direct comparison of absorbed SLE and NHS sera showed that SLE sera have markedly lower levels of antibodies specific for bacterial DNA than the levels found in NHS sera (Fig. 2).

These observations document a reduction in the specific response to bacterial DNA in NHS and raise two mechanistic possibilities. The first is that a shift in

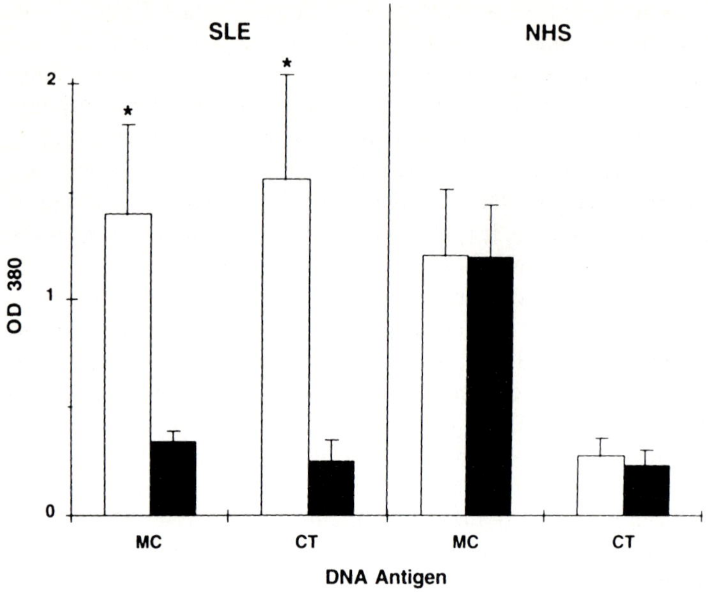


**Fig. 2.** Specificity analysis of antibodies in sera of normal human subjects (NHS) and patients with systemic lupus erythematosus (SLE). A panel of 20 NHS and 16 SLE sera were absorbed with either control cellulose (*open bar*) or calf thymus (CT) single-stranded-DNA cellulose (*closed bar*) and then tested for binding to either *Micrococcus lysodeikticus* or CT DNA. Asterisks indicates statistically significant differences ($P < 0.001$). Reproduced with permission from Proceedings of the Association of American Physicians

antibody specificity in SLE leads to the expression of antibodies to conserved rather than non-conserved determinants whether the inducing antigen is foreign or self. This shift could result from distortion in the B-cell repertoire because of impaired tolerance. The second interpretation is that SLE patients have an antibody-deficient state that prevents them from mounting a specific antibody response to bacterial DNA. In the absence of such specific antibodies, bacterial DNA, because of its content of ISS, could stimulate the immune system sufficiently to cause autoantibody production.

At present, there are insufficient data to distinguish these possibilities, although reference to other autoantibody responses suggests that an aberrant pattern of epitope recognition may mark the immune systems of SLE patients. In the response to protein autoantigens, SLE sera tend to react to conformational epitopes rather than sequential determinants of the kind recognized by antibodies induced by immunization in animals (Tan 1989; Yamanaka et al. 1989). The lack of binding to peptide antigens is striking and is contrary to predictions for antigenicity based on experimental immunization models. A failure to bind peptide determinants may reflect the process of antigen presentation, although skewing in the B-cell repertoire could also affect the epitopes recognized.

While bacterial DNA has the potential to drive anti-DNA production, its role in the setting of human disease is unknown at present. In other studies on anti-DNA production, chromatin has been implicated as the driving antigen, because antibodies to DNA frequently coexist with antibodies to histones and other chromatin components. Furthermore, in autoimmune mice, antibodies to chromatin may precede the expression of antibodies to isolated DNA, suggesting that antibodies to DNA may emerge from a response directed to a larger complex. The ability of nucleosomes to stimulate T cells is further evidence that chromatin functions as a key autoantigen to drive SLE autoantibody production (Burlingame et al. 1993; Mohan et al. 1993; Burlingame et al. 1994).

Even if chromatin is a major autoantigen, bacterial DNA may nevertheless have an important role in stimulating anti-DNA in SLE. As shown in longitudinal studies, levels of anti-DNA fluctuate dramatically over time and show patterns of temporal expression that differ from antibodies to other nuclear antigens, such as Sm (McCarty et al. 1982). If self-antigens were the exclusive driving antigens, more coordinate appearance of autoantibodies would be expected. However, a foreign antigen that periodically impinges on the immune system could lead to the flares of serological activity that characterize the course of SLE. We are currently performing experiments to address these possibilities.

## 7 Summary

Bacterial DNA has potent immunological properties that can stimulate the immune system in SLE in both specific and non-specific ways. As such, this molecule may play an important role in disease pathogenesis, because it can exert immunomodulatory activity and function as a molecular mimic. Future studies will hopefully both determine the role of foreign nucleic acids in the induction of autoantibodies and lead to strategies for their elimination.

## References

Braun RP, Lee JS (1988) Immunogenic duplex nucleic acids are nuclease resistant. J Immunol 141: 2084–2089

Burlingame RW, Boey ML, Starkebaum G, Rubin RL (1994) The central role of chromatin in autoimmune responses to histones and DNA in systemic lupus erythematosus. J Clin Invest 94:184–192

Burlingame RW, Rubin RL, Balderas RS, Theofilopoulos AN (1993) Genesis and evolution of antichromatin autoantibodies in murine lupus implicates T-dependent immunization with self antigen. J Clin Invest 91:1687–1696

Desai DD, Krishnan MR, Swindle JT, Marion TN (1993) Antigen-specific induction of antibodies against native mammalian DNA in nonautoimmune mice. J Immunol 151:1614–1626

Eilat D, Anderson WF (1994) Structure-function correlates of autoantibodies to nucleic acids. Lessons from immunochemical, genetic and structural studies. Mol Immunol 31:1377–1390

Emlen W, Mannik M (1984) Effect of DNA size and strandedness on the in vivo clearance and organ localization of DNA. Clin Exp Immunol 56:185–192

Frappier L, Price GB, Martin RG, Zannis-Hadjopoulos M (1989) Characterization of the binding specificity of two anticruciform DNA monoclonal antibodies. J Biol Chem 264:334–341

Gilkeson GS, Bloom DD, Pisetsky DS, Clarke SH (1993) Molecular characterization of anti-DNA antibodies induced in normal mice by immunization with bacterial DNA. Differences from spontaneous anti-DNA in the content and location of $V_H$ CDR3 arginines. J Immunol 151:1353–1364

Gilkeson GS, Grudier JP, Karounos DG, Pisetsky DS (1989a) Induction of anti-double stranded DNA antibodies in normal mice by immunization with bacterial DNA. J Immunol 142:1482–1486

Gilkeson GS, Grudier JP, Pisetsky DS (1989b) The antibody response of normal mice to immunization with single-stranded DNA of various species origin Clin Immunol Immunopath 51:362–371

Gilkeson GS, Pippen AMM, Pisetsky DS (1995) Induction of cross-reactive anti-dsDNA antibodies in preautoimmune NZB/NZW mice by immunization with bacterial DNA. J Clin Invest 95:1398–1402

Karounos DG, Grudier JP, Pisetsky DS (1988) Spontaneous expression of antibodies to DNA of various species origin in sera of normal subjects and patients with systemic lupus erythematosus. J Immunol 140:451–455

Krieg AM, Yi A-K, Matson S, Waldschmidt TJ, Bishop GA, Teasdale R, Koretzky GA, Klinman DM (1995) CpG motifs in bacterial DNA trigger direct B-cell activation. Nature 374:546–549

Krishnan MR, Marion TN (1993) Structural similarity of antibody variable regions from immune and autoimmune anti-DNA antibodies. J Immunol 150:4948–4957

Lafer EM, Moller A, Nordheim A, Stollar BD, Rich A (1981) Antibodies specific for left-handed Z-DNA. Proc Natl Acad Sci USA 78:3546–3550

Madaio MP, Hodder S, Schwartz RS, Stollar BD (1984) Responsiveness of autoimmune and normal mice to nucleic acid antigens. J Immunol 132:872–876

McCarty GA, Bembe MB, Rice JR, Pisetsky DS (1982) Independent expression of autoantibodies in systemic lupus erythematosus. J Rheumatol 9:691–695

Mohan C, Adams S, Stanik V, Datta SK (1993) Nucleosome: a major immunogen for pathogenic autoantibody-inducing T cells of lupus. J Exp Med 177:1367–1381

Pisetsky DS (1992) Anti-DNA antibodies in systemic lupus erythematosus. Rheum Dis Clin North Am 18:437–454

Pisetsky DS (1996a) The immunologic properties of DNA. J Immunol 156:421–423

Pisetsky DS (1996b) Immune activation by bacterial DNA: A new genetic code. Immunity 5:303–310

Pisetsky DS (1997) Immunostimulatory DNA: A clear and present danger? Nature Med 3:829–831

Pisetsky DS (1998) Antibody responses to DNA in normal immunity and aberrant immunity. Clin Diagn Lab Immunol 5:1–6

Pisetsky DS, Drayton DM (1997) Deficient expression of antibodies specific for bacterial DNA by patients with systemic lupus erythematosus. Proc Assoc Am Physicians 109:237–244

Radic MZ, Weigert M (1994) Genetic and structural evidence for antigen selection of anti-DNA antibodies. Annu Rev Immunol 12:487–520

Rekvig OP, Fredriksen K, Brannsether B, Moens U, Sundsfjord A, Traavik T (1992) Antibodies to eukaryotic, including autologous, native DNA are produced during BK virus infection, but not after immunization with non-infectious BK DNA. Scand J Immunol 36:487–495

Robertson CR, Gilkeson GS, Ward MM, Pisetsky DS (1992) Patterns of heavy and light chain utilization in the antibody response to single-stranded bacterial DNA in normal human subjects and patients with systemic lupus erythematosus. Clin Immunol Immunopath 62:25–32

Robertson CR, Pisetsky DS (1992) Specificity analysis of antibodies to single-stranded micrococcal DNA in the sera of normal human subjects and patients with systemic lupus erythematosus. Clin Exp Rheumatol 10:589–594

Stollar BD (1992) Immunochemical analysis of nucleic acids. In: Cohn WE, Moldave K (eds) Progress in nucleic acid research and molecular biology. Academic Press, San Diego, pp. 39–77

Stollar BD (1994) Molecular analysis of anti-DNA antibodies. FASEB J 8:337–342

Stollar D, Levine L, Marmur J (1962) Antibodies to denatured deoxyribonucleic acid in lupus erythematosus serum. II. Characterization of antibodies in several sera. Biochim Biophys Acta 61:7–18

Tan EM (1989) Antinuclear antibodies: diagnostic markers for autoimmune diseases and probes for cell biology. Adv Immunol 44:93–150

Wu Z-Q, Drayton D, Pisetsky DS (1997) Specificity and immunochemical properties of antibodies to bacterial DNA in sera of normal human subjects and patients with systemic lupus erythematosus (SLE). Clin Exp Immunol 109:27–31

Yamamoto S, Yamamoto T, Kataoka T, Kuramoto E, Yano O, Tokunaga T (1992a) Unique palindromic sequences in synthetic oligonucleotides are required to induce INF and augment INF-mediated natural killer activity. J Immunol 148:4072–4076

Yamamoto S, Yamamoto T, Shimada S, Kuramoto E, Yano O, Kataoka T, Tokunaga T (1992b) DNA from bacteria, but not from vertebrates, induces interferons, activates natural killer cells and inhibits tumor growth. Microbiol Immunol 36:983–997

Yamanaka H, Willis EH, Carson DA (1989) Human autoantibodies to poly(adenosine diphosphate ribose) polymerase recognize cross-reactive epitopes associated with the catalytic site of the enzyme. J Clin Invest 83:180–186

Yoshida M, Mahato RI, Kawabata K, Takakura Y, Hashida M (1996) Disposition characteristics of plasmid DNA in the single-pass rat liver perfusion system. Pharm Res 13:599–603

# CpG DNA in Cancer Immunotherapy

G.J. WEINER



## 1 Introduction

Anecdotal reports of tumor regression following systemic bacterial infection have been observed for centuries. The first reported attempt to explore this effect in a systematic fashion took place in the 1890s, when Dr. William Coley, a New York surgeon, performed a series of studies evaluating anti-tumor therapy with bacteria and bacterial products. Dr. Coley's original attempt to use bacteria as an anti-tumor agent involved the use of live cultures of streptococci injected directly into tumor masses (COLEY 1893). This resulted in tumor regression in some cases but proved to be toxic, with the first patient almost dying of erysipelas. Subsequent studies by Coley involved a mixture of heat-killed streptococci and serratia (then known as *Bacillus prodigiosus*) (COLEY 1894). This preparation was still quite toxic, but did result in some anti-tumor responses. It is this preparation that is now referred to as Coley's toxin. Much of the antitumor activity of Coley's toxin is attributed to endotoxin (WIEMANN and STARNES 1994). However, it is curious to note that Coley's original success was with an organism that does not produce

University of Iowa Cancer Center, and Department of Internal Medicine, 5970 JPP 200 Hawkins Drive, Iowa City, IA 52242, USA
E-mail: george-weiner@uiowa-edu

endotoxin. Thus, additional bacterial components, such as bacterial DNA, may well have played a role in the observed responses. Indeed, recent reports have confirmed that cancer can regress following bacterial infection (Mori et al. 1997).

Despite the fact that Coley performed his intriguing experiments over a century ago, immunotherapy still has little impact on the management of patients with cancer. The recent reawakening of interest in cancer immunotherapy results from a more complete understanding of the immune response in general and the interaction between the immune system and tumors in particular. We are currently making rapid progress in these areas. Nevertheless, we are still in the infancy of scientific cancer immunotherapy and are only now beginning to understand the true promise of specific agents, such as immunostimulatory DNA, as potentially useful agents for the treatment of cancer.

The complex immunologic effects of immunostimulatory DNA in general and CpG DNA in particular are outlined elsewhere in this volume and will not be discussed here. Instead, we will focus on a number of promising approaches to cancer immunotherapy and how CpG DNA might play a role in these approaches. Some convincing data is reviewed below; however, evaluation of CpG DNA as a cancer therapy is only beginning, and much of what follows is based on circumstantial evidence with little or no direct experimental data. Nevertheless, our basic understanding of tumor immunology, combined with an increasing knowledge of the immunologic effects of CpG DNA, is enough to warrant discussion of CpG DNA as a cancer therapy using these approaches. Approaches discussed below include:

1. CpG DNA as a single agent for the treatment of cancer
2. CpG DNA for the enhancement of the efficacy of antibody therapy
3. CpG DNA for the enhancement of the efficacy of immunization with tumor-associated antigens
4. CpG DNA for the enhancement of the efficacy of cellular immunotherapy
5. CpG DNA for the enhancement of the anti-tumor response following DNA immunization

## 2 CpG DNA as a Single Agent for the Treatment of Cancer

As outlined above, much of the effect of Coley's toxin has been attributed to endotoxin, which is known to induce production of a number of cytokines that have anti-tumor effects in vitro and in animal models. This has led, over the past several years, to the clinical evaluation of a number of recombinant cytokines as cancer treatments (Heaton and Grimm 1993). Some recombinant cytokines have been shown to have clear, although not dramatic, anti-tumor activity in clinical trials (Rosenberg et al. 1985; Bukowski 1997; Kantarjian et al. 1998). While such studies demonstrate that cancer immunotherapy can be effective, overall responses to such agents have been disappointing given the expectations in the

1960s and 1970s, when cytokines, such as the interferons (IFNs), were first being discovered and proposed as cancer cures.

In retrospect, the expectation that single cytokines might result in significant anti-tumor responses was quite naive. The immune response normally involves the integrated production of a variety of cytokines, which work in concert both locally and systemically. Agents that are able to orchestrate the production of cytokines by the host both temporally and spatially may be more effective at inducing an anti-tumor response compared with passive administration of recombinant cytokines. CpG is capable of inducing production of a number of cytokines that have been shown to have anti-tumor activity, including tumor necrosis factor $\alpha$ (TNF$\alpha$), interleukin 12 (IL-12) and IFN-$\gamma$ (Ballas et al. 1996; Cowdery et al. 1996; Lipford et al. 1997). In addition, CpG DNA has direct effects on immune cell subpopulations that play an important role in anti-tumor immunity such as B cells (Krieg et al. 1995), natural killer (NK) cells (Ballas et al. 1996), monocytes and macrophages (Chace et al. 1997; Sparwasser et al. 1997), dendritic cells (Jakob et al. 1998; Liu et al. 1998; Sparwasser et al. 1998), and possibly T cells (Lipford et al. 1997). The complex cascade of effects on the immune system induced by CpG DNA would seem to point towards the ability of CpG DNA to enhance non-specific anti-tumor immunity.

Limited data suggest that CpG DNA by itself can have anti-tumor effects. Smith et al. used a murine model of lymphoma to evaluate the anti-tumor effects of an antisense phosphorothioate DNA designed to block the c-*myc* oncogene. While their antisense DNA inhibited tumor growth, so did a control DNA that did not have antisense activity but was immunostimulatory (Smith and Wickstrom 1998).

Ballas has evaluated the effect of CpG DNA on the growth of B16 melanoma cells in syngeneic mice (manuscript in preparation). All mice treated with non-CpG DNA died within 24 days, while those treated with CpG DNA had 40% long-term survival. Surviving mice re-challenged with melanoma died similar to control, suggesting that there is no memory. Studies done in severe combined immunodeficiency mice gave identical results to immunocompetent mice. Taken together, these results suggest that innate immunity (probably NK cells) and not T cells are responsible for the therapeutic effect.

While toxicity from CpG DNA in animal models has been limited, toxicity from anti-cancer agents that work in a non-specific manner often limits their clinical efficacy. Clinical trials will be required before we can know whether CpG DNA has promise as a single agent for the treatment of cancer.

## 3 CpG DNA for the Enhancement of the Efficacy of Antibody Therapy

Kohler and Milstein's seminal technology provided monoclonal murine antibodies for both preclinical and clinical evaluation over 20 years ago (Kohler and

Milstein 1975). Such murine antibodies were effective in migrating to their designed target. However, unmodified murine antibodies were limited in their ability to manipulate the human immune system, had relatively short half-lives and proved to be immunogenic in themselves (Link and Weiner 1998). Since that time, protein chemists and molecular biologists have used a variety of approaches to "humanize" antibodies and so overcome some of the limitations of the original murine antibodies. Studies of antibodies with largely human frameworks demonstrate they are more effective at activating human effector mechanisms, have longer half-lives and are less immunogenic than their murine counterparts. Recent clinical trials with such agents have demonstrated monoclonal antibodies (mAbs) can have significant anti-tumor effects in patients (Maloney et al. 1997; Pegram et al. 1998), although there continues to be significant room for improvement. The mechanisms responsible for the clinical response to antibody therapy are not yet clear, although antibody-dependent cellular cytotoxicity (ADCC) mediated by NK cells and monocytes/macrophages likely plays a large role.

A number of investigators have explored approaches to enhancing the activation of NK cells with cytokines to increase ADCC in vitro and the efficacy of mAb therapy in vivo. For example, IL2 has been shown to increase the efficacy of mAbs in a mouse lymphoma model (Berinstein and Levy 1987). This approach is now being evaluated in clinical trials. CpG DNA activates both NK cells and macrophages and, therefore, could enhance the efficacy of antibody therapy by increasing effector cell killing of antibody-coated tumor cells. We found that, in vitro, CpG-DNA-activated splenocytes induced tumor cell lysis more effectively than did unactivated splenocytes. Such effector cells were also superior to unactivated splenocytes or cells activated with a control methylated DNA at inducing ADCC. In vivo, CpG DNA alone had no effect on survival of mice inoculated with lymphoma. However, a single injection of CpG DNA enhanced the anti-tumor response to anti-tumor mAb therapy. These anti-tumor effects were less pronounced when treatment consisted of an identical DNA containing methylated cytosines. A single dose of CpG DNA was more effective than multiple doses of IL-2 at inhibiting tumor growth when used along with anti-tumor mAbs (Wooldridge et al. 1997). We have found similar results in vitro using human effector cells and human tumor cells (manuscript in preparation). Thus, use of CpG DNA to enhance the efficacy of antibody therapy remains a real possibility. A clinical trial designed to assess this possibility is currently being planned.

## 4 CpG DNA for the Enhancement of the Efficacy of Immunization with Tumor-Associated Antigen

Vaccinations for a variety of deadly infectious diseases, such as small pox and diphtheria, have had a major impact on world-wide public health over the past century. However, cancer has proven to be less responsive to vaccination. Recent advances in the field of immunology are allowing us to understand why development of a successful vaccination strategy has been so difficult for cancer. In broad and simplified terms, the major problems include: finding an antigen that is adequately tumor specific yet expressed by all the malignant cells; breaking immune tolerance against that antigen; and inducing an immune response strong enough to induce tumor cell destruction (SPOONER et al. 1995; GILBOA 1996; HELLSTROM et al. 1997).

A number of tumor-associated antigens have been identified that appear to be acceptable, although not ideal, targets for cancer-immunization strategies. Examples include the idiotype (Id) expressed by B-cell malignancies (HSU et al. 1997) and a number of proteins associated with melanoma (MAEURER et al. 1996). Induction of an immune response against antigens that have been used for years as tumor markers, such as carcinoembryonic antigen (PERVIN et al. 1997) and prostate-specific antigen (CORREALE et al. 1997; HODGE et al. 1995), is also showing early promise.

The question of whether a humoral or a cellular anti-tumor response is most desirable is still debated and likely depends on the tumor type and the antigen being targeted. Many cancer-vaccine studies use antibody production as an indication of a successful immune response. This is based in large part on the relative ease of quantitating the antibody response compared to the cellular response. There is also evidence that passive antibody therapy can have anti-tumor activity and that the ability to induce a humoral response following immunization in some cancer-vaccine trials correlates with clinical outcome (LIVINGSTON 1998). Obviously, an antibody response can only be effective if the target antigen is a surface molecule. In addition, numerous solid-tumor animal models demonstrate that a cellular immune response is required if a significant anti-tumor response is desired. Overall, there is now general agreement that induction of a potent anti-tumor cellular response is required if cancer immunotherapy is to be effective against most tumor types.

Unfortunately, the immune adjuvants used in most cancer-vaccination trials to date enhance the T helper 2 (Th2) response, and so have a sub-optimal effect on cellular immunity. Aluminum hydroxide, which is used in many commercial vaccine preparations, has been shown to selectively block activation of $CD8^+$ cytotoxic T lymphocytes (CTLs) in mice immunized against hepatitis-B surface antigen (SCHIRMBECK et al. 1994). Other adjuvants that induce a more extensive Th1 response are currently being evaluated in both pre-clinical and clinical studies (BALDRIDGE and WARD 1997), and are reviewed elsewhere in this volume. These include threonyl-muryl dipeptide (HSU et al. 1993, 1997), a variety of attenuated or killed bacteria (CHEN et al. 1993; BALDRIDGE and WARD 1997) and bacterial derivatives (JOHNSTON and BYSTRYN 1991), BCG (MASTRANGELO et al. 1996) and *Quillaja saponaria* 21 (LIVINGSTON et al. 1994). While each of these has shown some efficacy, none are ideal due to toxicity (including both systemic toxicity and local inflammation after repeated immunization), minimal efficacy at stimulating a cellular response or difficulties associated with production. Thus, new adjuvants are needed for vaccine approaches designed to enhance the Th1 response.

Synthetic DNA containing immunostimulatory sequences are particularly attractive as adjuvants for tumor-antigen immunization because of their ease of production and potent ability to induce a Th1 response. In a number of murine systems, CpG DNA have been evaluated as immune adjuvants and shown to be effective. The studies outlined below were not all performed in tumor models, but the results point to the promise of CpG DNA as an adjuvant in tumor-antigen immunization. Chu et al. used hen-egg lysozyme (HEL) as a model antigen and found that immunization with HEL in incomplete Freund's adjuvant (IFA) resulted in Th2-dominated immune responses characterized by HEL-specific secretion of Th2 cytokines (IL-5 but not IFN-$\gamma$). In contrast, the addition of CpG DNA to immunization with IFA–HEL switched the immune response to a Th1-dominated cytokine pattern (high levels of IFN-$\gamma$ and decreased IL-5) (Chu et al. 1997). CpG DNA also enhanced production of anti–HEL immunoglobulin G2a (IgG2a, a Th1-associated isotype) when compared with IFA–HEL alone. This Th1 response was more marked than that seen with complete Freund's adjuvant (CFA). Using hepatitis-virus-B surface antigen as an immunogen, Davis et al. demonstrated a marked increase in antigen-specific IgG and IgG2a (Davis et al. 1998). Importantly, they also found development of an antigen-specific cytotoxic-T-cell response. A similar humoral and cellular response was found by Lipford et al. using ovalbumin as the target antigen (Lipford et al. 1997). In addition, these investigators detected an enhanced CTL response to both unprocessed protein antigen and to major histocompatibility complex (MHC) class-I-restricted peptide.

We utilized the Id from the 38C13 murine lymphoma model as the target antigen. The Id served as the tumor antigen. CpG DNA was as effective as CFA at inducing an antigen-specific antibody response, but was associated with less toxicity. CpG DNA induced a higher titer of antigen-specific and tumor-specific IgG2a than did CFA. Mice immunized with CpG DNA as an adjuvant were protected from tumor challenge to a degree similar to that seen in mice immunized with CFA but with less toxicity (Weiner et al. 1997). We also explored synergy with granulocyte–monocyte colony-stimulating factor (GM-CSF). Immunization using antigen (Id), CpG DNA and soluble GM-CSF enhanced production of antigen-specific antibody and shifted production towards the IgG2a isotype. This effect was most pronounced after repeat immunizations with CpG DNA and antigen/GM-CSF fusion protein. A single immunization with CpG DNA and antigen/GM-CSF fusion protein 3 days prior to tumor inoculation prevented tumor growth, whereas other approaches to vaccination in this model are ineffective under those conditions (Liu et al. 1998).

Studies in these varied systems indicate that CpG DNA enhances an antigen-specific Th1 response after immunization with soluble antigen and can lead to development of antigen-specific cellular immunity. CpG DNA have been well tolerated and have been equal or superior to standard adjuvants, such as CFA, at inducing an antibody response. Perhaps most importantly, CpG DNA has also been superior at inducing a therapeutic effect. Additional studies that compare CpG DNA to the other adjuvants outlined above are needed in order to determine whether synergy exists between these adjuvants and to determine the efficacy of CpG DNA as an adjuvant in humans.

Classic immunologic teaching states that intracellular proteins are processed and presented in class-I molecules, and this leads to a cellular immune response. In contrast, extracellular antigens are presented in class-II molecules, which leads to a humoral response. There is increasing evidence that there is also cross-talk between the class-I and class-II pathways. Some extracellular antigens are taken up by antigen-presenting cells (APCs) and processed in a manner that leads to presentation in class-I molecules and development of a CTL response termed "cross-priming" (Rock 1996). The results outlined above support the concept that an effective cellular response can be induced by immunization with an intact tumor antigen plus an adjuvant, such as CpG DNA, that activates cells that can process exogenous antigen and present peptides in class-I MHC (Davis 1998).

It is important to note the CpG DNA have yet to be evaluated as immune adjuvants in humans. Although individual CpG DNA differ somewhat in their ability to activate various immune cell populations and induce cytokine production in human and murine systems, it is now clear that both human and murine leukocytes respond to this novel pathway of immune activation (Krieg et al. 1996). It is, therefore, likely that sequences that induce activation of human antigen presenting cells in vitro will be effective adjuvants in vivo, as is observed in the mouse models.

## 5 CpG DNA for the Enhancement of the Efficacy of Cellular Immunotherapy

The 1980s saw the development of recombinant cytokines, such as IL-2, that allowed for the ex vivo expansion of various populations of lymphocytes. Cells expanded in this manner were called "lymphokine-activated killer" or LAK cells and had significant in vitro tumoricidal activity. Initial clinical trials of "adoptive immunotherapy" using LAK cells were promising (Lotze and Rosenberg 1986). However, it soon became apparent that LAK cells were not migrating well to tumor (Basse 1995). In addition, large doses of cytokines were needed to keep the LAK cells functional once they were re-infused into the patient. This resulted in significant toxicity. Finally, LAK cells expanded from the peripheral blood were not tumor specific. "Tumor-infiltrating lymphocytes", or TILs, were obtained from the tumor mass itself and expanded in a similar manner in an attempt to enhance the number of tumor-specific cells. Despite enhanced specificity in vitro, clinical trials of TILs demonstrated little anti-tumor activity (Schwartzentruber et al. 1994). Although the lymphocytes were harvested from the tumor mass, many may have been bystander cells and not truly specific for the tumor. There is also evidence that the signaling pathways in TILs are abnormal and that this may limit the cytotoxic capabilities of TILs (Finke et al. 1993). Adoptive immunotherapy studies using cytotoxic lymphocytes are continuing in both the laboratory and the clinic, but problems related to target specificity and migration of lymphocytes to the target remain.

If these problems can be overcome, CpG DNA might have a role in adoptive immunotherapy approaches. Animal models suggest much of the anti-tumor activity observed with LAK and TIL cells may actually be related to NK activity. CpG DNA is known to enhance NK function in vitro and in vivo (BALLAS et al. 1996). CpG DNA might, therefore, be useful in expanding and activating NK cells intended for adoptive immunotherapy. The effect of CpG DNA on T-cells remains controversal. In studies using highly purified T cells, Lipford et al. found that CpG DNA enhances proliferation of T-cells when the T-cell receptor is engaged, suggesting that CpG DNA is capable of supplying a co-stimulatory signal directly to the T cell (LIPFORD et al. 1997). In contrast, we have been unable to demonstrate a direct co-stimulatory effect of CpG DNA on T-cell activation or proliferation. Further studies are needed to determine whether CpG DNA does indeed have a co-stimulatory effect on T-cell activation and whether such an effect can be used to enhance T-cell-mediated lysis of tumors.

While it is unclear whether CpG DNA can impact directly on T-cell activation, there is no doubt that CpG DNA can enhance T-cell activity indirectly by improving antigen presentation. This effect fits in well with the recent focus of adoptive-transfer studies on the use of APCs to induce an active immune response in the host. Among the most promising recent cancer-immunotherapy studies are those involving dendritic cells (DCs), which are extremely potent APCs (GILBOA et al. 1998; LOTZE et al. 1997; MORSE and LYERLY 1998). Of particular interest is the ability of DCs to induce a cellular immune response (HAMBLIN 1996; STEINMAN 1996; GIROLOMONI and RICCIARDICASTAGNOLI 1997; MCCANN 1997). Levy and colleagues have demonstrated induction of an antigen-specific response following treatment with antigen-pulsed DCs in both an animal model (TIMMERMAN and LEVY 1997) and humans (HSU et al. 1996). In a small clinical trial, four of four treated patients demonstrated a cellular immune response to Id following immunization with Id-pulsed DCs, and clinical regression of tumor was seen, to some degree, in all patients. Interestingly, no humoral response was seen. This was in stark contrast to the studies, reported by the same group, which demonstrated that immunization of patients with Id-keyhole limpet hemocyanin leads to an intense humoral response (HSU et al. 1993; NELSON et al. 1996). Of particular interest is the suggestion of prolonged survival in the small number of patients treated with Id-pulsed DCs (HSU et al. 1997).

DCs can be generated using a variety of cytokines and from a variety of organs, including the bone marrow, lymph nodes and skin. The resulting DCs have varied phenotypes and functions depending upon their site of origin and the cytokines that are used to expand and activate them in vitro. The ideal source of DCs or approach to in vitro expansion, activation, exposure to antigen and reinfusion has yet to be determined.

CpG DNA has effects on DCs; these effects enhance the ability of the DCs to present antigen and induce an antigen-specific cellular response. Sparwasser et al.

have shown that CpG DNA induces maturation of immature DC obtained from murine bone marrow and activates mature DC to produce cytokines, including IL-12, IL-6 and TNF-$\alpha$ (Sparwasser et al. 1998). Jakob et al. found that treatment of DCs derived from murine fetal skin decreased E-cadherin-mediated adhesion, upregulated MHC class-II and co-stimulatory molecules and enhanced accessory cell activity. Injection of CpG DNA into murine dermis led to enhanced expression of MHC class II and CD86 by the Langerhans' cells (Jakob et al. 1998).

We also found that CpG DNA markedly enhances the production of cytokines, including IL-12, from bone-marrow-derived DCs (Liu et al. 1998). In addition, the effect of CpG DNA on DCs was synergistic with that induced by GM-CSF. Recent studies in our laboratory suggest that the effect of CpG DNA on cytokine production by DCs is largely responsible for their enhanced antigen-presenting capabilities. In particular, the effect of CpG DNA is due largely, but not fully, to enhanced production of IL-12 by the DCs. We have also found that CpG DNA can enhance the survival, maturation and differentiation of human DCs in vitro (Harmann et al. 1999).

Given the increasing interest in DC-based cancer immunotherapy and the potent effect of CpG DNA on DCs, it is clear that the combination of CpG DNA and DCs requires further evaluation. However, before a rational clinical trial can begin, we need to have a better understanding of the cellular response of DCs to CpG DNA so that the optimal sequence of DC generation, antigen pulsing and CpG-DNA administration can be used.

## 6 CpG DNA for the Enhancement of the Anti-Tumor Response Following DNA Immunization

Another area of intense interest in the field of cancer immunotherapy is the use of DNA immunization using naked DNA constructs containing sequences that encode for the antigen of interest (Spooner et al. 1995; Ulmer et al. 1996; Cohen et al. 1998). The intent of such therapy is to have host cells take up the DNA, produce protein based on the infused DNA, and express peptides derived from that protein in host class I, thereby inducing a cellular immune response directed towards that antigen. It is unclear what cell type is responsible for uptake of the DNA, production of the protein and presentation of peptide in class I following DNA immunization. Although initial thoughts were that any cells (such as myocytes) could perform these functions, there is now evidence that professional APCs are involved (Doe et al. 1996; Fu et al. 1997). Irrespective of which cells are responsible, this approach has been evaluated with some success in tumor models (Syrengelas et al. 1996).

The importance of professional APCs in DNA immunization is consistent with recent studies demonstrating that the presence of CpG motifs in DNA-vaccine constructs impacts on their effectiveness as vaccines. Sato et al. found that human monocytes transfected with plasmid DNA or double-stranded oligonucleotides containing CpG sequences transcribed larger amounts of IFN-α, IFN-β, and IL-12 when compared with cells transfected with DNA that did not contain such sequences (Sato et al. 1996). Krieg and colleagues have also found that modifying the CpG content of vectors intended for DNA immunization can have a significant impact on their ability to induce development of a cellular response (Brazolot Millan et al. 1998). The ability to construct vectors that encode for a specific protein and enhance a Th1 response to peptides derived from that protein would have clear implications in the area of tumor vaccination and surely will be evaluated extensively in the years ahead.

## 7 Conclusion

Advances in our understanding of the relationship between the immune system and malignancy have resulted in a reawakening of interest in the field of cancer immunotherapy (Table 1). Recent recognition of the potent immunostimulatory effects of CpG DNA suggest that such agents may well be important if cancer immunotherapy is to play a major role in our treatment and prevention of malignancy. Preliminary studies suggest that CpG DNA can be effective, alone or in combination with mAbs, at inducing tumor regression. Perhaps even more prom-

**Table 1.** Summary of potential uses of CpG DNA in cancer immunotherapy

| | CpG as a single agent | CpG to enhance effect of antibody therapy | CpG as an adjuvant with protein immunization | CpG to enhance efficacy of adoptive immunotherapy with dendritic cells | CpG as an important component of DNA vaccines |
|---|---|---|---|---|---|
| Most likely mechanism | Enhanced NK activation | Enhanced ADCC by monocytes and NK cells | Enhanced production of antibody and improved activation of cellular immunity | Increased production of Th1 cytokines, including IL-12 | Enhanced presentation of antigen by antigen-presenting cells |
| Evidence | Enhanced in vitro killing and in vivo efficacy in a mouse melanoma model | Enhanced ADCC in vitro and improved therapeutic effects in a mouse lymphoma model | Effective at inducing a cellular responses and improving efficacy in mouse models | Enhancement of dendritic cell function | CpG content of vectors impacts on efficacy of DNA immunization |

ADCC, antibody-dependent cellular cytotoxicity; IL, interleukin.

ising is the potential use of CpG DNA to enhance development of an active cellular anti-tumor response in the host through its use as a vaccine adjuvant, as an agent capable of stimulating APCs or as part of DNA immunization strategy. It remains to be determined whether CpG DNA has a future as a single agent in the treatment of NK-sensitive malignancies, such as melanoma or renal cell cancer, as an agent in combination with antibodies currently being used to treat lymphoma and breast cancer, such as Rituximab and Trastuzumab, or as a part of vaccine therapy for a variety of malignancies. Further work in both the laboratory and the clinic is needed before we can know the true promise of this exciting new class of agents.

## References


Baldridge JR, Ward JR (1997) Effective adjuvants for the induction of antigen-specific delayed type hypersensitivity. Vaccine 15:395–401

Ballas ZK, Rasmussen WL, Krieg AM (1996) Induction of NK activity in murine and human cells by CpG motifs in oligodeoxynucleotides and bacterial DNA. J Immunol 157:1840–5

Basse PH (1995) Tissue distribution and tumor localization of effector cells in adoptive immunotherapy of cancer. APMIS Suppl 55:1–28

Berinstein N, Levy R (1987) Treatment of a murine B-cell lymphoma with monoclonal antibodies and IL-2. J Immunol 139:971–976

Brazolot Millan CL, Weeratna R, Krieg AM, Siegrist CA, Davis HL (1998) CpG DNA can induce strong Th1 humoral and cell-mediated immune responses against hepatitis B surface antigen in young mice. Proc Natl Acad Sci USA, 15553–8

Bukowski RM (1997) Natural history and therapy of metastatic renal cell carcinoma: the role of interleukin-2. Cancer 80:1198–220

Chace JH, Hooker NA, Mildenstein KL, Krieg AM, Cowdery JS (1997) Bacterial DNA-induced NK cell IFN-gamma production is dependent on macrophage secretion of IL-12. Clin Immunol Immunopathology 84:185–93

Chen HY, Wu SL, Yeh MY, Chen CF, Mikami Y, Wu JS (1993) Antimetastatic activity induced by *Clostridium butyricum* and characterization of effector cells. Anticancer Research 13:107–11

Chu RS, Targoni OS, Krieg AM, Lehmann PV, Harding CV (1997) CpG oligodeoxynucleotides act as adjuvants that switch on T helper 1 (Th1) immunity. Journal of Experimental Medicine 186: 1623–1631

Cohen AD, Boyer JD, Weiner DB (1998) Modulating the immune response to genetic immunization. FASEB J 12:1611–26

Coley WB (1893) The treatment of malignant tumors by repeated inoculations of *Erysipelas* with a report of ten original cases. Am J Med Sci 105:487–511

Coley WB (1894) Treatment of inoperable malignant tumors with the toxins of *Erysipelas* and the bacillus *Prodigiosus*. Am J Med Sci 108:183–212

Correale P, Walmsley K, Nieroda C, Zaremba S, Zhu M, Schlom J, Tsang KY (1997) In vitro generation of human cytotoxic T lymphocytes specific for peptides derived from prostate-specific antigen. J Natl Cancer Inst 89:293–300

Cowdery JS, Chace JH, Yi AK, Krieg AM (1996) Bacterial DNA induces NK cells to produce IFN-$\gamma$ in vivo and increases the toxicity of lipopolysaccharides. J Immunol 156:4570–5

Davis HL, Weeratna R, Waldschmidt TJ, Tygrett L, Schorr J, Krieg AM (1998) CpG DNA is a potent adjuvant in mice immunized with recombinant hepatitis B surface antigen. J Immunol 160:870–876

Doe B, Selby M, Barnett S, Baenziger J, Walker CM (1996) Induction of cytotoxic T lymphocytes by intramuscular immunization with plasmid DNA is facilitated by bone marrow-derived cells. Proc Natl Acad Sci USA 93:8578–83

Finke JH, Zea AH, Stanley J, Longo DL, Mizoguchi H, Tubbs RR, Wiltrout RH, Oshea JJ, Kudoh S, Klein E, Bukowski RM, Ochoa AC (1993) Loss of T-cell receptor $\zeta$-chain and p56(lck) in T-cells infiltrating human renal cell carcinoma. Cancer Res 53:5613–5616

Fu TM, Ulmer JB, Caulfield MJ, Deck RR, Friedman A, Wang S, Liu X, Donnelly JJ, Liu MA (1997) Priming of cytotoxic T lymphocytes by DNA vaccines: requirement for professional antigen presenting cells and evidence for antigen transfer from myocytes. Mol Med 3:362–71

Gilboa E (1996) Immunotherapy of cancer with genetically modified tumor vaccines. Semin Oncol 23:101–107

Gilboa E, Nair SK, Lyerly HK (1998) Immunotherapy of cancer with dendritic-cell-based vaccines. Cancer Immunol Immunother 46:82–7

Girolomoni G, Ricciardicastagnoli P (1997) Dendritic cells hold promise for immunotherapy. Immunology Today 18:102–104

Hamblin TJ (1996) From dendritic cells to tumour vaccines. Lancet 34:7705–706

Hartmann G, Weiner GJ, Krieg AM (1999) CpG DNA: A potent signal for growth, activation and maturation of human dendritic cells. Proc Natl Acad Sci 96:9305–9310

Heaton KM, Grimm EA (1993) Cytokine combinations in immunotherapy for solid tumors – a review. Cancer Immunol Immunother 37:213–219

Hellstrom KE, Gladstone P, Hellstrom I (1997) Cancer vaccines – challenges and potential solutions. Molecular Medicine Today 3:286–290

Hodge JW, Schlom J, Donohue SJ, Tomaszewski JE, Wheeler CW, Levine BS, Gritz L, Panicali D, Kantor JA (1995) A recombinant vaccinia virus expressing human prostate-specific antigen (PSA): safety and immunogenicity in a non-human primate. Int J Cancer 63:231–7

Hsu FJ, Benike C, Fagnoni F, Liles TM, Czerwinski D, Taidi B, Engleman EG, Levy R (1996) Vaccination of patients with B-cell lymphoma using autologous antigen-pulsed dendritic cells. Nature Med 2:52–58

Hsu FJ, Caspar CB, Czerwinski D, Kwak LW, Liles T, Syrengelas A, Taidi-Laskowski A, Levy R (1997) Tumor-specific idiotype vaccines in the treatment of patients with B-cell lymphoma – long term results of a clinical trial. Blood 89:3129–3135

Hsu FJ, Kwak L, Campbell M, Liles T, Czerwinski D, Hart S, Syrengelas A, Miller R, Levy R (1993) Clinical trials of idiotype-specific vaccine in B-cell lymphomas. Annals of the New York Academy of Sciences 690:385–7

Jakob T, Walker PS, Krieg AM, Udey MC, Vogel JC (1998) Activation of cutaneous dendritic cells by CpG-containing oligodeoxynucleotides: a role for dendritic cells in the augmentation of Th1 responses by immunostimulatory DNA. J Immunol 161:3042–9

Johnston D, Bystryn JC (1991) Effect of cell wall skeleton and monophosphoryl lipid A adjuvant on the immunogenicity of a murine B16 melanoma vaccine. J Natl Cancer Inst 83:1240–5

Kantarjian HM, Giles FJ, O'Brien SM, Talpaz M (1998) Clinical course and therapy of chronic myelogenous leukemia with interferon-alpha and chemotherapy. Hematol Oncol Clin North Am 12: 31–80

Kohler G, Milstein C (1975) Continuous cultures of fused cells secreting antibody of predefined specificity. Nature 256:495

Krieg AM, Matson S, Fisher E (1996) Oligodeoxynucleotide modifications determine the magnitude of B cell stimulation by CpG motifs. Antisense and Nucletic Acid Drug Development 6:133–9

Krieg AM, Yi AK, Matson S, Waldschmidt TJ, Bishop GA, Teasdale R, Koretzky GA, Klinman DM (1995) CpG motifs in bacterial DNA trigger direct B-cell activation. Nature 374:546–9

Link BK, Weiner GJ (1998) Monoclonal antibodies in the treatment of human B-cell malignancies. Leuk Lymphoma 31:237–49

Lipford GB, Bauer M, Blank C, Reiter R, Wagner H, Heeg K (1997) CpG-containing synthetic oligonucleotides promote B and cytotoxic T cell responses to protein antigen: a new class of vaccine adjuvants. Eur J Immunol 27:2340–4

Lipford GB, Sparwasser T, Bauer M, Zimmermann S, Koch ES, Heeg K, Wagner H (1997) Immunostimulatory DNA: sequence-dependent production of potentially harmful or useful cytokines. Eur J Immunol 27:3420–6

Liu HM, Newbrough SE, Bhatia SK, Dahle CE, Krieg AM, Weiner GJ (1998) Immunostimulatory CpG oligodeoxynucleotides enhance the immune response to vaccine strategies involving granulocyte-macrophage colony- stimulating factor. Blood 92:3730–6

Livingston P (1998) Ganglioside vaccines with emphasis on GM2. Semin Oncol 25:636–45

Livingston PO, Adluri S, Helling F, Yao TJ, Kensil CR, Newman MJ, Marciani D (1994) Phase 1 trial of immunological adjuvant QS-21 with a GM2 ganglioside-keyhole limpet haemocyanin conjugate vaccine in patients with malignant melanoma. Vaccine 12:1275–80

Lotze MT, Rosenberg SA (1986) Results of clinical trials with the administration of interleukin 2 and adoptive immunotherapy with activated cells in patients with cancer. Immunobiology 172:420–37

Lotze MT, Shurin M, Davis I, Amoscato A, Storkus WJ (1997) Dendritic cell based therapy of cancer. Adv Exp Med Biol 417:551–69

Maeurer MJ, Storkus WJ, Kirkwood JM, Lotze MT (1996) New treatment options for patients with melanoma: review of melanoma-derived T-cell epitope-based peptide vaccines. Melanoma Research 6:11–24

Maloney DG, Grillolopez AJ, White CA, Bodkin D, Schilder RJ, Neidhart JA, Janakiraman N, Foon KA, Liles TM, Dallaire BK, Wey K, Royston I, Davis T, Levy R (1997) Idec-C2b8 (Rituximab) anti-Cd20 monoclonal antibody therapy patients with relapsed low-grade non-Hodgkin's lymphoma. Blood 90:2188–2195

Mastrangelo MJ, Maguire HC Jr, Sato T, Nathan FE, Berd D (1996) Active specific immunization in the treatment of patients with melanoma. Semin Oncol 23:773–81

McCann J (1997) Immunotherapy using dendritic cells picks up steam. Journal of the National Cancer Institute 89:541–542

Mori Y, Tsuchiya H, Tsuchida T, Asada N, Nojima T, Tomita K (1997) Disappearance of Ewings sarcoma following bacterial infection – a case report. Anticancer Research 17:1391–1397

Morse MA, Lyerly HK (1998) Immunotherapy of cancer using dendritic cells. Cytokines Cell Mol Ther 4:35–44

Nelson EL, Li XB, Hsu FJ, Kwak LW, Levy R, Clayberger C, Krensky AM (1996) Tumor-specific, cytotoxic T-lymphocyte response after idiotype vaccination for B-cell, non-Hodgkin's lymphoma. Blood 88:580–589

Pegram MD, Lipton A, Hayes DF, Weber BL, Baselga JM, Tripathy D, Baly D, Baughman SA, Twaddell T, Glaspy JA, Slamon DJ (1998) Phase II study of receptor-enhanced chemosensitivity using recombinant humanized anti-p185HER2/neu monoclonal antibody plus cisplatin in patients with HER2/neu-overexpressing metastatic breast cancer refractory to chemotherapy treatment. J Clin Oncol 16:2659–71

Pervin S, Chakraborty M, Bhattacharya-Chatterjee M, Zeytin H, Foon KA, Chatterjee SK (1997) Induction of antitumor immunity by an anti-idiotype antibody mimicking carcinoembryonic antigen. Cancer Research 57:728–34

Rock KL (1996) A new foreign policy: MHC class I molecules monitor the outside world. Immunology Today 17:131–7

Rosenberg SA, Mule JJ, Spiess PJ, Reichert CM, Schwarz SL (1985) Regression of established pulmonary metastases and subcutaneous tumor mediated by the systemic administration of high-dose recombinant interleukin 2. J Exp Med 161:1169–88

Sato Y, Roman M, Tighe H, Lee D, Corr M, Nguyen MD, Silverman GJ, Lotz M, Carson DA, Raz E (1996) Immunostimulatory DNA sequences necessary for effective intradermal gene immunization. Science 273:352–4

Schirmbeck R, Melber K, Kuhrober A, Janowicz ZA, Reimann J (1994) Immunization with soluble hepatitis B virus surface protein elicits murine H-2 class I-restricted CD8+ cytotoxic T-lymphocyte responses in vivo. J Immunol 152:1110–9

Schwartzentruber DJ, Hom SS, Dadmarz R, White DE, Yannelli JR, Steinberg SM, Rosenberg SA, Topalian SL (1994) In vitro predictors of therapeutic response in melanoma patients receiving tumor-infiltrating lymphocytes and interleukin-2. J Clin Oncol 12:1475–83

Smith JB, Wickstrom E (1998) Antisense c-myc and immunostimulatory oligonucleotide inhibition of tumorigenesis in a murine B-cell lymphoma transplant model. J Natl Cancer Inst 90:1146–54

Sparwasser T, Koch ES, Vabulas RM, Heeg K, Lipford GB, Ellwart JW, Wagner H (1998) Bacterial DNA and immunostimulatory CpG oligonucleotides trigger maturation and activation of murine dendritic cells. Eur J Immunol 28:2045–54

Sparwasser T, Miethke T, Lipford G, Erdmann A, Hacker H, Heeg K, Wagner H (1997) Macrophages sense pathogens via DNA motifs: induction of tumor necrosis factor-alpha-mediated shock. Eur J Immunol 27:1671–9

Spooner RA, Deonarain MP, Epenetos AA (1995) DNA vaccination for cancer treatment. Gene Therapy 2:173–80

Steinman RM (1996) Dendritic cells and immune-based therapies. Exp Hematol 24:859–862

Syrengelas AD, Chen TT, Levy R (1996) DNA immunization induces protective immunity against B-cell lymphoma. Nature Med 2:1038–1041

Timmerman J, Levy R (1997) Enhanced immunogenicity of tumor-specific immunoglobulin-pulsed dendritic cells using a chimeric idiotype-GM-CSF fusion protein. Proc Am Assoc Cancer Res 38:616

Ulmer JB, Donnelly JJ, Liu MA (1996) Toward the development of DNA vaccines. Current Opinion in Biotechnology 7:653–8

Weiner GJ, Liu HM, Wooldridge JE, Dahle CE, Krieg AM (1997) Immunostimulatory oligodeoxynucleotides containing the CpG motif are effective as immune adjuvants in tumor antigen immunization. Proc Natl Acad Sci USA 94:10833–10837

Wiemann B, Starnes CO (1994) Coley's toxins, tumor necrosis factor and cancer research: a historical perspective. Pharm Ther 64:529–564

Wooldridge JE, Ballas Z, Krieg AM, Weiner GJ (1997) Immunostimulatory oligodeoxynucleotides containing CpG motifs enhance the efficacy of monoclonal antibody therapy of lymphoma. Blood 89:2994–2998

# Use of CpG DNA for Enhancing Specific Immune Responses

H.L. Davis



## 1 Introduction

### 1.1 CpG DNA and the Vertebrate Immune System

Bacterial DNA, in contrast to vertebrate DNA, has the capacity to act as a potent immune stimulator (Tokunaga et al. 1984; Messina et al. 1991; Yamamoto et al. 1992b). This is now known to be due to unmethylated CpG dinucleotides that are immunostimulatory when present within the context of certain flanking bases (CpG-S motifs) (Krieg et al. 1995). CpG dinucleotides are present at the expected frequency of 1/16 bases in bacterial DNA but are under-represented (1/50–1/60 bases) in vertebrate DNA, where they are almost always methylated (Bird et al. 1987). CpG-S motifs are virtually absent from vertebrate DNA, and the rapid immune activation in response to CpG DNA appears to be an evolutionary adaptation whereby CpG-S motifs act as a "danger" signal that the vertebrate innate immune defense mechanisms can recognize and respond to (Krieg et al. 1996).

Loeb Health Research Institute, 725 Parkdale Avenue, Ottawa, ON, K1Y 4E9, Canada
E-mail: hdavis@LRI.ca

CpG DNA can induce proliferation of almost all (>95%) B cells and further protects B cells from apoptosis. It also triggers polyclonal immunoglobulin (Ig), interleukin 6 (IL-6) and IL-12 secretion from B cells and can drive isotype shifting in vitro (Krieg et al. 1995; Klinman et al. 1996; Yi et al. 1996a,b). This B-cell activation by CpG DNA is T-cell independent and antigen non-specific although, as discussed below, these effects have strong synergy with antigen-specific responses. In addition to its direct effects on B cells, CpG DNA also directly activates monocytes, macrophages and dendritic cells to secrete interferon α/β (IFN-α/β), IL-6, IL-12, granulocyte–monocyte colony-stimulating factor (GM-CSF), chemokines and tumor necrosis factor α (TNF-α) (Cowdery et al. 1996; Halpern et al. 1996; Klinman et al. 1996; Schwartz et al. 1997). These cytokines stimulate natural killer (NK) cells to secrete IFN-γ and have increased lytic activity (Yamamoto et al. 1992a; Ballas et al. 1996; Cowdery et al. 1996; Klinman et al. 1996; Yi et al. 1996; Chace et al. 1997). Overall, CpG induces a T-helper type 1 (Th1)-like pattern of cytokine production dominated by IL-12 and IFN-γ with little secretion of Th2 cytokines (Klinman et al. 1996). CpG ODN also induces co-stimulatory molecule expression in vitro and in vivo (Davis et al. 1998) and upregulates expression of major histocompatibility complex (MHC) molecules. Similar effects to those induced by bacterial DNA may be obtained with synthetic oligodeoxynucleotides (ODN) containing CpG-S motifs (CpG ODN).

Not all DNA sequences containing unmethylated CpG dinucleotides are stimulatory, and there is some species specificity to such effects (see below). CpG-S motifs are typically preceded on the 5′ side by an ApA, GpA, or GpT dinucleotide and followed on the 3′ side by two pyrimidines, especially TpT. Many sequences containing CpG dinucleotides that are non-stimulatory for a given species appear to have no effect on the immune system; however, some DNA sequences containing CpG dinucleotides may actually counteract the stimulatory effects of CpG-S motifs. Such "neutralizing" CpG motifs (CpG-N), which are over-represented in certain adenoviral genomes (types 2 and 5), typically include CCG, CGG, or CG-repeats in their sequences (Krieg et al. 1998). Such sequences are typical of the so-called "CpG islands" in vertebrate DNA, described by Bird et al. (1987), which are usually unmethylated but nevertheless non-stimulatory.

## 1.2 Adjuvant Properties of CpG DNA

In addition to the activation of innate immune responses, CpG DNA can serve as a potent adjuvant for the enhancement of antigen-specific responses. Enhancement of humoral responses appears to be due to the strong synergy between the direct activation of B cells by CpG DNA and the signals, delivered through the B-cell antigen receptor, for both B-cell proliferation and Ig secretion (Krieg et al. 1995). In addition, antigen-specific humoral responses are likely enhanced by the induction of cytokines that could have indirect effects on B cells via T-helper pathways. Overall, CpG DNA induces a Th1-like pattern of cytokine production dominated by IL-12 and IFN-γ, with little secretion of Th2 cytokines (Klinman et al. 1996).

Th1-type immune responses are desirable because (i) they are associated with the production of predominantly IgG2a antibodies, which are thought to have better neutralizing capabilities than the Th2-associated IgG1 antibodies, (ii) antigen-specific cell-mediated immunity (CMI) is enhanced by better antigen presentation with the enhanced expression of MHC and co-stimulatory molecules (DAVIS et al. 1998), and (iii) the Th1-type cytokines can aid the maturation of cytotoxic T lymphocytes (CTL).

CpG DNA appears to enhance immune responses against any type of antigen and, as such, it could be used to improve the efficacy of any prophylactic vaccines against infectious disease. The strong Th1 bias of CpG DNA and the ability to induce strong CMI also opens the possibility for use of CpG DNA in therapeutic vaccines against, for example, chronic infections or cancer.

For the purpose of in vivo enhancement of antigen-specific responses in humans, it is preferable to use synthetic short ODN, which can be well characterized, rather than bacterial DNA. As well, for in vivo use, the ODN should be made with a synthetic phosphorothioate backbone to render them more nuclease-resistant without losing the immune-stimulatory effects (KRIEG et al. 1996; ZHAO et al. 1993, 1996; BOGGS et al. 1997). In vivo adjuvant effects of CpG ODN with a native phosphodiester backbone (O-ODN) or a chimeric backbone composed of a phosphodiester center and phosphorothioate ends (SOS-ODN) are diminished or lost (Davis, unpublished observation), presumably due to more rapid destruction by lymphocyte nucleases (ZHAO et al. 1993). In practice, we have found 18–24-mer ODN to be optimally effective but of a manageable size for easy and inexpensive manufacturing.

CpG-S motifs may also be used to enhance immune responses with DNA vaccines. In this case, for reasons discussed below, it is best to clone the sequences into the backbones of the plasmid vectors.

## 1.3 Species-Specificity of CpG DNA

CpG-S motifs are species-specific, to a certain extent. For example, our optimal mouse motif with a single CpG dinucleotide in a phosphorothioate ODN (TGACGTT) does not activate human cells. It is possible to induce more potent immune activation by having more than one CpG motif in the ODN, in which case the spacing between motifs is critical (Davis and Krieg, unpublished observations). Furthermore, the same CpG sequence in phosphodiester or phosphorothioate backbones may not show the same stimulatory effects or work at the same optimal concentration. Thus, factors that seem to influence the stimulatory effect of a particular CpG ODN in a given species include the backbone of the DNA and the sequence, number, and spacing of the CpG motifs.

Mouse cells will respond to a wide variety of CpG motifs, including human motifs, whereas human cells are stimulated by a much more restricted subset of CpG-S motifs. Therefore, it is not possible to predict the in vivo effects of CpG-S motifs in humans from in vivo results in mice. Furthermore, it may not even be possible to make such predictions from primate studies. For example, we have found that immune cells from such closely related species as the human and the chimpanzee don't necessarily respond to the same CpG-S ODN, even when the sequence is very similar to that of another ODN that does activate cells from both species (Krieg and Davis, unpublished results).

While the mechanism of the surprisingly complex role of the bases flanking CpG dinucleotides has yet to be elucidated, it is important to take them into consideration when designing in vivo studies to evaluate CpG ODN as a vaccine adjuvant. In particular, it is absolutely necessary to use immune cells from the desired target species and test them with a panel of in vitro tests (i.e. B-cell proliferation, cytokine secretion, NK activity) to determine which are immunostimulatory for that species before attempting to use them as vaccine adjuvants in vivo.

## 2 Experience with CpG DNA and Antigen-Based Immunization

CpG ODN have been shown to be potent adjuvants for augmenting both humoral and cell-mediated immune responses against a wide variety of antigens. These include protein antigens of a diverse nature and polysaccharide (PS) antigens. Furthermore, CpG ODN has been delivered with antigens both parenterally as well as mucosally.

The CpG ODN is best given mixed together with the antigen. While this may not be absolutely essential, we have found that enhancement of the humoral response is greatly abrogated if the CpG ODN is given at a different time (i.e. one day before or after) or in the contralateral leg (Davis and Krieg, unpublished observations). This effect may be partly antigen dependent, since others have reported adjuvant effects when the CpG ODN was given up to 30 days prior to a model antigen (Lipford and Wagner, personal communication) or when it was given in the contralateral leg to the site of injection of whole killed influenza vaccine (B. Wahren, personal communication).

### 2.1 CpG DNA with Protein Antigens

The greatest amount of experience using CpG ODN to augment antigen-specific immune responses has been with protein antigens. Many of these are infectious-disease-related antigens, including live attenuated measles virus (Kovarik et al. 1999), whole killed influenza virus (Moldoveanu et al. 1997; Roman et al. 1997), hepatitis B surface antigen (HBsAg, a virus-like particle) (Brazolot Millan et al. 1998; Davis et al. 1998), and a peptide derived from tetanus toxoid (Kovarik et al. 1999). Enhanced immune responses have also been demonstrated using CpG ODN with model antigens, such as ovalbumin (Klinman et al. 1997; Lipford et al. 1997a,b), hen-egg lysozyme (Chu et al. 1997), β-galactosidase (Roman et al. 1997;

Horner et al. 1998) and fowl γ-globulin (Sun et al. 1998). Finally, CpG ODN has been shown to be highly effective for both the enhancement of Th1-like immune responses against a tumor antigen, the idiotype of the surface IgM of 38C13 mouse lymphoma cells, and for protection against subsequent tumor challenge (Weiner et al. 1997; Liu et al. 1998). The ability to use CpG ODN with virtually any type of antigen is particularly attractive, because alum, the only adjuvant currently licensed for human use in most countries, cannot be used with live attenuated pathogens and has not been found to be particularly useful with other vaccines, such as those containing whole killed influenza virus.

The findings of the above studies have shown that, in virtually all cases, CpG ODN causes significant enhancement of antigen-specific antibody titers. These can be as much as 2–3 log higher than titers obtained with antigen alone or with a non-CpG control ODN. In several studies, strong Th1 bias of the CpG-induced response is also shown by the induction of potent CTL responses (Lipford et al. 1997a,b; Brazolot Millan et al. 1998; Davis et al. 1998), preferential secretion of Th1-type cytokines, such as IFN-γ or IL-12 (Chu et al. 1997; Horner et al. 1998; Kovarik et al. 1999), and increased relative proportion of IgG2a antibodies (Chu et al. 1997; Roman et al. 1997; Brazolot Millan et al. 1998; Davis et al. 1998; Sun et al. 1998).

CpG ODN, which is water soluble, has a strong synergy with other adjuvants, such as alum (Davis et al. 1998), GM-CSF cholera toxin (CT) (McCluskie and Davis 1998), and incomplete Freund's adjuvant (IFA). Indeed, these combinations give responses equal or superior to those with complete Freund's adjuvant (CFA), with much less toxicity. The synergy may be related to different mechanisms of the immune modulators (GM-CSF, CT) or in the case of alum and IFA, the CpG ODN may benefit from the depot effect. We have not detected any signs of toxicity in mice, even with doses 50 times higher than those required for strong adjuvant effects.

The strong adjuvant effects of CpG ODN may also allow the use of reduced doses of antigen. For example, reducing the dose of HBsAg (alone or with alum) by tenfold likewise reduces the antibody titers in mice by tenfold but, when given with CpG ODN, there is no loss of the high titers of anti-HBs antibodies (Davis, unpublished results).

## 2.2 CpG DNA for Enhancing Immune Responses in Early Life

Vaccination of infants is often delayed, because the neonatal immune system is considered too immature to respond appropriately (Hunt et al. 1994). In particular, B-cell responses are weak and preferentially generate IgM/IgG1 antibody isotypes, and cytotoxic responses are poor (Kovarik et al. 1998). Nevertheless, newborns are at risk for exposure to many infectious diseases, and it would be highly desirable to have vaccines that are effective even with an immature immune system. In particular, the neonatal immune system is strongly Th2 biased, and this results in very poor CMI, which is crucial for the prevention or control of many diseases that neonates are at risk of developing.

Young mice are useful models in which to test immunization strategies for newborn humans, because their responses to protein antigens have similar limitations (Barrios et al. 1996). CpG ODN has been found to be effective in very young mice for inducing or enhancing antibodies and CTLs against HBsAg (Brazolot Millan et al. 1998), live attenuated measles virus, and a tetanus-toxoid-derived peptide (Kovarik et al. 1999). Not only does CpG ODN allow the induction of immune responses at younger ages than does antigen alone or antigen with alum, but these responses are more Th1-like. This is particularly important, because the immature immune system is strongly Th2 biased.

## 2.3 CpG DNA as a Mucosal Adjuvant

Parenteral immunization can induce systemic (i.e. circulating antibodies and CTLs), but only rarely mucosal, immunity (i.e. secretory IgA antibodies). In contrast, antigens delivered at a mucosal surface can trigger both mucosal immunity at local and distant sites and systemic responses (Haneberg et al.1994; Gallichan and Rosenthal, 1995). While live attenuated organisms have proven effective for mucosal immunization, results have been disappointing with subunit vaccines, in large part due to the lack of a safe and effective adjuvant.

CT and the closely related *Escherichia coli* heat-labile enterotoxin (LT) are effective mucosal adjuvants in animal models (Lycke et al. 1992; Spangler et al. 1992); however, they are highly toxic, especially in humans. In efforts to produce effective but safe mucosal adjuvants for use in humans, two strategies have been adopted to render CT and LT less toxic: (i) the use of the non-toxic B subunits that lack enzymatic activity (Holmgren et al. 1993; Verweij et al. 1998) or (ii) genetically detoxified mutants that have little or no enzymatic activity. Such compounds, which have greatly reduced toxicities, retain some adjuvanticity in animal models (Fontana et al. 1995; Rappuoli et al. 1995; Di Tommaso et al. 1996; Yamamoto et al. 1997a,b, 1998; Chong et al. 1998; Douce et al. 1998; Guiliani et al. 1998; Komase et al. 1998).

CpG ODN has been found to be a highly effective mucosal adjuvant when administered intranasally to mice. For example, both systemic (IgG) and mucosal (secretory IgA) humoral responses against whole killed influenza virus were significantly enhanced by the addition of CpG ODN (Moldoveanu et al. 1997). Furthermore, CpG ODN was able to induce antibodies and CTL against HBsAg, which by itself did not induce any detectable responses even with high doses and boosting (McCluskie and Davis 1998). In fact, this study showed CpG ODN to be as effective as CT, and to have a strong synergy with CT when the two adjuvants were co-administered. In addition, CpG ODN gave a more Th1-like response than CT, which gave a Th2-type response.

Effective doses of CpG ODN for intranasal delivery (1–10μg) are lower than those required for parenteral immunization (10–50μg). In addition, CpG ODN appears to be safe and well tolerated with intranasal delivery to mice, even at very high doses (up to 500μg). Mice inhaling CpG showed no short-term signs of distress over those receiving HBsAg alone, and all recovered quickly, with no apparent long-lasting effects. In contrast, mice receiving high doses of CT (>10μg) show signs of toxicity, such as ruffling of fur and diarrhea. CT is even more toxic in humans, where a dose as low as 1–5μg can cause diarrhea (JERTBORN et al. 1992).

Thus, CpG ODN appears to be a promising new mucosal adjuvant that could be administered with a variety of antigens. It might also be used in combination with mucosal toxins to obtain better immune responses with a lower dose and less adjuvant-related toxicity. The ability to induce mucosal immunity that can prevent entry of pathogens into the body is highly desirable, but equally important is the ability to deliver vaccines in an easy and needle-free fashion. This could radically change the prospect of delivering vaccines to less developed areas of the world and, even in more developed nations, avoidance of the risk of needle-stick injury is highly desirable.

## 2.4 Th1 Influence of CpG DNA

The strong Th1-bias of CpG ODN has been demonstrated with in vitro experiments and in vivo delivery of antigens by parenteral and mucosal routes to adult and young mice. The best marker of this is the type of cytokine induced (high IFN-γ and IL-12 and low IL-4 and IL-5); however, it is also indicated by presence of CTL and a preponderance of IgG2a relative to IgG1 antibodies.

For vaccination against infectious diseases, a Th1-like response is generally desirable. Furthermore, CTLs appear to be essential for protection against certain diseases, especially those where the pathogen resides intracellularly. It is generally accepted that induction of CTL is important for immunotherapy of chronic infections, such as hepatitis B virus (HBV), hepatitis C virus, *Mycobacterium* (tuberculosis), and human immunodeficiency virus. In addition, the Th1 cytokines, such as IFN-γ and TNF-α, have been shown to be important for non-lytic control of chronic HBV infection (CHISARI 1997).

While strong Th1 CMI may not be necessary to protect against all infectious diseases, it is clearly a requirement for cancer vaccines where the intent is to induce CTL that can attack and kill cancer cells, for example those remaining after surgical removal of a solid tumor or after chemotherapy. In this regard, CpG ODN has been found to be a more potent Th1 adjuvant than many other adjuvants, including CFA (Davis, unpublished data).

A Th1-like response may be preferential in the lungs, because allergic asthma is generally associated with Th2-like responses (KAY 1996; HOGG 1997). Asthma has become an increasingly prevalent disease in the developed world, and this has been has been linked to the reduced incidence of bacterial infections owing to better hygiene and widespread use of antibiotics (COOKSON and MOFFATT 1997). It is also possible that the strong Th2 bias of alum, which is currently used in many pediatric vaccines, may contribute to the rising incidence of asthma. The ability to induce

Th1 responses in babies and young children with CpG ODN may help reduce the incidence of asthma. Indeed, CpG ODN has recently been shown, in mice, to prevent allergen-induced asthmatic responses, which include airway eosinophilia, Th2 cytokine induction, IgE production, and bronchial hyperactivity. Furthermore, CpG ODN can even "cure" pre-established asthmatic responses in mice through redirection of the immune response towards a more Th1-like profile (Broide et al. 1998; Kline et al. 1998).

## 2.5 Polysaccharide Antigens

PS antigens are particularly difficult, because immune responses against them are largely T-cell independent. As such, there is no CMI, antibody responses are short lived, and there is no memory response. One published report found that CpG ODN actually reduced the level of antibodies against a high-molecular-weight PS in mice (Threadgill et al. 1997). However, others have found modest to good increases in antibodies, including those recognized as being T dependent (IgG2a), such as CpG ODN and pneumococcal PS (Siegrist, personal communication), meningitis type-C PS (Westerink, personal communication), and trinitrophenyl-Ficoll (Waldschmidt, personal communication), although some of these studies failed to demonstrate induction of memory responses. The poor results in the first study may be related to the dose of 500μg CpG ODN, which is greatly in excess of an optimal dose for mice (about 10μg). Nevertheless, results obtained with CpG ODN and PS antigens have been less dramatic relative to those obtained with protein antigens. This is likely related the T-independent nature of anti-PS responses, since many of the effects of CpG-S motifs are involved with antigen presentation to T-cells or induction of cytokines that serve T-helper functions.

## 2.6 CpG DNA with Antigen–Antibody Complexes

Antigen–antibody complexes have been shown to induce significantly better immune responses than antigen alone in a number of different animal models (Pokric et al. 1993; Hanke et al. 1994; Haddad et al. 1997). It has also recently been found that CpG ODN can significantly enhance these responses in mice when administered parenterally with complexes of hepatitis B surface antigen (HBsAg) mixed with antibodies against HBsAg (anti-HBs; HBsAg/Ab complexes) (Wen, personal communication).

We have also shown that HBsAg/Ab complexes can induce strong immune responses when given by a mucosal route (McCluskie et al. 1999). Although CpG ODN is a highly effective mucosal adjuvant with some antigens (discussed below), in the case of antigen–antibody complexes, CpG ODN does not enhance antibody titers, although it can shift the humoral response to be more Th1-like.

## 3 Role of CpG Motifs in DNA Vaccines

DNA vaccination has emerged as an exciting new approach to vaccination that promises significant advantages over current vaccination strategies. Numerous groups have demonstrated the induction of potent and long-lasting immune responses, including induction of CTL, against viral, bacterial, or parasitic antigens encoded in DNA plasmids (DAVIS 1998; DONNELLY et al. 1998). The strength of the immune response with DNA vaccines is impressive considering the very small amount of antigen produced. This appears to be due to the in vivo synthesis of antigens (and, thus, efficient MHC presentation of antigens) over an extended period of time, as well as to the CpG-S motifs inherent in the backbone of the plasmid DNA vectors, such as those commonly used for DNA-based immunization. For example, removal or methylation of potent CpG-S motifs has been found to reduce or abolish immune responses against encoded antigens (SATO et al. 1996; KLINMAN et al. 1997)

Other investigators working in the field of DNA vaccines have also observed that the DNA backbone itself provides a Th1-type adjuvant effect when administered with a protein antigen (LECLERC et al. 1997). This adjuvant effect of the DNA backbone appears to require unmethylated CpG motifs, because it is lost if the DNA vector is treated with CpG methylase (KLINMAN et al. 1997). Moreover, a non-expressing plasmid enriched in CpG motifs is a more effective adjuvant for protein vaccines than a non-optimized plasmid (SATO et al. 1996; ROMAN et al. 1997).

In order to use CpG-S motifs to enhance immune responses with DNA vaccines, it is really essential to clone them into the vector backbone. We have shown that it is not possible to directly mix S-ODN with plasmid DNA, because this will result in an ODN-dose-dependent reduction in gene expression from the plasmid (WEERATNA et al. 1998), possibly due to competitive interference at binding sites on the surfaces of target cells (ZHAO et al. 1993). While ODN with a phosphorothioate-phosphodiester chimeric backbone (SOS-ODN) do not adversely affect the level of gene expression (except those with poly-G sequences, which have greatly increased binding to cell membranes) (HUGHES et al. 1994), this is not useful, because SOS-ODN are apparently not sufficiently nuclease resistant to exert a strong CpG-adjuvant effect. It is also not possible to augment responses to DNA vaccines by administering the CpG S-ODN at a different time or site than the plasmid DNA (Weeratna and Davis, unpublished results). Thus, at least for the present, it appears necessary to clone CpG motifs into DNA vaccine vectors in order to take advantage of their adjuvant effects.

Optimization of the CpG content of DNA vaccines is further complicated by the presence of numerous CpG-N motifs, which can neutralize the effects of the CpG-S motifs. We have shown that it is possible to improve the efficacy of a DNA vaccine in mice by using site-directed mutagenesis to remove many of the CpG-N motifs. Further improvement was then obtained by cloning additional CpG-S motifs into the non-coding backbone (KRIEG et al. 1998). Such CpG-optimized

DNA-vaccine vectors may be particularly advantageous for use in humans, where the more restricted subset of effective CpG-S motifs will negatively skew the balance of neutralizing and stimulatory motifs.

## 4 Summary

CpG ODN, owing to its wide range of immunostimulatory effects has been found to be a potent Th1-type adjuvant that is effective with virtually any type of antigen, although responses are less impressive with PS than protein antigens. The use of CpG ODN as an adjuvant may allow the development of vaccines against a wider range of diseases, which could include therapeutic vaccines for chronic infections or cancer, effective pediatric vaccines for newborns, and easily delivered mucosal vaccines.

## References

Ballas ZK, Rasmussen WL, Krieg AM (1996) Induction of natural killer activity in murine and human cells by CpG motifs in oligodeoxynucleotides and bacterial DNA. J Immunol 157:1840–1845

Barrios C, Brawand P, Berney M, Brandt C, Lambert PH, Siegrist CA (1996) Neonatal and early life immune responses to various forms of vaccine antigens qualitatively differ from adult responses: predominance of a Th2-biased pattern which persists after adult boosting. Eur J Immunol 26: 1489–1496

Bird AP (1987) CpG islands as gene markers in the vertebrate nucleus. Trends in Genetics 3:342

Boggs RT, McGraw K, Condon T, Flournoy S, Villiet P, Bennett CF, Monia BP (1997) Characterization and modulation of immune stimulation by modified oligonucleotides. Antisense Nucl Acid Drug Develop 7:461–471

Brazolot Millan CL, Weeratna R, Krieg AM, Siegrist CA, Davis HL (1998) CpG DNA can induce strong Th1 humoral and cell-mediated immune responses against hepatitis B surface antigen in young mice. Proc Natl Acad Sci USA (In press)

Broide D, Schwarze J, Tighe H, Gifford T, Nguyen MD, Malek S, Van Uden J, Martin-Orozco E, Gelfand EW, Raz E (1998) Immunostimulatory DNA sequences inhibit IL-5, eosinophilic inflammation, and airway hyper-responsiveness in mice. J Immunol 161:7054–7062

Chace JH, Hooker NA, Mildenstein KL, Krieg AM, Cowdery JS (1997) Bacterial DNA-induced NK cell IFN-γ production is dependent on macrophage secretion of IL-12. Clin Immunol Immunopath 84:185–193

Chisari FV (1997) Cytotoxic T cells and viral hepatitis. J Clin Invest 99:1472–1477

Chong C, Friberg M, Clements JD (1998) LT(R192G), a non-toxic mutant of the heat-labile enterotoxin of *Escherichia coli*, elicits enhanced humoral and cellular immune responses associated with protection against lethal oral challenge with *Salmonella* spp. Vaccine 16:732–740

Chu RS, Targoni OS, Krieg AM, Lehmann PV, Harding CV (1997) CpG oligodeoxynucleotides act as adjuvants that switch on T helper 1 (Th1) immunity. J Exp Med 186:1623–1631

Cookson OCM, Moffatt MF (1997) Asthma: An epidemic in the absence of infection? Science 275:41–42

Cowdery JS, Chace JH, Yi A-K, Krieg AM (1996) Bacterial DNA induces NK cells to produce interferon-γ in vivo and increases the toxicity of lipopolysaccharide. J Immunol 156:4570–4575

Davis HL, Weeratna R, Waldschmidt TJ, Schorr J, Krieg AM (1998) CpG DNA is a potent enhancer of specific Immunity in mice immunized with recombinant hepatitis B surface antigen. J Immunol 160:870–876

Di Tommaso A, Saletti G, Pizza M, Rappuoli R, Dougan G, Abrignani S, Douce G, De Magistris MT (1996) Induction of antigen-specific antibodies in vaginal secretions by using a nontoxic mutant of heat-labile enterotoxin as a mucosal adjuvant. Infect Immun 64:974–979

Donnelly JJ, Ulmer JB, Liu MA (1997) DNA vaccines. Life Sciences 60:163–172

Douce G, Giuliani MM, Giannelli V, Pizza MG, Rappuoli R, Dougan G (1998) Mucosal immunogenicity of genetically detoxified derivatives of heat labile toxin from *Escherichia coli*. Vaccine 16: 1065–1073

Fontana MR, Manetti R, Giannelli V, Magagnoli C, Marchini A, Olivieri R, Domenighini M, Rappuoli R, Pizza M (1995) Construction of nontoxic derivatives of cholera toxin and characterization of the immunological response against the A subunit. Infect Immun 63:2356–2360

Gallichan WS, Rosenthal KL (1995) Specific secretory immune responses in the female genital tract following intranasal immunization with a recombinant adenovirus expressing glycoprotein B of herpes simplex virus. Vaccine 13:1589–1595

Giuliani MM, Del Giudice G, Giannelli V, Dougan G, Douce G, Rappuoli R, Pizza M (1998) Mucosal adjuvanticity and immunogenicity of LTR72, a novel mutant of *Escherichia coli* heat-labile enterotoxin with partial knockout of ADP-ribosyltransferase activity. J Exp Med 187:1123–1132

Haddad EE, Whitfill CE, Avakian AP, Ricks CA, Andrews PD, Thoma JA, Wakenell PS (1997) Efficacy of a novel infectious bursal disease virus immune complex vaccine in broiler chickens. Avian Dis 41:882–889

Halpern MD, Kurlander RJ, Pisetsky DS (1996) Bacterial DNA induces murine interferon-$\gamma$ production by stimulation of interleukin-12 and tumor necrosis factor-$\alpha$. Cell Immunol 167:72–78

Haneberg B, Kendall D, Amerongen HM, Apter FM, Kraehenbuhl JP, Neutra MR (1994) Induction of specific immunoglobulin A in the small intestine, colon-rectum, and vagina measured by a new method of collection of secretions from local mucosal surfaces. Infection and Immunity 62:15–23

Hanke T, Botting C, Green EA, Szawlowski PW, Rud E, Randall RE (1994) Expression and purification of nonglycosylated SIV proteins, and their use in induction and detection of SIV-specific immune responses. AIDS Res Hum Retroviruses 10:665–674

Hogg JC (1997) The pathology of asthma. APMIS 105:735–745

Holmgren J, Lycke N, Czerkinsky C (1993) Cholera toxin and cholera B subunit as oral-mucosal adjuvant and antigen vector systems. Vaccine 11:1179–1184

Horner AA, Ronaghy A, Cheng PM, Nguyen MD, Cho HJ, Broide D, Raz E (1998) Immunostimulatory DNA is a potent mucosal adjuvant. Cell Immunol 190:77–82

Hughes JA, Awrutskaya AV, Juliano RL (1994) Influence of base composition cellular uptake of 10-mer phosphorothioate oligonucleotides in Chinese hamster ovary (CHRC5) cells. Antisense Res Develop 4:211–215

Hunt DW, Huppertz HI, Jiang HJ, Petty RE (1994) Studies of human cord blood dendritic cells: evidence for functional immaturity. Blood 84:4333–4343

Jertborn M, Svennerholm AM, Holmgren J (1992) Safety and immunogenicity of an oral recombinant cholera B subunit- whole cell vaccine in Swedish volunteers. Vaccine 10:130–132

Kay AB (1996) TH2-type cytokines in asthma. Ann NY Acad Sci 796:1–8

Kline JN, Businga TR, Waldschmidt TJ, Weinstock JV, Krieg AM (1998) Modulation of airway inflammation by CpG oligodeoxynucleotides in a murine model of asthma. J Immunol 160: 2555–2559

Klinman DM, Yi A, Beaucage SL, Conover J, Krieg AM (1996) CpG motifs present in bacterial DNA rapidly induce lymphocytes to secrete interleukin 6, interleukin 12 and interferon $\gamma$. Proc Natl Acad Sci USA 93:2879–2883

Klinman DM, Yamshchikov G, Ishigatsubo Y (1997) Contribution of CpG motifs to the immunogenicity of DNA vaccines. J Immunol 158:3635–3639

Komase K, Tamura S, Matsuo K, Watanabe K, Hattori N, Odaka A, Suzuki Y, Kurata T, Aizawa C (1998) Mutants of *Escherichia coli* heat-labile enterotoxin as an adjuvant for nasal influenza vaccine. Vaccine 16:248–254

Kovarik J, Siegrist CA (1998) Immunity in early life. Immunol Today 19:150–152

Kovarik J, Bozzotti P, Love-Homan L, Pihlgren M, Davis HL, Lambert P-H, Krieg AM, Siegrist CA (1999) CpG oligonucleotides can circumvent the TH2 polarization of neonatal responses to vaccines but fail to fully redirect TH2 responses established by neonatal priming. J Immunol (in press)

Krieg AM, Yi A-K, Matson S, Waldschmidt TJ, Bishop GA, Teasdale R, Koretzky GA, Klinman DM (1995) CpG motifs in bacterial DNA trigger direct B-cell activation. Nature 374:546–549

Krieg AM, Matson S, Fisher E (1996) Oligodeoxynucleotide modifications determine the magnitude of immune stimulation by CpG motifs. Antisense Res Dev 6:133–139

Krieg AM, Wu T, Weeratna R, Efler SM, Love-Homan L, Yang L, Yi AK, Short D, Davis HL (1998) Sequence motifs in adenoviral DNA block immune activation by stimulatory CpG motifs. Proc Natl Acad Sci USA 95:12631–12636

Leclerc C, Deriaud E, Rojas M, Whalen RG (1997) The preferential induction of a Th1 immune response by DNA-based immuniaztion is mediated by the immunostimulatory effect of plasmid DNA. Cell Immunol 179:97–106

Lipford GB, Bauer M, Blank C, Reiter R, Wagner H, Heeg K (1997a) CpG-containing synthetic oligonucleotides promote B and cytotoxic T cell responses to protein antigen: a new class of vaccine adjuvants. Eur J Immunol 27:2340–2344

Lipford GB, Sparwasser T, Bauer M, Zimmermann S, Koch E-S, Heeg K, Wagner H (1997b) Immunostimulatory DNA: sequence-dependent production of potentially harmful or useful cytokines. Eur J Immunol 27:3420–3426

Liu SJ, Sher YP, Ting CC, Liao KW, Yu CP, Tao MH (1998) Treatment of B-cell lymphoma with chimeric IgG and single-chain Fv antibody-interleukin-2 fusion proteins. Blood 92:2103–2112

Lycke N, Tsuji T, Holmgren J (1992) The adjuvant effect of *Vibrio cholerae* and *Escherichia coli* heat-labile enterotoxins is linked to their ADP-ribosyltransferase activity. European Journal of Immunology 22:2277–2281

McCluskie MJ, Davis HL (1998) CpG DNA is a potent enhancer of systemic and mucosal immune responses against hepatitis B surface antigen with intranasal administration to mice. J Immunol 161:4463–4466

McCluskie MJ, Di Q, Wen Y-M, Davis HL (1999) Mucosal immunization against hepatitis B virus with antigen-antibody complexes: induction of Th1 immune responses. J Infec Dis (in press)

Messina JP, Gilkeson GS, Pisetsky DS (1991) Stimulation of in vitro murine lymphocyte proliferation by bacterial DNA. J Immunol 147:1759–1764

Moldoveanu Z, Love-Homan L, Huang WQ, Krieg AM (1998) CpG DNA: a novel adjuvant for systemic and mucosal immunization with influenza virus. Vaccine 16:1216–1224

Pokric B, Sladic D, Juros S, Cajavec S (1993) Application of the immune complex for immune protection against viral disease. Vaccine 11:655–659

Rappuoli R, Douce G, Dougan G, Pizza M (1995) Genetic detoxification of bacterial toxins: a new approach to vaccine development. Int Arch Allergy Immunol 108:327–333

Roman M, Martinorozco E, Goodman S, Nguyen MD, Sato Y, Ronaghy A, Kornbluth RS, Richman DD, Carson DA, Raz E (1997) Immunostimulatory DNA sequences function as T helper-1-promoting adjuvants. Nature Med 3:849–854

Sato Y, Roman M, Tighe H, Lee D, Corr M, Nguyen M-D, Silverman GJ, Lotz M, Carson DA, Raz E (1996) Immunostimulatory DNA sequences necessary for effective intradermal gene immunization. Science 273:352–354

Schwartz D, Quinn TJ, Thorne PS, Sayeed S, Yi A-K, Krieg AM (1997) CpG motifs in bacterial DNA cause inflammation in the lower respiratory tract. J Clin Invest 100:68–73

Spangler BD (1992) Structure and function of cholera toxin and the related *Escherichia coli* heat-labile enterotoxin. Microbiological Rev 56:622–647

Sun S, Kishimoto H, Sprent J (1998) DNA as an adjuvant: capacity of insect DNA and synthetic oligodeoxynucleotides to augment T cell responses to specific antigen. J Exp Med 187:1145–1150

Threadgill DS, McCormick LL, McCool TL, Greenspan NS, Schreiber JR (1998) Mitogenic synthetic polynucleotides suppress the antibody response to a bacterial polysaccharide. Vaccine 16:76–82

Tokunaga T, Yamamoto H, Shimada S, Abe H, Fukuda T, Fujisawa Y, Furutani Y, Yano O, Kataoka T, Sudo T, Makiguchi N, Suganuma T (1998) Antitumor activity of deoxyribonucleic acid fraction from *Mycobacterium bovis* GCG. I. Isolation, physicochemical characterization and antitumor activity. JNCI 72:955–962

Verweij WR, de Haan L, Holtrop M, Agsteribbe E, Brands R, van Scharrenburg GJ, Wilschut J (1998) Musosal immunoadjuvant activity of recombinant *Escherichia coli* heat- labile enterotoxin and its B subunit: induction of systemic IgG and secretory IgA responses in mice by intranasal immunization with influenza virus surface antigen. Vaccine 16:2069–2076

Weeratna R, CL Bm, Krieg AM, Davis HL (1998) Reduction of antigen expression from DNA vaccines by co-administered oligonucleotides. Antisense Nucleic Acid Drug Devel 8:351–356

Weiner GJ, Liu H-M, Wooldridge JE, Dahle CE, Krieg AM (1997) Immunostimulatory oligodeoxynucleotides containing the CpG motif are effective as immune adjuvants in tumor antigen immunization. Proc Natl Acad Sci USA 94:10833–10837

Yamamoto S, Yamamoto Y, Kataoka T, Kuramoto E, Yano O, Tokunaga T (1992) Unique palindromic sequences in synthetic oligonucleotides are required to induce INF and augment INF-mediated natural killer activity. J Immunol 148:4072–4076

Yamamoto S, Yamamoto Y, Kataoka T, Kuramoto E, Yano O, Tokunaga T (1992) DNA from bacteria, but not from vertebrates, induces interferons, activates natural killer cells and inhibits tumor growth. Microbiol Immunol 36:983–997

Yamamoto S, Kiyono H, Yamamoto M, Imaoka K, Fujihashi K, Van Ginkel FW, Noda M, Takeda Y, McGhee JR (1997) A nontoxic mutant of cholera toxin elicits Th2-type responses for enhanced mucosal immunity. Proceedings of the National Academy of Sciences of the United States of America 94:5267–5272

Yamamoto S, Takeda Y, Yamamoto M, Kurazono H, Imaoka K, Yamamoto M, Fujihashi K, Noda M, Kiyono H, McGhee JR (1997) Mutants in the ADP-ribosyltransferase cleft of cholera toxin lack diarrheagenicity but retain adjuvanticity. Journal of Experimental Medicine 185:1203–1210

Yamamoto M, Briles DE, Yamamoto S, Ohmura M, Kiyono H, McGhee JR (1998) A nontoxic adjuvant for mucosal immunity to pneumococcal surface protein A. J Immunol 161:4115–4121

Yi A-K, Klinman DM, Martin TL, Matson S, A.M. K (1996) Rapid immune activation by CpG motifs in bacterial DNA: Systemic induction of IL-6 transcription through an antioxidant-sensitive pathway. J Immunol 157:5394–5402

Yi A-K, Hornbeck P, Lafrenz DE, Krieg AM (1996) CpG DNA rescue of murine B lymphoma cells from anti-IgM induced growth arrest and programmed cell death is associated with increased expression of c-myc and bcl-XL. J Immunol 157:4918–4925

Zhao Q, Matson S, Herrera CJ, Fisher E, Yu H, Krieg AM (1993) Comparison of cellular binding and uptake of antisense phosphodiester, phosphorothioate and mixed phosphorothioate and methylphosphonate oligonucleotides. Antisense Res Develop 3:53–66

# Immunostimulatory-Sequence DNA is an Effective Mucosal Adjuvant

A.A. Horner and E. Raz



## 1 Introduction

The respiratory, gastrointestinal, vaginal, and rectal mucosa are major portals of entry for infectious agents and foreign proteins (Czerkinsky and Holmgren 1995; Staats and McGhee 1996). Locally produced and secreted immunoglobulin A (IgA) and cytotoxic T lymphocytes (CTL) within the mucosal tissue and draining lymph nodes protect against microbial infection (Gallichan et al. 1993; Czerkinsky and Holmgren 1995; Staats and McGhee 1996; Ada and McElrath 1997). Because the mucosal immune system serves a front-line role in protecting us from our environment, there is a great deal of interest in developing strategies for stimulating protective mucosal immune responses. Unfortunately, traditional vaccination methods are not effective and, at present, there is no simple, safe, and generally applicable approach for the induction of mucosal immune responses.

Mucosal delivery of replicating immunogens, such as live attenuated vaccines, elicits robust and long-lasting immune responses, including the production of mucosal IgA, serum antibody, and A CTL response (Czerkinsky and Holmgren

Department of Medicine and The Sam and Rose Stein Institute for Aging, University of California, San Diego, 9500 Gilman Drive, La Jolla, CA 92093-0663, USA

1995; STAATS and McGHEE 1996). However, simple protein antigens are often used for research purposes and clinical applications, and they are much less immunogenic. Systemic [i.e. intradermal (i.d.)] vaccination with monomeric proteins elicits serum antibody production, but many arms of the immune response, such as mucosal IgA synthesis and CTL activity are not elicited. However, mucosal vaccination with protein alone generally will not elicit any immune response and may even induce tolerance (CZERKINSKY and HOLMGREN 1995; GUPTA and SIBER 1995; STAATS and McGHEE 1996). In general, to elicit a mucosal IgA response, monomeric protein antigens need to be delivered to mucosal surfaces with an adjuvant (CZERKINSKY and HOLMGREN 1995; STAATS and McGHEE 1996). Cholera toxin (CT) is the best-studied and most potent mucosal adjuvant identified to date. Unfortunately, like many adjuvants, it has unacceptable toxicity for use in humans (GUPTA and SIBER 1995). In addition, CT biases systemic immune responses toward a T-helper 2 ($Th_2$) phenotype, potentially leading to IgE production and allergic hypersensitivity toward the antigen (SNIDER et al. 1994; MARINARO et al. 1995).

Immunostimulatory-sequence DNA contained within plasmids and oligodeoxynucleotides (ISS-ODN) has previously been shown to have potent adjuvant activity for systemic immune responses to protein antigens. The immune responses seen after systemic vaccination with simple proteins and immunostimulatory-sequence DNA shares many features with the immune responses seen after viral infections. This immune response characteristically has a $Th_1$ bias, is robust and long lasting, and generally includes CTL activity (SATO et al. 1996; ROMAN et al. 1997; DAVIS et al. 1998; ZIMMERMANN et al. 1998). Here we review our recent observations, which lead us to conclude that ISS-ODN is a potent mucosal adjuvant that induces both secretory IgA, and a strong $Th_1$-biased systemic immune response following intranasal (i.n.) delivery with antigen (HORNER et al. 1998).

## 2 ISS-ODN is an Adjuvant for $Th_1$-Biased Systemic Immune Responses

ISS-ODN containing the hexamer 5′-pyrimidine-pyrimadine-CpG-purine-purine-3′ have been shown to provide effective adjuvant activity for the induction of systemic $Th_1$-biased immune responses toward protein antigens co-administered via i.d. and intramuscular routes. The immune response includes the induction of a $Th_1$-cytokine profile [interferon $\gamma$ (IFN$\gamma$) but not interleukin 4 (IL-4)], the production of high IgG2a and low IgG1 titers (IFN$\gamma$ and IL-4-dependent isotypes respectively), and a CTL response. In contrast, immunization with protein alone leads to a relatively $Th_2$-biased immune response without CTL activity (ROMAN et al. 1997; DAVIS et al. 1998; HORNER et al. 1998; ZIMMERMANN et al. 1998).

### 2.1 Insights into How ISS-ODN Functions as an Adjuvant

The mechanisms underlying the $Th_1$-biased adjuvant activity of ISS-ODN are not well understood at the molecular level. However, at the cellular level, we do have some mechanistic insights. ISS-ODN can induce the production of IFN $\alpha+\beta$, IL-12, and IL-18, from macrophages and IFN$\gamma$ from natural killer (NK) cells in an antigen-independent manner (YAMAMOTO et al. 1992; KLINMAN et al. 1996; PISETSKY 1996; ROMAN et al. 1997). These cytokines are known to promote $Th_1$-biased immune responses (COFFMAN and MOSMANN 1988; MOSMANN and COFFMAN 1989; ROMAN et al. 1997). In addition, ISS-ODN stimulates IL-6 production from B cells and IL-10 production from macrophages (KLINMAN et al. 1996; ANITESCU et al. 1997; personal observations). IL-6 stimulates polytypic antibody production, while IL-10 promotes IgA synthesis specifically (MURAGUCHI et al. 1988; DEFRANCE et al. 1992). However, neither IL-6 nor IL-10 are considered to be characteristic $Th_1$ cytokines (COFFMAN and MOSMANN 1988; MOSMANN and COFFMAN 1989).

Along with the induction of an innate cytokine response, ISS-ODN induces or increases, in an antigen-independent manner, the expression of a variety of cell-surface proteins important for productive immune responses. Incubation of naive splenocytes with ISS-ODN leads to expression of B7.1 and B7.2 on B cells and macrophages. These molecules provide important co-activation signals to T cells via CD 28 engagement (JAKOB et al. 1998; SPARWASSER et al. 1998; SHAHINIAN et al. 1993; Martin-Orozco et al. manuscript submitted). In addition, ISS-ODN increases CD40 expression on B cells and macrophages. CD40 ligation by CD154 (CD40 ligand) stimulates isotype switching by B cells and activates antigen-presenting cells (APCs) (SPARWASSER et al. 1998; FULEIHAN et al. 1993; Martin-Orozco et al. manuscript submitted). ISS-ODN also increase B-cell and macrophage expression of class-I and class-II molecules, further increasing their capacity to function as APCs (JAKOB et al. 1998; SPARWASSER et al. 1998; KARLSSON et al. 1996; Martin-Orozco et al. manuscript submitted). Lastly, cytokine-receptor expression (IL2R and IFN$\gamma$R) is increased on B cells and macrophages, increasing their potential to respond to cytokines in their local environments (Martin-Orozco et al. manuscript submitted).

A central dogma in immunology is that T cells direct the Th bias of antigen-specific immune responses. However, while cytokine production and expression of cell-surface molecules on B cells and macrophages are promoted by incubation with ISS-ODN, purified T cells do not demonstrate these responses when incubated with ISS-ODN (Martin-Orozco et al. manuscript submitted). We believe that the antigen-independent innate responses of B cells, NK cells, macrophages, and other APCs toward ISS-ODN provide important upstream signals, which both promote and bias subsequent antigen-dependent and -specific T-cell and B-cell immune responses. If correct, this model implies that the $Th_1$-biased immune response which develops toward antigens delivered with ISS-ODN is initiated and shaped by B cells, macrophages, and possibly other APCs, but not T cells.

## 3 The Mucosal Immune System

Mucosal immunity clearly plays an important front-line role in protecting man from his/her environment. However, adjuvants and mucosal delivery are needed for the induction of mucosal immune responses to many antigens, including most monomeric proteins (Czerkinsky and Holmgren 1995; Staats and McGhee 1996). Research suggests that lymphocytes involved in mucosal immune responses express a unique set of surface proteins, such as integrin $\alpha_4\beta_7$, which direct these cells to mucosal organs. These homing receptors allow lymphocytes from the site of primary contact with antigen to traffic to multiple mucosal sites, leading to both local and distal mucosal immunity (Berlin et al. 1993; Quiding-Jarbrink 1997). Therefore, the mucosal immune system can be considered to be made up of a unique population of lymphocytes that can function semi-independently of lymphocytes involved in systemic immune responses.

### 3.1 Mucosal Immunization with ISS-ODN Induces a Secretory IgA Response

To further characterize the adjuvant potential and clinical utility of ISS-ODN, we assessed its ability to function as a mucosal adjuvant. CT is the best-studied and most potent known mucosal adjuvant for the induction of secretory IgA. Therefore, the IgA response of mice mucosally immunized with protein and ISS-ODN was compared to the IgA response when mice were mucosally immunized with protein and CT. Both ISS-ODN and CT were found to be effective mucosal adjuvants for a variety of antigens and mouse strains. In Fig. 1, the mucosal IgA response of BALB/c mice i.n. immunized with β-galactosidase (β-gal) is presented (Horner et al. 1998). While direct comparison of the IgA levels from different

**Fig. 1A–D.** Immunoglobulin A (IgA) responses. Mice received a single immunization with β-galactosidase (β-gal; 50μg) alone, with oligodeoxynucleotide immunostimulatory sequence (ISS-ODN; 50μg), mutated oligodeoxynucleotide (50μg), or cholera toxin (CT; 10μg) via intranasal (i.n.) or intradermal routes. Immunostimulatory sequences and mutated oligodeoxynucleotides used in these experiments have the sequences 5′-TGACTGTGAACGTTCGAGATGA-3′, and 5′-TGACTGTGAACCTTAGA-GATGA-3′, respectively. Results were obtained by enzyme-linked immunosorbent assay and represent mean values for four mice per group; *error bars* reflect the standard errors of the means. Results are representative of three similar and independent experiments. **A** Bronchoalveolar lavage fluid (BALF) IgA. Bronchial lavage was carried out at sacrifice during week 7 with 800μl of phosphate-buffered saline (PBS). There was no significant difference in BALF anti-β-gal IgA levels between i.n. β-gal/ISS-ODN- and β-gal/CT-immunized groups. **B** Vaginal IgA. Vaginal lavage with 50μl of PBS was carried out at sacrifice. There was no significant difference in vaginal anti-β-gal IgA levels between i.n. β-gal/ISS-ODN- and β-gal/CT-immunized groups. **C** Fecal IgA. Feces were collected at 2, 4, and 7 weeks, and IgA was extracted as previously described (Haneberg et al. 1994). There was no significant difference in fecal anti-β-gal IgA levels between the i.n. β-gal/ISS-ODN and β-gal/CT groups except at 2 weeks ($P = 0.03$). **D** BALF, fecal, vaginal, and serum IgA. Serum was obtained at sacrifice during week 7, and IgA content was compared to week-7 BALF, fecal, and vaginal IgA. In i.n. β-gal/ISS-ODN- and β-gal/CT-immunized mice, the mucosal compartments demonstrated consistently higher anti-β-gal IgA levels then the serum

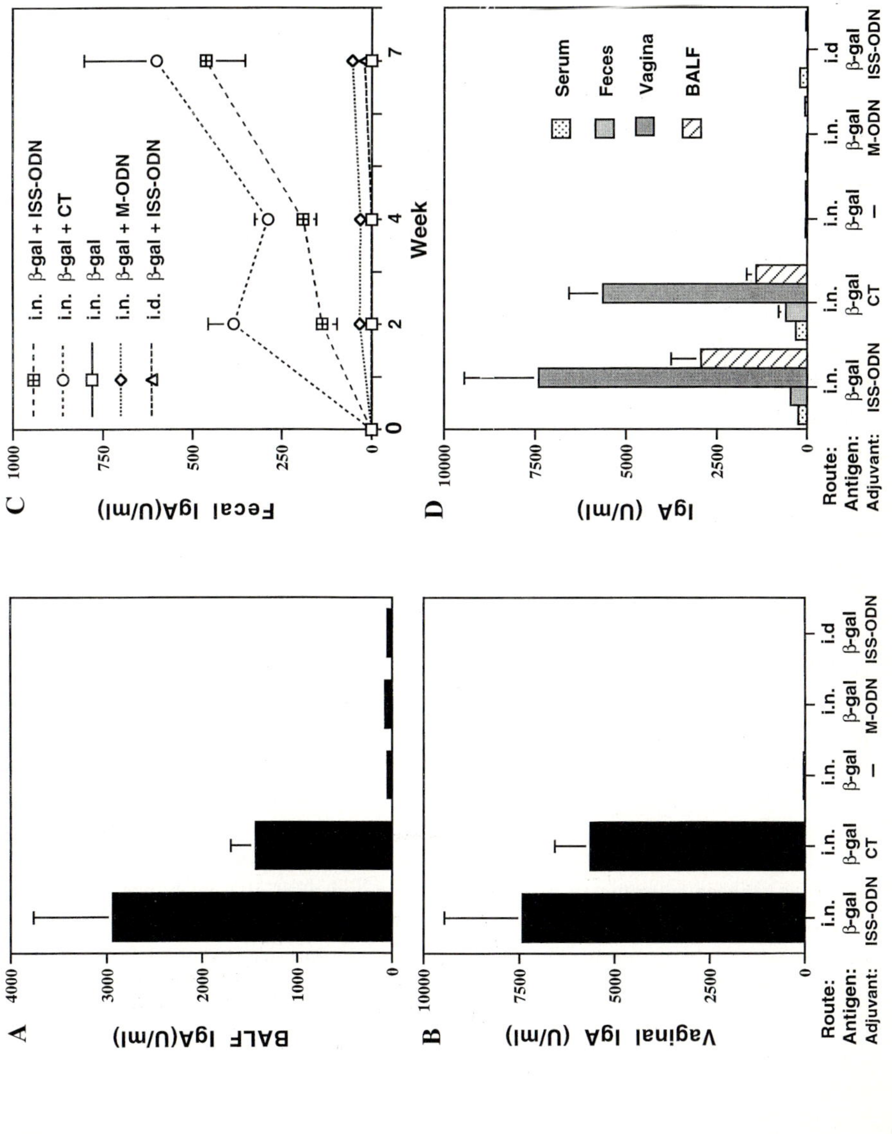
A
BALF IgA(U/ml)
B
Vaginal IgA (U/ml)
C
Fecal IgA(U/ml)
Week
i.n. β-gal + ISS-ODN
i.n. β-gal + CT
i.n. β-gal
i.n. β-gal + M-ODN
i.d. β-gal + ISS-ODN
D
IgA (U/ml)
Serum
Feces
Vagina
BALF
Route:
Antigen:
Adjuvant:
i.n.
β-gal
ISS-ODN
CT
—
M-ODN
i.d

mucosal sites can not be made due to differences in sample-collection techniques, the data clearly demonstrate that β-gal/ISS-ODN- and β-gal/CT-immunized mice had equivalent antigen-specific IgA levels in bronchial, vaginal, and intestinal secretions. To establish that a mucosal adjuvant was needed for the induction of mucosal IgA, mice were i.n. immunized with β-gal alone or with a mutated oligodeoxynucleotide (M-ODN). However, i.n. vaccination without mucosal adjuvant resulted in no detectable IgA. To evaluate whether contact with the respiratory mucosa was required for ISS-ODN to have mucosal adjuvant activity, mice were immunized with β-gal and ISS-ODN via intragastric (i.g.) and i.d. routes. These routes of immunization did not lead to measurable IgA in mucosal secretions (data for i.g. immunization not shown). To confirm that the IgA detected in bronchoalveolar lavage fluid (BALF), fecal material, and vaginal washes of immunized mice was actively secreted by mucosal tissue and did not passively diffuse from serum, anti-β-gal IgA levels in serum, feces, vaginal washings, and BALF were compared. It should be noted that initial acquisition of BALF, fecal samples, and vaginal washes requires an unmeasurable dilution of the IgA contained in the material, which does not occur when obtaining serum. Despite this fact, i.n. β-gal/ISS-ODN- and i.n. β-gal/CT-immunized mice produced higher levels of anti-β-gal IgA in feces, vaginal washes, and BALF than in serum, demonstrating the active secretion of antigen-specific IgA from the mucosal surfaces of these mice.

These results show that ISS-ODN and CT have equivalent mucosal adjuvant activity with a test antigen that has no capacity to induce mucosal IgA production when delivered alone. In addition, mucosal delivery of antigen with ISS-ODN is shown to lead to a secretory IgA response, while i.d. delivery does not. Taken together, these findings demonstrate that ISS-ODN is an excellent adjuvant for the induction of both local and distal mucosal immunity when co-delivered with antigen via the nose.

## 3.2 Mucosal Immunization with ISS-ODN Induces a $Th_1$-Biased Systemic Immune Response

Although systemic immune responses can occur in isolation, immune responses involving mucosal lymphocytes generally occur with concomitant stimulation of the systemic immune system (Czerkinsky and Holmgren 1995; Staats and McGhee 1996). Systemic-immune responses can be subdivided into $Th_1$- and $Th_2$-biased responses based on the specific cytokine and antibody profile elicited. $Th_1$-biased immune responses are seen following viral infection and are generally associated with a CTL response (Mosmann and Coffman 1988, 1989). In contrast, $Th_2$-biased immune responses are associated with allergic hypersensitivities and the synthesis of IgE (Snider et al. 1994). The adjuvants ISS-ODN and CT are considered to promote $Th_1$- and $Th_2$-biased immune responses, respectively (Snider et al. 1994; Marinaro et al. 1995; Roman et al. 1997; Davis et al. 1998; McCluskie and Davis 1998; Horner et al. 1998). Therefore, the magnitude and phenotype of the systemic immune response induced by i.n. β-gal/ISS-ODN and i.n. β-gal/CT immunization was evaluated. Splenocytes from immunized mice were incubated with β-gal, and culture supernatants were assayed for the production of IFNγ and IL-4, cytokines classically associated with $Th_1$ and $Th_2$ immunity, respectively (Fig. 2) (MOSMANN and COFFMANN 1989). Splenocytes

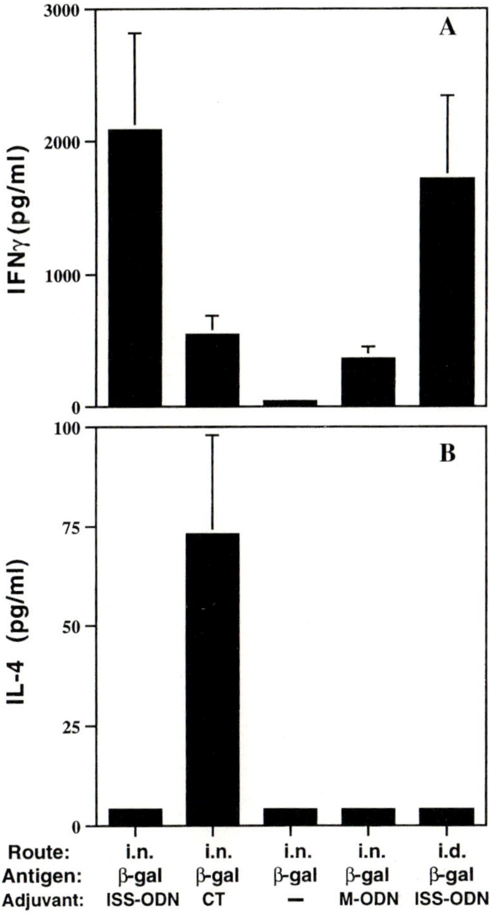


**Fig. 2A,B.** Splenocyte antigen-induced cytokine profiles. Mice received a single immunization with β-galactosidase (β-gal; 50μg) alone, with oligodeoxynucleotide immunostimulatory sequence (ISS-ODN; 50μg), mutated oligodeoxynucleotide (50μg), or cholera toxin (CT; 10μg) via intranasal (i.n.) or intradermal (i.d.) routes. The sequences of the oligodeoxynucleotides are provided in the Fig. 1 legend. Splenocytes were harvested from sacrificed mice during week 7, cultured in media with or without β-gal (10μg/ml), and 72-h supernatants were assayed by enzyme-linked immunosorbent assay. Splenocytes cultured without β-gal produced no detectable interferon γ (IFNγ) or interleukin 4 (IL-4; data not shown). Results represent the mean for four mice in each group, and similar results were obtained in two other independent experiments. *Error bars* reflect standard errors of the means. **A** IFNγ levels. IFNγ levels were equivalent in i.n. and i.d. β-gal/ISS-ODN-immunized mice but statistically higher then in other immunization groups ($P = 0.05$ for i.n. β-gal/ISS-ODN versus β-gal/CT mice). **B** IL-4 levels. IL-4 levels above background were detected only in mice immunized with i.n. β-gal/CT ($P = 0.04$ versus i.n. β-gal/ISS-ODN)

from mice immunized with β-gal and ISS-ODN via i.n. and i.d. routes produced similarly high levels of IFNγ but no detectable IL-4. In contrast, i.n. vaccination with β-gal and CT led to low-level splenocyte production of IFNγ but significant levels of IL4. Intranasal immunization with β-gal alone or with M-ODN led to

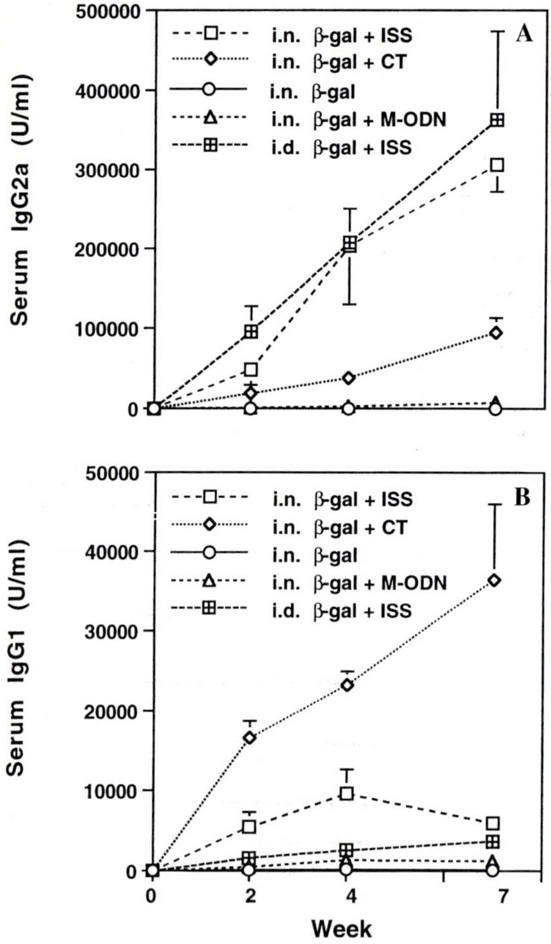


**Fig. 3A,B.** Serum immunoglobulin G (IgG) subclass profiles. Mice received a single immunization with β-galactosidase (β-gal; 50μg) alone, with oligodeoxynucleotide immunostimulatory sequence (ISS-ODN; 50μg), mutated oligodeoxynucleotide (50μg), or cholera toxin (CT; 10μg) via intranasal (i.n.) or intradermal (i.d.) routes. The sequences of the oligodeoxynucleotides are provided in the Fig. 1 legend. Serum was collected at 2, 4, and 7 weeks from immunized mice and assayed by enzyme-linked immunosorbent assay. Results represent mean values for four mice per group, and *error bars* reflect standard errors of the means. Results are representative of three similar and independent experiments. **A** Serum IgG2a. Serum IgG2a levels were equivalent in i.n. and i.d. β-gal/ISS-ODN but statistically higher than in other immunization groups at 7 weeks ($P = 0.005$ for i.n. β-gal/ISS-ODN versus β-gal/CT mice). **B** Serum IgG1. Serum IgG1 levels were equivalent in i.n. and i.d. β-gal/ISS-ODN- but statistically lower than in i.n. β-gal/CT-immunized mice at all time points ($P = 0.003$, $P = 0.02$, and $P = 0.02$ for i.n. β-gal/ISS-ODN versus β-gal/CT mice at 2, 4, and 7 weeks, respectively)

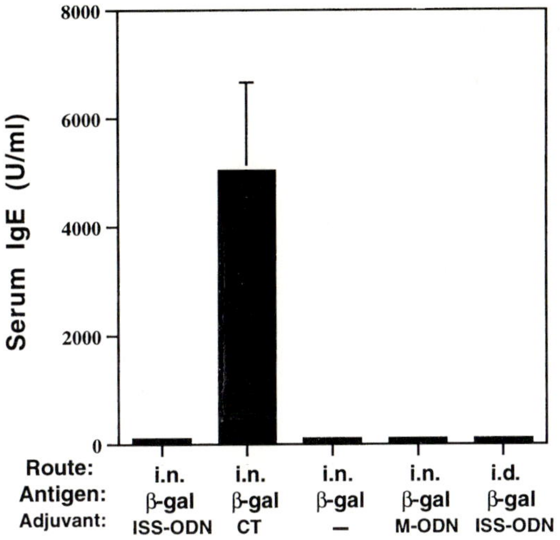


**Fig. 4.** Serum immunoglobulin E (IgE). Mice received a single immunization with β-galactosidase (β-gal; 50μg) alone, with oligodeoxynucleotide immunostimulatory sequence (ISS-ODN; 50μg), mutated oligodeoxynucleotide (50μg), or cholera toxin (CT; 10μg) via intranasal (i.n.) or intradermal routes. The sequences of the oligodeoxynucleotides are provided in the Fig. 1 legend. Serum was collected from mice during week 7 and anti-β-gal IgE levels were assayed by enzyme-linked immunosorbent assay. Results represent mean values for four mice per group, and *error bars* reflect standard errors of the means. Results are representative of three similar and independent experiments. Serum IgE was only detectable in i.n. β-gal/CT immunized mice ($P = 0.004$ for i.n. β-gal/CT- versus i.n. β-gal/ISS-ODN-immunized mice)

much lower or undetectable cytokine production in splenocytes (Horner et al. 1998).

IFNγ is a switch factor for IgG2a production, while IL-4 is a switch factor for IgG1 and IgE (Coffman and Mosmann 1988). Given the previously noted splenic cytokine profiles, it would, therefore, be predicted that i.n. β-gal/ISS-ODN co-administration would lead to higher IgG2a and lower IgG1 and IgE levels than i.n. β-gal/CT immunization. Indeed, i.n. and i.d. β-gal/ISS-ODN-immunized mice produced equivalent $Th_1$-biased serum antibody responses, while i.n. β-gal/CT vaccination led to a $Th_2$-biased antibody profile. As can be seen in Fig. 3, both i.n. and i.d. vaccination with ISS-ODN and β-gal led to high levels of IgG2a and low levels of IgG1 when compared to i.n. β-gal/CT vaccination (Horner et al. 1998). In addition, IgE production in i.n. β-gal/ISS-ODN- and i.n. β-gal/CT-immunized mice was evaluated. As expected and seen in Fig. 4, only β-gal/CT-immunized mice produced IgE (previously unpublished data).

Cumulatively, these observations demonstrate that, under the conditions employed in this series of experiments, i.n. and i.d. delivery of antigens with ISS-ODN leads to equivalent systemic $Th_1$-biased cytokine and antibody profiles, while i.n. β-gal/CT co-administration leads to a $Th_2$-biased immune response that includes the production of IgE. Considered in conjunction with the IgA data previously presented, the data further demonstrate that induction of mucosal IgA can occur in the context of both $Th_1$- and $Th_2$-biased systemic immune responses.

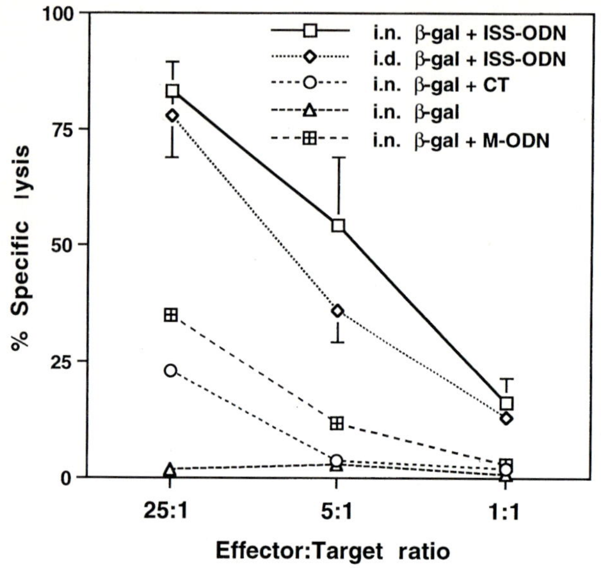


**Fig. 5.** Splenic cytotoxic T-lymphocyte (CTL) responses. Mice received a single immunization with β-galactosidase (β-gal; 50μg) alone, with oligodeoxynucleotide immunostimulatory sequence (ISS-ODN; 50μg), mutated oligodeoxynucleotide (50μg), or with cholera toxin (CT; 10μg) via intranasal (i.n.) or intradermal (i.d.) routes. The sequences of the oligodeoxynucleotides are provided in the Fig. 1 legend. Splenocytes were harvested from mice at week 7. They were cultured for 5 days with interleukin 2, β-gal peptide, and mitomycin-C-treated splenocytes from naïve mice, and CTL responses were subsequently determined. Results represent mean values for four mice per group, and *error bars* reflect standard errors of the means. Results are representative of three similar and independent experiments. CTL responses were equivalent in i.n. and i.d. β-gal/ISS-ODN immunized mice at all enzyme:target (E:T) ratios but were statistically higher then in i.n. β-gal/CT-immunized mice ($P = 0.005$ and $P = 0.05$ for i.n. β-gal/ISS-ODN- versus β-gal/CT-immunized mice at E:T ratios of 25:1 and 5:1, respectively)

## 3.3 Mucosal Immunization with ISS-ODN Induces a Splenic CTL Response

Although development of antigenspecific CTL activity is associated with $Th_1$-biased immunity, not all $Th_1$-biased immune responses include the development of cytotoxic T cells (Staats and McGhee 1996; Ada and McElrath 1997). Therefore, the ability of i.n. co-delivery of β-gal and ISS-ODN to induce a CTL response was evaluated. As demonstrated in Fig. 5, mice immunized with β-gal and ISS-ODN by either the i.n. or i.d. route displayed vigorous splenic CTL activity. However, i.n. β-gal/CT immunization resulted in a poor CTL response. Likewise, i.n. immunization with β-gal alone or with M-ODN led to poor or undetectable CTL responses (Horner et al. 1998). Considered in conjunction with the splenic cytokine and serum antibody responses previously presented, this data confirms that i.d. and i.n. vaccination with β-gal and ISS-ODN leads to $Th_1$-biased systemic immune responses of equivalent magnitude. In addition, the CTL assay results further demonstrate the dichotomy between the $Th_1$- and $Th_2$-biased systemic immune responses seen when β-gal is co-delivered i.n. with ISS-ODN and CT, respectively.

## 4 Discussion

In summary, these results demonstrate that ISS-ODN is a potent mucosal adjuvant. Intranasal delivery of antigen with either ISS-ODN or CT leads to an equivalent and vigorous mucosal IgA response, while i.d. co-delivery of antigen with ISS-ODN does not lead to mucosal IgA production. However, under the experimental conditions employed, i.d. and i.n. vaccination with target antigen and ISS-ODN induces equivalent systemic $Th_1$-biased immune responses characterized by high levels of antigen specific IFNγ but no IL-4 production from cultured splenocytes, high IgG2a and low IgG1 and IgE serum concentrations, and vigorous CTL responses. In contrast, i.n. co-delivery of antigen with CT leads to a $Th_2$-biased systemic immune response characterized by low IFNγ but substantial IL-4 production from in vitro antigen-stimulated splenocytes, high IgG1 and IgE and low IgG2a serum concentrations, and a poor CTL response. While initial work utilized BALB/c mice and β-gal as antigen, subsequent studies have demonstrated the generalizability of this phenomenon to other mouse strains and protein antigens (unpublished data).

Consistent with our findings, Dr. Moldoveanu and colleagues have shown that ISS-ODN is an effective mucosal adjuvant for IgA production against a formalin-inactivated influenza vaccine (Moldoveanu et al. 1998). Interestingly, under the experimental conditions utilized in these studies, the inactivated influenza vaccine stimulated secretory IgA production even without an adjuvant. The multimeric nature of the antigen preparation and the epithelial-binding properties of the influenza virus may help explain this observation (Smith 1998). These investigators were unable to elicit a CTL response against influenza virus. Formalin inactivation leads to cross-linking of protein. This, in turn, could limit peptide processing and presentation by major histocompatibility complex molecules, leading to relatively weak T-cell responses, as suggested by the authors (Moldoveanu et al. 1998). McCluskie and colleagues have also recently shown that ISS-ODN serves as an effective mucosal adjuvant for the induction of an IgA response, using the hepatitis-B surface antigen (McCluskie and Davis 1998). Unlike the influenza vaccine, but consistent with antigens used in our studies, hepatitis B surface antigen did not induce an IgA response when mucosally delivered without adjuvant. However, in contrast with our findings, these investigators were also unable to elicit a significant CTL response if ISS-ODN was the only adjuvant used to immunize mice. Our experience suggests that the low doses of antigen and ISS-ODN used in these experiments may have played a role in the poor CTL responses reported.

The fact that equivalent mucosal IgA levels can develop in the context of $Th_1$- and $Th_2$-biased systemic immune responses with i.n. β-gal/ISS-ODN and β-gal/CT immunizations, respectively, is consistent with other published results. Marinaro and colleagues recently demonstrated that oral delivery of tetanus toxoid with CT led to mucosal IgA production in conjunction with a $Th_2$-biased immune profile. However, when tetanus toxoid and CT were co-administered with oral IL-12, the systemic immune response was skewed toward a $Th_1$ phenotype, while mucosal IgA production was unaffected (MARINARO et al. 1997). Taken together, these findings document that synthesis of mucosal IgA can occur in the context of both $Th_1$- and $Th_2$-biased systemic immunity.

Mucosal IgA and CTL responses are known to provide protection against a number of infectious agents. Human immunodeficiency virus is but one important example (GALLICHAN et al. 1993; CZERKINSKY and HOLMGREN 1995; STAATS and MCGHEE 1996; ADA and MCELRATH 1997; LETVIN 1998; VANCOTT et al. 1998). There are a number of strategies available for the development of vaccines that induce these immune parameters. However, none appear to be globally applicable. Live attenuated vaccines produce robust immune responses, including mucosal IgA synthesis and CTL activity. Unfortunately, difficulty in attenuating many pathogens and the risk of iatrogenic disease limits the use and development of live attenuated vaccines (CZERKINSKY and HOLMGREN 1995; GUPTA and SIBER 1995; STAATS and MCGHEE 1996; LETVIN 1998). However, recombinant proteins from infectious agents are generally safe but induce relatively poor immune responses and generally induce none when delivered to mucosal surfaces. However, mucosal adjuvants can improve immune responses towards co-administered protein antigens substantially (CZERKINSKY and HOLMGREN 1995; GUPTA and SIBER 1995; STAATS and MCGHEE 1996).

CT is an extremely potent mucosal adjuvant but is inherently toxic and induces a $Th_2$-biased immune response that can include the development of IgE and consequent allergic sensitization toward the target antigen (SNIDER et al. 1994; MARINARO et al. 1995; personal observations). At present, such toxicity and other technical problems have kept many adjuvants from becoming available for use in humans (GUPTA and SIBER 1995). Alum is essentially the only adjuvant in clinical use today. It is relatively weak, does not work with a number of antigens, does not induce CTL activity, and because it must be delivered systemically, does not induce mucosal IgA (GUPTA and SIBER 1995). A safe and effective mucosal adjuvant would be of great value in the development of better vaccines. ISS-ODN is a potent adjuvant which works with a wide range of protein antigens and generally induces a $Th_1$-biased immune response with CTL activity (SATO et al. 1996; ROMAN et al. 1997; DAVIS et al. 1998; ZIMMERMANN et al. 1998). Intranasal and i.d. administration of protein with ISS-ODN leads to vigorous and equivalent $Th_1$-biased systemic immune responses, while only i.n. delivery induces a mucosal immune response (HORNER et al. 1998). Therefore, i.n. delivery of relevant antigens with ISS-ODN may well prove superior to i.d. delivery for the induction of protective immunity against mucosal pathogens.

## 5 Future Prospects

ISS-ODN are easy to manufacture, stable, and lack identified toxicity at immunogenic doses in mice and primates (unpublished observations). Additionally, use of antisense ODN in monkeys and human clinical trials have demonstrated no significant toxicity with daily doses of up to fivefold more per kilogram then those used in the present study (WEBB et al. 1997). Moreover, we and others have shown that human and mouse immunocytes display similar immunologic responses to ISS-ODN (BALLAS et al. 1996; ROMAN et al. 1997). Because ISS-ODN can be utilized as an adjuvant for both mucosal and systemic $Th_1$-biased immune responses toward simple monomeric protein antigens, it may well prove to be a valuable reagent for clinical applications. Thus, future investigations in primates and humans are needed to establish whether ISS-ODN has utility in the development of vaccines that induce protective immune responses against mucosal pathogens.

*Acknowledgements.* This work was supported in part by NIH grants AI01490 and AI40682 and a grant from Dynaway Technologies.

## References

Ada GL, McElrath MJ (1997) HIV type 1 vaccine induced cytotoxic T cell responses: potential role in vaccine efficacy. AIDS Res Hum Retroviruses 13:205–210

Anitescu M, Chace JH, Tuetken R, Yi AK, Berg DJ, Krieg AM, Cowdery JS (1997) Interleukin-10 functions in vitro and in vivo to inhibit bacterial DNA induced secretion of interleukin-12. J Interferon Cytokine Res 17:781–788

Ballas ZK, Rasmussen WL, Krieg AM (1996) Induction of natural killer cell activity in murine and human cells by CpG motifs in oligodeoxynucleotides and bacterial DNA. J Immunol 157:1840–1845

Berlin C, Berg EL, Briskin MJ, Andrew DP, Kilshaw PJ, Holzmann B, Weissman IL, Hamann A, Butcher EC (1993) $\alpha_4\beta_7$ Integrin mediates lymphocyte binding to the mucosal vascular addressin MAdCAM-1. Cell 74:185–195

Coffman RL, Mosmann TR (1988) Isotype regulation by helper T cells and lymphokines. Monogr Allergy 24:96–103

Czerkinsky C, Holmgren J (1995) The mucosal immune system and prospects for anti-infectious and anti-inflammatory vaccines. Immunologist 3:97–103

Davis HL, Weeranta R, Waldschmidt TJ, Tygrett L, Schorr J, Krieg AM (1998) CpG DNA is a potent enhancer of specific immunity in mice immunized with recombinant hepatitis B surface antigen. J Immunol 160:870–876

Defrance T, Vanbervliet B, Briere F, Durand I, Rousett F, Banchereau J (1992) Iinterleukin-10 and transforming growth factor β cooperate to induce anti-CD40 –activated na human B cells to secrete immunoglobulin A. J Exp Med 175:671–682

Fuleihan R, Ramesh N, Loh R, Jabara H, Rosen FS, Chatila T, Fu SM, Stamenkovic I, Geha RS (1993) Defective expression of the CD40 ligand in X-chromosome-linked immunoglobulin deficiency with normal or elevated IgM. Proc Natl Acad Sci USA 90:2170–2173

Gallichan WS, Johnson DC, Graham FL, Rosenthal KL (1993) Mucosal immunity and protection after intranasal immunization with recombinant adenovirus expressing herpes simplex virus glycoprotein. J Infect Dis 168:622–629

Gutpa RK, Siber GR (1995) Adjuvants for human vaccines-current status, problems and future prospects. Vaccine 13:1263–1276

Haneberg B, Kendell D, Amerongen HM, Apter FM, Kraehenbuhl JP, Neutra MR (1994) Induction of specific immunoglobulin A in the small intestine, colon-rectum, and vagina measured by a new method for collection of secretions from local mucosal surfaces. Infect Immun 62:15–23

Horner AA, Ronaghy A, Cheng PM, Nguyen MD, Cho HJ, Broide D, Raz E (1998) Immunostimulatory DNA is a potent mucosal adjuvant. Cell Immunol 190:77–82

Jakob T, Walker PS, Krieg AM, Udey MC, Vogel JC (1998) Activation of cutaneous dendritic cells by CpG-containing oligodeoxynucleotides: a role for dentritic cells in the augmentation of $Th_1$ responses by immunostimulatory DNA. J Immunol 161:3042–3049

Karlsson L, Castano AR, Peterson PA (1996) Principles of antigen processing and presentation. In: Kagnoff MF, Kiyono H (eds) Essentials of mucosal immunology. Academic Press, San Diego

Klinman DM, Yi AK, Beaucage SL, Conover J, Krieg AM (1996) CpG motifs present in bacterial DNA rapidly induce lymphocytes to secrete interleukin 6, interleukin 12, and interferon-$\gamma$. Proc Natl Acad Sci USA 93:2879–2883

Letvin NL (1998) Progress in the development of an HIV-1 Vaccine. Science 280:1875–1880

Marinaro M, Staats HF, Hiroi T, Jackson RJ, Coste M, Boyaka PN, Okahashi N, Yamamoto M, Kiyono H, Bluethmann H, Fujihashi K, McGhee JR (1995) Mucosal adjuvant effect of cholera toxin in mice results in the production of T helper 2 ($Th_2$) cells and IL-4. J Immunol 155:4621–4629

Marinaro M, Boyaka PN, Finkelman FD, Kiyono H, Jackson RJ, Jirillo E, McGhee JR (1997) Oral but not parenteral interleukin (IL)-12 redirects T helper 2 ($Th_2$)-type responses to an oral vaccine without altering mucosal IgA responses. J Exp Med 185:415–427

McCluskie MJ, Davis HL (1998) CpG DNA is a potent enhancer of systemic and mucosal immune responses against hepatitis B surface antigen with intranasal administration in mice. J Immunol 161:4463–4466

Moldoveanu Z, Love-Homan L, Huang WQ, Kreig AM (1998) CpG DNA, a novel immune enhancer for systemic and mucosal immunization with influenza virus. Vaccine 16:1216–1221

Mosmann TR, Coffmann RL (1989) $Th_1$ and $Th_2$ cells: Differential patterns of lymphokine secretion lead to different functional properties. Annu Rev Immunol 7:145–173

Muraguchi A, Hirano T, Tang B, Matsuda T, Horii Y, Nakajima K, Kishimoto T (1988) The essential role of B cell stimulatory factor 2 (BSF-2/IL-6) for the terminal differentiation of B cells. J Exp Med 167:332–344

Pisetsky DS (1996) Immune activation by bacterial DNA: a new genetic code. Immunity 5:303–310

Quiding-Jarbrink M, Nordstrom I, Granstrom G, Kilander A, Jertborn M, Butcher EC, Lazarovits AI, Holmgren J, Czerkinsky C (1997) Differential expression of Tissue-specific adhesion molicules on human circulating antibody-forming cells after systemic, enteric, and nasal immunizations. J Clin Invest 99:1281–1286

Roman M, Martin-Orozco E, Goodman JS, Nguyen MD, Sato Y, Ronaghy A, Kornbluth R, Richman DD, Carson DA, Raz E (1997) Immunostimulatory DNA sequences function as T helper-1-promoting adjuvants. Nature Med 3:849–854

Sato Y, Roman M, Tighe H, Lee D, Corr M, Nguyen MD, Silverman GJ, Lotz M, Carson DA, Raz E (1996) Immunostimulatory DNA sequences necessary for effective intradermal gene immunization. Science 273:352–354

Shahinian A, Pfeffer K, Lee KP, Kundig TM, Kishihara K, Wakeham A, Kawai K, Ohashi PS, Thompson CB, Mak TW (1993) Differential T cell costimulatory requirements in CD28-deficient mice. Science 261:609–612

Smith CB (1998) Influenza viruses. In: Gornbach SL, Bartlett JG, Blacklow NR (eds) Infectious diseases. Saunders, Philadelphia

Snider DP, Marshal JS, Perdue MH, Liang H (1994) Production of IgE antibody and allergic sensitization of intestinal and peripheral tissues after oral immunization with protein antigen and cholera toxin. J Immunol 153:647–657

Sparwasser T, Koch ES, Vabulas RM, Heeg K, Lipford GB, Ellwart JW, Wagner H (1998) Bacterial DNA and immunostimulatory CpG oligonucleotides trigger maturation and activation of murine dendritic cells. Euro J Immunol 28:2045–2054

Staats HF, McGhee JR (1996) Application of basic principles of mucosal immunity to vaccine development. In: Kiyono H, Ogra PL, McGhee JR (eds) Mucosal vaccines. Academic, San Diego

VanCott TC, Kaminski RW, Mascola JR, Kalyanaraman VS, Wassef NM, Alving CR, Ulrich JT, Lowell GH, Birx DR (1998) HIV-1 neutralizing antibodies in the genital and respiratory tracts of mice intranasally immunized with oligomeric gp120. J Immunol 160:2000–2012

Webb A, Cunningham D, Cotter F, Clarke PA, di Stefano F, Ross P, Corbo M, Dziewanowska Z (1997) BCL-2 antisense therapy in patients with non-Hodgkin lymphoma. Lancet 349:1137–1141

Yamamoto S, Yamamoto T, Kataoka T, Kuramoto E, Yano O, Tokunaga T (1992) Unique palindromic sequences in synthetic nucleotides required to induce IFN and augment IFN-mediated natural killer cell activity. J Immunol 148:4072–4076

Zimmermann S, Egeter O, Hausmann S, Lipford GB, Rocken M, Wagner H, Heeg K (1998) CpG oligodeoxynucleotides trigger protective and curative $Th_1$ responses in lethal murine leishmaniasis. J Immunol 160:3627–3630

# CpG DNA Switches on Th1 Immunity and Modulates Antigen-Presenting Cell Function

R.S. Chu, D. Askew, and C.V. Harding



## 1 Introduction

DNA containing CpG motifs, which are present in bacterial DNA but suppressed in mammalian DNA, can stimulate immune cells and modulate immune responses. A CpG motif consists of an unmethylated CpG dinucleotide flanked by two 5′ purines and two 3′ pyrimidines (Krieg et al. 1995). CpG motifs are similar to palindromic sequences identified earlier in bacille Calmette-Guerin DNA, which increase the cytolytic function of natural killer (NK) cells (Yamamoto et al. 1992), and immunostimulatory sequences, which increase the immunogenicity of DNA vaccines (Sato et al. 1996). Bacterial DNA or oligodeoxynucleotides (ODN) containing CpG motifs can stimulate B cells (Krieg et al. 1995; Sun et al. 1997), macrophages (Stacey et al. 1996; Sparwasser et al. 1997b), dendritic cells (Jakob et al. 1998; Sparwasser et al. 1998), and NK cells (Yamamoto et al. 1992; Ballas et al. 1996).

ODN containing CpG motifs (CpG ODN) have important potential utility as vaccine adjuvants due to their ability to stimulate cytokines (including Th1-associated cytokines), which may enhance vaccine efficacy. In vitro, CpG ODN stimulate splenocytes to produce IL-6, IL-12, and IFN-$\gamma$ (Klinman et al. 1996). In particular, macrophages produce IL-12 in response to CpG ODN, and production of IFN-$\gamma$ by NK cells is increased by CpG ODN both directly and indirectly via macrophage-derived IL-12 produced in response to CpG ODN (Cowdery et al. 1996; Halpern et al. 1996; Chace et al. 1997). These observations suggest that CpG ODN could affect the differentiation of T-cell responses by controlling the

Institute of Pathology, Case Western Reserve University, Cleveland, OH 44106 cytokine milieu present during the generation of T-cell responses. Experiments that address this hypothesis are discussed below, in Sect. 3.

In addition to their effects via modulation of the cytokine milieu, CpG DNA may influence T-cell responses by altering antigen processing and presentation functions of APCs, such as dendritic cells, macrophages, and B cells. Stimulation of antigen presenting cells (APCs) with CpG DNA can alter the expression of MHC molecules and co-stimulator molecules (Davis et al. 1998; Jakob et al. 1998; Sparwasser et al. 1998; Askew et al. 1999; Chu et al. 1999). Our recent studies of the effects of CpG ODN on antigen-processing functions of macrophages and dendritic cells are discussed below, in Sect. 4.

## 2 Th Subsets and Their Importance in Disease States

The CD4 Th cell plays a crucial role in the adaptive immune response, providing activation and differentiation signals in the form of cytokines and direct cell–cell contact to B cells, $CD8^+$ T cells, and dendritic cells. The ability of Th cells to influence multiple aspects of immune responses is facilitated by the existence of different Th subsets with particular effector functions. One current delineation of different Th subsets is the Th1/Th2 model, which divides mature Th cells by the types of cytokines that they secrete (Mosmann and Coffman 1989; Abbas et al. 1996). Specifically, Th1 cells secrete cytokines, such as IFN-γ, IL-2, and lymphotoxin, while Th2 cells secrete IL-4, IL-5, IL-6, IL-10, and IL-13. Certain immune responses and disease states may involve predominance of Th1 or Th2 effects. Much evidence has been generated to support this Th-cell dichotomy, although it is clear that this is a simplified model, and in many situations a mixture of T-cell subsets may occur, with secretion of cytokines associated with both Th1 and Th2 cells (Kelso 1995).

Th1 cells play an important role in responses to microbes. The secretion of IFN-γ by Th1 cells allows this subset to activate and increase the microbicidal activity of macrophages and to recruit other inflammatory cells. IFN-γ also promotes the development of cytotoxic-T-lymphocyte responses and Ig class switching to the IgG2a and IgG3 isotypes (in mice). These isotypes are effective in complement fixation and opsonization, thus facilitating the antimicrobial response. In contrast, IL-4-secreting Th2 cells are important mediators of B-cell differentiation and cause Ig class switching to the IgG1 isotype (in mice). IgG1, while less effective in complement fixation, is the predominant isotype in antibody responses against protein antigens. The production of IL-4 by Th2 cells also promotes secretion of IgE, while IL-5 from Th2 cells causes eosinophil activation. Thus, Th2 responses promote hypersensitivity responses. Also, some of the cytokines produced by Th2 cells, such as IL-10, are important in negatively regulating inflammatory responses by suppressing macrophage activation.

The development of Th cells towards either subset is mainly dependent upon the cytokine environment at the time of activation of uncommitted Th cells (also called Th0 cells). Other factors, such as antigen dose and particular co-stimulatory molecules involved in Th activation, may also play a role, but the overall importance of these factors is uncertain. The presence of IL-12 at the time of antigen presentation, however, has been clearly shown to stimulate the differentiation of Th1 cells, while IL-4 is the predominant cytokine which promotes the differentiation of Th2 cells (O'Garra 1998). The cytokines produced by one subset are often inhibitory toward the development or activity of the other. For example, IL-4 and IL-10 are capable of inhibiting IL-12 production by APCs and thus inhibit the development of Th1 cells (O'Garra 1998).

Because the effector functions of Th1 and Th2 cells are varied and often antagonistic toward those of the opposite subset, it would be desirable, in many disease states, to be able to direct immune responses to differentiate toward one subset versus the other (immune deviation). For example, in many infections involving intracellular microbes, a Th1-dominated response is crucial for resistance. In contrast, a Th2-dominated response may be more protective during infections with some parasites and helminths. In addition, various autoimmune diseases, such as multiple sclerosis and insulin-dependent diabetes mellitus, have been proposed to be a result of Th1-driven autoimmune responses, and patients with such diseases may benefit from therapy that promotes development of opposing Th2 cells (Liblau et al. 1995). Th2 cells are also implicated in disease states, specifically atopic disease and asthma. Clinically, immune deviation towards Th1 responses might be useful in alleviating allergic disease (Finkelman 1995).

## 3 CpG ODN Act as Th1-Directing Adjuvants

Adjuvants are used to enhance immune responses to immunogens. Different adjuvants have been found to direct Th-subset differentiation in different ways. Alum, an adjuvant that is used for clinical purposes in humans, has been noted to induce weak Th2 responses in mice (Raz et al. 1996; Ada and Ramsay 1997). Immunization with antigen in incomplete Freund's adjuvant (IFA) also generates Th2-type responses, with antigen-specific memory T cells secreting IL-5, while immunization with complete Freund's adjuvant (CFA) generates Th1 responses with T cells that secrete IFN-$\gamma$ (Forsthuber et al. 1996). IL-12 has also been used as an adjuvant for Th1 responses in experimental systems (Bliss et al. 1996). Some of these adjuvants (such as IFA and CFA), however, are difficult to use in human vaccines because of their toxic potential.

In the interest of anti-microbial vaccine design, there is a need for Th1-directing adjuvants that can elicit protective cellular immunity while being safe for human use. CpG ODN are potential vaccine adjuvants. CpG ODN are potent activators of macrophages and dendritic cells, causing to them to secrete IL-12 and other cytokines (see above and other chapters in this volume). The ability of CpG ODN to induce IL-12 suggested the hypothesis that CpG ODN would act as Th1-

directing adjuvants in vivo, and a murine experimental system was developed to test this hypothesis.

To test the ability of CpG ODN to direct specific Th1 responses in vivo, BALB/c mice were injected with hen-egg lysozyme (HEL) in IFA with or without the addition of CpG ODN or non-CpG ODN (see Table 1 for ODN sequences). Figure 1 shows that immunization with HEL in IFA alone produced a Th2 response against HEL, characterized by a high level of IL-5 secretion with little or no IFN-γ production by splenocytes upon in vitro antigen challenge (Chu et al. 1997), consistent with prior observations (Forsthuber et al. 1996). However, the addition of CpG ODN 1826 to the HEL/IFA immunization caused a marked increase in antigen-specific IFN-γ production by splenocytes and a decrease in IL-5 secretion compared to the response induced by HEL/IFA alone (Chu et al. 1997) (Fig. 1). The in vitro splenocyte response to HEL was shown to be $CD4^+$ T-cell dependent, as addition of anti-CD4 blocking antibodies abrogated cytokine production. In addition to CpG ODN 1826, two other CpG ODN 1760 and 1585, had similar effects. In

**Table 1.** Sequences of synthetic oligodeoxynucleotides (ODNs)

| ODN | Sequence[a] | Motif |
|---|---|---|
| 1826 | TCCATGACGTTCCTGACGTT | CpG |
| 1745 | TCCAATGAGCTTCCTGAGTCT | non-CpG |
| 1760 | ATAATCGACGTTCAAGCAAG | CpG |
| 1908 | ATAATAGAGCTTCAAGCAAG | non-CpG |
| 1585 | GGGGTCAACGTTGAGGGGGG | CpG |
| 1972 | GGGGTCTGTGCTTTTGGGGGG | non-CpG |

[a] The CpG motifs or corresponding non-CpG motifs are underlined.

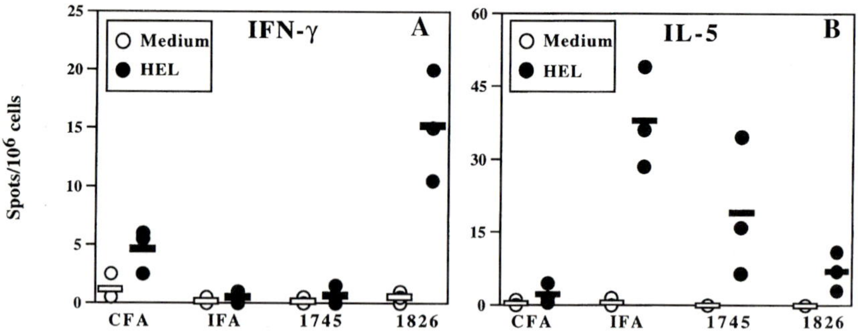


**Fig. 1A,B.** Immunization with CpG oligodeoxynucleotide (ODN) as an adjuvant induces a T helper cell 1 (Th1)-dominated cytokine profile, with increased interferon γ (IFN-γ) and decreased interleukin 5 (IL-5). BALB/c mice were injected i.p. with complete Freund's adjuvant/hen-egg lysozyme (HEL) (a control for a Th1-dominated response), incomplete Freund's adjuvant/HEL (a control for a Th2-dominated response), or IFA/HEL with 100μg (**A**) or 30μg (**B**) of ODN (non-CpG ODN 1745 or CpG ODN 1826). After 3 weeks, splenocytes were isolated and incubated with HEL (*closed circles*) or medium alone (*open circles*). Cytokine production by individual cells was assessed by an enzyme-linked immunosorbent assay spot assay for IFN-γ (**A**) or IL-5 (**B**). Each *point* represents one mouse; *horizontal bars* indicate the mean of points for each group of mice. Figure adapted from Chu et al. (Chu et al. 1997) and reproduced with permission from the Journal of Experimental Medicine contrast, addition of control non-CpG ODN (1982 1908, or 1972; Table 1) to the HEL/IFA immunization did not alter the cytokine profile of HEL-specific memory cells, indicating the importance of the CpG sequence in generating a Th1 response.

These immunization studies showed that CpG ODN are very potent Th1 adjuvants. Optimal induction of Th1 responses was seen with 30μg of CpG ODN per mouse, and doses of as little as 3–10μg of ODN per mouse were also effective at inducing Th1 differentiation. The levels of antigen-specific IFN-γ secretion generated by CpG ODN were relatively high and even exceeded those induced by immunization with HEL in CFA.

Immunization with CpG ODN also induced changes in humoral responses consistent with enhanced levels of antigen-specific IFN-γ production. Immunization with HEL in IFA, IFA plus non-CpG ODN, and IFA plus CpG ODN all produced similar total levels of HEL-specific antibodies (Fig. 2A). In contrast, immunization with HEL in IFA plus CpG ODN induced production of HEL-specific IgG2a, an isotype associated with Th1 responses, whereas immunization with HEL in IFA alone or in IFA plus non-CpG ODN did not induce HEL-specific IgG2a (Fig. 2B). The levels of IgG2a, similar to the in vitro splenocyte IFN-γ responses, were consistently higher with CpG ODN than with CFA as an adjuvant (Chu et al. 1997).

Thus, CpG ODN act as immunomodulators and produce strong Th1 responses against co-administered antigens, manifested by both cytokine and antibody isotype profiles. The addition of non-viable *Mycobacterium tuberculosis* to IFA, producing CFA, is an established method used to induce Th1 responses. In our

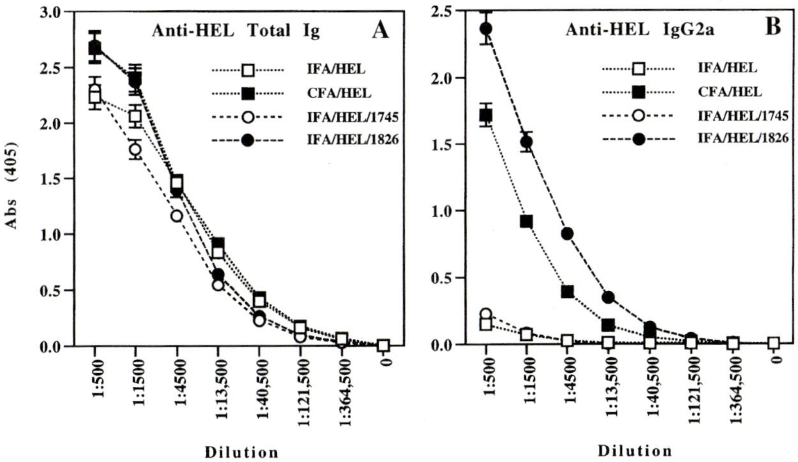


**Fig. 2A,B.** T helper cell 1 (Th1)-associated antigen-specific immunoglobulin G2a (IgG2a) responses are induced by immunization of BALB/c mice with incomplete Freund's adjuvant (IFA)/hen-egg lysozyme (HEL)/CpG oligodeoxynucleotide (ODN) but not IFA/HEL/non-CpG ODN. Mice were immunized as in Fig. 1. Sera were collected from mice 15–18 days after injection and assayed by enzyme-linked immunosorbent assay for: anti-HEL total antibody response (**A**) and anti-HEL IgG2a, an isotype associated with Th1-dominated responses (**B**). Figure adapted from Chu et al. (Chu et al. 1997) and reproduced with permission from the Journal of Experimental Medicine studies, preparations of IFA plus CpG ODN caused more effective induction of Th1 responses than did CFA. Strong Th1 responses were seen after a single immunization with CpG ODN. Furthermore, the Th1 adjuvant activity of CpG ODN was observed in both BALB/c mice, which have a Th2 bias in some experimental systems, and B10.D2 mice, which show stronger Th1 responses.

In addition to the use of immunomodulatory substances, many adjuvants, such as alum and IFA, create an antigen depot that is critical for retention of antigens and immunomodulators, allowing for persistent stimulation of the immune system. The IFA in these studies provided an important depot function that enhanced the adjuvant effect. In comparison, an injection of antigen plus CpG ODN in saline generated a weaker Th1 response, as measured by antigen-specific IFN-$\gamma$ secretion in vitro, than immunization with antigen plus CpG ODN in IFA, although both adjuvant preparations similarly induced antigen-specific IgG2a (R.S. Chu, unpublished data). Future vaccines using CpG ODN as Th1-directing adjuvants may include less toxic components, such as biodegradable oils or alum, which can generate an antigen depot in vivo. It has already been shown that the use of alum in conjunction with CpG ODN is effective in generating strong humoral responses against hepatitis-B surface antigen, including induction of specific IgG2a (Davis et al. 1998). Alternatively, multiple injections of CpG ODN may obviate the need for an additional depot-forming adjuvant. Such an approach was used by Roman et al. in the generation of Th1-biased humoral and cellular responses against proteins encoded by DNA vaccines (Roman et al. 1997).

One concern for vaccine adjuvants is toxic potential. Repeated administration of large doses of CpG ODN (150mg/kg) can cause mortality in mice (Sarmiento et al. 1994), and the induction of tumor necrosis factor $\alpha$ (TNF-$\alpha$) secretion by CpG ODN has been shown to cause fatal shock in mice that have been previously sensitized with D-galactosamine (Sparwasser et al. 1997a). However, in our studies using low doses of ODN (30–100μg/mouse), we did not observe any significant toxicity in mice treated with CpG ODN. There were no observable changes in mouse appearance, behavior, or body weight. Administration of CpG ODN alone, at the doses used in our studies, produced mild splenomegaly and hyperplasia of draining lymph nodes that were reversible within 10–14 days, while a single injection of up to 1mg of CpG ODN does not alter mouse feeding, grooming, physical activity, or behavior (A.M. Krieg, personal communication). Thus, CpG ODN provide potent adjuvant activities at doses that do not produce significant toxicity.

Another concern regarding CpG ODN as Th1-directing adjuvants is the potential induction of Th1-dominated autoimmunity, which has been observed with other manipulations that enhance Th1 responses. For example, lipopolysaccharide (LPS)- or CpG-DNA-induced IL-12 production causes in vitro differentiation of antigen-specific lymph-node cells that produce IFN-$\gamma$. When injected into normally experimental allergic encephalitis (EAE)-resistant mice, these cells are subsequently able to cause EAE, a disease associated with autoreactive Th1 effector cells (Segal et al. 1997). When CpG ODN are used in humans, their potent Th1 adjuvant activity will warrant careful monitoring of patients prone to Th1-dominated autoimmune diseases.

There are many exciting potential clinical applications for CpG ODN. The use of CpG ODN to produce protective Th1 responses has been explored in disease models involving Th1 or Th2 subsets, such as asthma (Kline et al. 1998) and murine leishmaniasis (Zimmerman et al. 1998), as discussed elsewhere in this volume. In addition to promoting Th1 differentiation, CpG ODN have general adjuvant effects in other systems, increasing total specific antibody levels (Davis et al. 1998; Sun et al. 1998) and augmenting cytotoxic-T-lymphocyte responses after immunization with protein alone or with protein encapsulated in liposomes (Lipford et al. 1997; Davis et al. 1998).

## 4 Modulation of MHC-II Antigen Processing and Presentation by CpG ODN

The adjuvant activities of CpG DNA include activation of cells via both direct and indirect mechanisms. The direct effects of CpG DNA result from signals transduced in cells, such as macrophages, dendritic cells, B cells, and NK cells, which directly recognize CpG DNA (Yamamoto et al. 1992; Stacey et al. 1996; Sparwasser et al. 1997b, 1998; Jakob et al. 1998; Yi and Krieg 1998). The direct effects of CpG DNA include the induction of cytokines, including IL-6, IL-12, and IFN-$\gamma$, which are expressed by macrophages, dendritic cells, B cells, and NK cells. These cytokines then act on many cell types to promote the indirect effects of CpG DNA, such as enhancement of T-cell responses and Th1 differentiation, as discussed above. Cytokine-induced effects may occur via signaling in T cells to alter T-cell differentiation or via activation of APCs to enhance antigen presentation and T-cell responses.

Many questions about the impact of CpG DNA on antigen presentation remain. In vivo, CpG DNA may indirectly influence antigen presentation by inducing cytokines (IFN-$\gamma$) that are known to activate APCs. In addition, CpG DNA may directly signal APCs to alter antigen processing or presentation. Induction of an increase in processing and presentation of exogenous antigen may contribute to the adjuvant activity of CpG DNA. Such an upregulation of antigen presentation would benefit the host in the context of bacterial infection, allowing more bacterial epitopes to be presented to T cells. However, the direct effects of CpG DNA on antigen processing and presentation remain poorly characterized, and this topic has been the focus of our recent studies.

In order to examine the direct effects of CpG DNA on intracellular antigen processing, we studied the in vitro effects of CpG ODN on processing and presentation of exogenous antigens by isolated macrophages. Macrophages were isolated from peritoneal exudate cells by adherence and were treated with CpG ODN overnight (18–24h). The macrophages were then pulsed briefly (1–2h) with protein antigen and fixed with paraformaldehyde. Antigen processing and presentation were measured using T-cell hybridomas specific for HEL or bovine ribonuclease (RNase), including T hybridomas that recognize the following complexes: RNase(42–56):I-$A^k$, RNase(90–105):I-$E^k$, HEL(34–45):I-$A^k$, and HEL(48–61): I-$A^k$.

After overnight treatment with CpG ODN (1 μg/ml), macrophage processing of all four epitopes was actually inhibited to varying degrees rather than upregulated, as measured by IL-2 production by the antigen-specific T-cell hybridomas (CHU et al. 1999). Interestingly, similar results were seen when the macrophages were treated with another bacterial product, LPS (100ng/ml), but not with non-CpG-ODN controls.

Additional studies examined the effects of CpG ODN on binding and presentation of exogenous peptides, which do not require intracellular protein antigen processing. The presentation of exogenous peptides by CpG ODN- or LPS-treated macrophages was also decreased compared to untreated cells, but to a lesser extent than presentation of peptides derived from processed whole antigen. These results suggested that CpG ODN and LPS downregulate expression of major histocompatibility complex class II (MHC-II) molecules. Surface I-$A^k$ levels were analyzed by flow cytometry and were found to be decreased after treatment of macrophages with CpG ODN or LPS, but not after treatment with non-CpG ODN. Northern analysis revealed that the level of I-$A^k$ mRNA was lower in macrophages that were treated with CpG ODN than in untreated macrophages. Thus, the downregulation of MHC-II expression mediated by CpG ODN occurred via transcriptional regulation (or possibly by regulation of mRNA stability). In contrast, the endocytic function of macrophages, as measured by uptake of fluoresceinated dextran, was not specifically affected by CpG ODN (CHU et al. 1999). In summary, CpG ODN and LPS caused a downregulation in macrophage antigen processing and presentation, which was primarily mediated by a decrease in the synthesis of MHC-II molecules. These results are reminiscent of previous studies demonstrating an inhibitory effect of LPS on MHC-II expression by IFN-γ-stimulated macrophages (STEEG et al. 1982; KOERNER et al. 1987; SICHER et al. 1994). The effects of LPS on MHC-II synthesis were proposed to be a result of increased levels of prostaglandin (STEEG et al. 1982) or nitric oxide (SICHER et al. 1994). These substances may also be involved in the effects on MHC-II synthesis mediated by CpG ODN.

It is unclear how the downregulation of MHC-II expression and antigen processing produced by CpG ODN in vitro relates to either the actual in vivo effects of bacterial DNA in the physiological context of an infection or the in vivo effects of CpG DNA as a vaccine adjuvant. There is a significant delay between the exposure to CpG DNA and the onset of inhibition of antigen processing, and bacterial antigens could be effectively processed during this interval. In addition, our in vitro system examines the direct effects of CpG ODN on processing and presentation functions of isolated macrophages. In vivo, the interactions of multiple cell types must be considered, and the in vivo effects of CpG DNA may include indirect effects on macrophages via stimulation of other cell types, e.g. NK cells, to produce IFN-γ. For example, the ability of CpG DNA to upregulate transcription of inducible nitric oxide synthase in macrophages is detected only when the macrophages are first treated with IFN-γ (Stacey et al. 1996). Similarly, other indirect effects may occur in vivo to counteract the direct inhibitory effects of CpG DNA on antigen processing. Alternatively, an inhibitory effect may be the predominant effect of CpG ODN on macrophage antigen processing in vivo. As such, the macrophage response to bacterial DNA may have evolved to produce a macrophage activation state characterized by other beneficial anti-microbial functions, such as cytokine secretion and microbicidal activity, rather than an upregulation in antigen presentation.

Even if macrophage antigen processing is not enhanced by CpG ODN, it is necessary to consider the multiplicity of APC types available in vivo. Some of the adjuvant properties of CpG ODN may still be explained by enhanced antigen processing and presentation by other types of APCs. In fact, previous studies have shown that CpG ODN can increase expression of co-stimulator molecules on B cells (Davis et al. 1998) and increase expression of both MHC-II and co-stimulator molecules on dendritic cells (Jakob et al. 1998; Sparwasser et al. 1998). We have recently investigated the effects of CpG ODN on the antigen processing and presentation functions of dendritic cells, and these studies have shown effects quite different from those we have observed in macrophages.

Dendritic cells exhibit extensive changes in their ability to process and present antigen as a function of their maturation. Immature dendritic cells reside in the periphery, express a high level of endocytic activity; however, they express lower levels of MHC-II than do mature dendritic cells, and they lack co-stimulatory molecules required for antigen presentation to primary T cells (Cella et al. 1997a,b; Pierre et al. 1997; Banchereau and Steinman 1998). Upon introduction of a maturation signal, such as that initiated by bacteria or bacterial products (LPS or CpG DNA) (Cella et al. 1997a), dendritic cells migrate to lymphoid tissues and mature. Maturation of dendritic cells involves decreased endocytic activity, decreased expression of the mannose receptor, FcR (CD32 and CD23), CD14, and intracellular MHC-II molecules, increased expression of cell surface MHC-I and MHC-II molecules, and increased expression of co-stimulatory molecules, such as CD40, CD80, and CD86 (Sallusto et al. 1995; Henderson et al. 1997; Winzler et al. 1997; Jakob et al. 1998; Rescigno et al. 1998; Sparwasser et al. 1998). Mature dendritic cells also have increased production of cytokines, such as IL-1, IL-6, IL-10, IL-12, and TNF (Henderson et al. 1997; Sparwasser et al. 1997a; Jakob et al. 1998; Rescigno et al. 1998).

Recent studies by Askew et al. (manuscript in preparation) have examined the effects of CpG ODN on the maturation of bone-marrow-derived dendritic cells and their ability to process and present protein antigens. Murine bone-marrow-derived dendritic cells were cultured in a medium containing GM-CSF with or without the addition of CpG ODN. CpG ODN induced the differentiation of cells with dendritic morphology and increased expression of MHC-II, CD80, CD86, and CD40. In addition, CpG ODN enhanced the processing of HEL or RNase when the antigen was added at the same time as the CpG ODN. After exposure to CpG ODN for 24–48h, however, dendritic cells displayed decreased antigen-processing function. More detailed antigen-processing studies suggest that the long-term inhibitory effects on dendritic cell antigen processing particularly involve an inhibition of processing mechanisms that use newly synthesized MHC-II molecules, whereas processing mechanisms that use recycling MHC-II appear to be maintained in mature dendritic cells. In summary, CpG ODN induced dendritic cell maturation that was accompanied by a transient increase in antigen-processing function, followed by a decrease in certain antigen-processing activities. However, the mature dendritic cells maintained high levels of expression of MHC-II and co-stimulator molecules and had the ability to present antigens that were processed earlier. Thus, the ability of CpG ODN to transiently increase antigen-processing activity by dendritic cells may contribute to the adjuvant activity of CpG ODN.

## 5 Conclusion

Cpg ODN activate immune cells and induce production of IL-6, IL-12, and IFN-g by splenocytes in vitro. As development of Th1 cells is dependent upon the presence of IL-12 at the time of antigenic stimulation, we investigated whether CpG ODN could act as Th1-directing adjuvants in vivo. Administration of CpG ODN with protein antigen caused a Th1-dominated cytokine response by splenocytes, with increased production of IFN-g and decreased production of IL-5 upon antigenic recall stimulation in vitro. In addition, CpG ODN induced production of antigen-specific lgG2a, a Th1-associated isotype. These data indicate that CpG ODN direct Th1 responses against simultaneously delivered antigen.

One mechanism for the adjuvant activity of CpG ODN may be modulation of antigen processing and presentation by antigen presenting cells such as macrophages, dendritic cells, or B cells. We investigated the effects of CpG ODN on macrophages and dendritic cells. Overnight treatment of macrophages with CpG ODN resulted in an inhibition of antigen processing and presentation that was due in part to a decrease in synthesis of MHC-II molecules, and no enhancement of antigen processing by macrophages was detexted at any time point. In contrast, treatment of dendritic cells with CpG ODN induced dendritic cell maturation and a transient increase in antigen processing, followed by a decline in antigen processing with maintained presentation of previously processed antigen.

CpG ODN have been shown to act as powerful adjuvants for the development of Th1 responses. They are, therefore, potentially useful in vaccine design for the induction of protective cellular immunity or in clinical situations where IFN-$\gamma$-dominated Th responses are desirable (atopic disease). Future studies in humans will determine the usefulness of these agents in altering human disease states and will, at the same time, provoke few toxic or harmful effects. Additional studies are also needed to elucidate the mechanisms of the Th1 and general adjuvant effects of CpG ODN. In addition to the modulation of cytokine secretion, one of these mechanisms may be the upregulation of antigen processing and presentation by dendritic cells.

## References


Abbas AK, Murphy KM, Sher A (1996) Functional diversity of helper T lymphocytes. Nature 383: 787–793

Ada G, Ramsay A (1997) Immunopotentiation and the selective induction of immune responses. In: Ada G, Ramsay A (eds), Vaccines, vaccination and the immune response. Lippincott-Raven, Philadelphia pp. 122–136

Askew DA, Chu RS, Krieg AM, Harding CV (1999) CpG DNA activates dendritic cells with distinct effects on nascent and recycling MHC-II antigen processing. In preparation

Ballas ZK, Rasmussen WL, Krieg AM (1996) Induction of NK activity in murine and human cells by CpG motifs in oligodeoxynucleotides and bacterial DNA. J Immunol 157:1840–1845

Banchereau J, Steinman RM (1998) Dendritic cells and the control of immunity. Nature 392:245–252

Bliss J, Van Cleave V, Murray K, Wiencis A, Ketchum M, Maylor R, Haire T, Resmini C, Abbas AK, Wolf SF (1996) IL-12, as an adjuvant, promotes a T helper 1 cell, but does not suppress a T helper 2 cell recall response. Journal of Immunology 156:887–894

Cella M, Engering A, Pinet V, Peiters J, Lanzavecchia A (1997a) Inflammatory stimuli induce accumulation of MHC class-II complexes on dendritic cells. Nature 388:782–787

Cella M, Sallusto F, Lanzavecchia A (1997b) Origin, maturation and antigen presenting function of dendritic cells. Current Opinion in Immunology 9:10–16

Chace JH, Hooker NA, Mildenstein KL, Krieg AM, Cowdery JS (1997) Bacterial DNA-induced NK cell IFN-$\gamma$ production is dependent on macrophage secretion of IL-12. Clin Immunol Immunopathology 84:185–193

Chu RS, Askew D, Noss EH, Tobian A, Krieg AM, Harding CV (1999) CpG oligodeoxynucleotides downregulate macrophage class-II MHC antigen processing. J Immunol 163:1188–1194

Chu RS, Targoni OT, Krieg AM, Lehmann PV, Harding CV (1997) CpG oligodeoxynucleotides act as adjuvants that switch on T helper 1 (Th1) immunity. Journal of Experimental Medicine 186:1623–1631

Cowdery JS, Chace JH, Yi A-K, Krieg AM (1996) Bacterial DNA induces NK cells to produce IFN$\gamma$ in vivo and increases the toxicity of lipopolysaccharides. J Immunol 156:4570–4575

Davis HL, Weeranta R, Waldschmidt T, Tygrett L, Schorr J, Krieg AM (1998) CpG DNA is a potent enhancer of specific immunity in mice immunized with recombinant hepatitis B surface antigen. J Immunol 160:870–876

Finkelman FD (1995) Relationships among antigen presentation, cytokines, immune deviation, and autoimmune disease. J Exp Med 182:279–282

Forsthuber T, Yip HC, Lehmann PV (1996) Induction of $T_H1$ and $T_H2$ immunity in neonatal mice. Science 271:1728–1730

Halpern MD, Kurlander RJ, Pisetsky DS (1996) Bacterial DNA induces murine interferon-$\gamma$ production by stimulation of interleukin-12 and tumor necrosis factor-$\alpha$. Cell Immunol 167:72–78

Henderson RA, Watkins SC, Flynn JL (1997) Activation of human dendritic cells following infection with *Mycobacterium tuberculosis*. Journal of Immunology 159:635–643

Jakob T, Walker PS, Krieg AM, Udey MC, Vogel JC (1998) Activation of cutaneous dendritic cells by CpG-containing oligodeoxynucleotides: a role for dendritic cells in the augmentation of Th1 responses by immunostimulatory DNA. Journal of Immunology 161:3042–3049

Kelso A (1995) Th1 and Th2 subsets: Paradigms lost? Immunology Today 16:374–379

Kline JN, Waldschmidt TJ, Businga TR, Lemish JE, Weinstock JV, Thorne PS, Krieg AM (1998) Modulation of airway inflammation by CpG oligodeoxynucleotides in a murine model of asthma. Journal of Immunology 160:2555–2559

Klinman DM, Yi A-K, Beaucage SL, Conover J, Krieg AM (1996) CpG motifs present in bacterial DNA rapidly induce lymphocytes to secrete interleukin 6, interleukin 12, and interferon $\gamma$. Proc Natl Acad Sci USA 93:2879–2883

Koerner TJ, Hamilton TA, Adams DO (1987) Suppressed expression of surface Ia on macrophages by lipopolysaccaride: evidence for regulation at the level of accumulation of mRNA. Journal of Immunology 139:239–243

Krieg AM, Yi A-K, Matson S, Waldschmidt TJ, Bishop GA, Teasdale R, Koretzky GA, Klinman DM (1995) CpG motifs in bacterial DNA trigger direct B cell activation. Nature 374:546–549

Liblau RS, Singer SM, McDevitt HO (1995) Th1 and Th2 $CD4^+$ T cells in the pathogenesis of organ-specific autoimmune diseases. Immunology Today 16:34–38

Lipford GB, Bauer M, Blank C, Reiter R, Wagner H, Heeg K (1997) CpG-containing synthetic oligonucleotides promote B and cytotoxic T cell responses to protein antigen: a new class of vaccine adjuvants. European Journal of Immunology 27:2340–2344

Mosmann TR, Coffman RL (1989) TH1 and TH2 cells: different patterns of lymphokine secretion lead to different functional properties. Annual Review of Immunology 7:145–173

O'Garra A (1998) Cytokines induce the development of functionally heterogeneous T helper cell subsets. Immunity 8:275–283

Pierre P, Turley SJ, Gatti E, Hull M, Meltzer J, Mirza A, Inaba K, Steinman RM, Mellman I (1997) Developmental regulation of MHC class II transport in mouse dendritic cells. Nature 388:787–792

Raz E, Tighe H, Sato Y, Dudler JA, Roman M, Swain SL, Spiegelberg HL, Carson DA (1996) Preferential induction of a $Th_1$ immune response and inhibition of specific IgE antibody formation by plasmid DNA immunization. Proc Natl Acad Sci USA 93:5141–5145

Rescigno M, Citterio S, Thery C, Rittig M, Medaglini D, Pozzi G, Amigorena S, Ricciardi-Castagnoli P (1998) Bacteria-induced neo-biosynthesis, stabilization, and surface expression of functional class I molecules in mouse dendritic cells. Proc Natl Acad Sci USA 95:5229–5234

Roman M, Martin-Orozco E, Goodman JS, Nguyen M-D, Sato Y, Ronaghy A, Kornbluth RS, Richman DD, Carson DA, Raz E (1997) Immunostimulatory DNA sequences function as T helper-1-promoting adjuvants. Nature Med 3:849–854

Sallusto F, Cella M, Danieli C, Lanzavecchia A (1995) Dendritic cells use macropinocytosis and the mannose receptor to concentrate macromolecules in the major histocompatibility complex class-II compartment: downregulation by cytokines and bacterial products. Journal of Experimental Medicine 182:389–400

Sarmiento UM, Perez JR, Becker JM, Narayanan R (1994) In vivo toxicological effects of rel A antisense phosphorothioates in CD-1 mice. Antisense research development 4:99–107

Sato Y, Roman M, Tighe H, Lee D, Corr M, Nguyen M-D, Silverman GJ, Lotz M, Carson DA, Raz E (1996) Immunostimulatory DNA sequences necessary for effective intradermal gene immunization. Science 273:352–354

Segal BM, Klinman DM, Shevach EM (1997) Microbial products induce autoimmune disease by an IL-12-dependent pathway. Journal of Immunology 158:5087–5090

Sicher SC, Vazquez MA, Lu CY (1994) Inhibition of macrophage Ia expression by nitric oxide. Journal of Immunology 153:1293–1300

Sparwasser T, Koch E-S, Vabulas RM, Heeg K, Lipford GB, Ellwart JW, H. W (1998) Bacterial DNA and immunostimulatory CpG oligonucleotides trigger maturation and activation of murine dendritic cells. Eur J Immunol 28:2045–2054

Sparwasser T, Miethke T, Lipford G, Borschert K, Hcker H, Heeg K, Wagner H (1997a) Bacterial DNA causes septic shock. Nature 386:336–337

Sparwasser T, Miethke T, Lipford G, Erdmann A, Haecker H, Heeg K, Wagner H (1997b) Macrophage sense pathogens via DNA motifs: induction of tumor-necrosis factor-α-mediated shock. European Journal of Immunology 27

Stacey KJ, Sweet MJ, Hume DA (1996) Macrophages ingest and are activated by bacterial DNA. J Immunol 157:2116–2122

Steeg PS, Johnson HM, Oppenheim JJ (1982) Regulation of murine macrophage Ia antigen expression by an immune interferon-like lymphokine: inhibitory effect of endotoxin. Journal of Immunology 129:2402–2406

Sun S, Beard C, Jaenisch R, Jones P, Sprent J (1997) Mitogenicity of DNA from different organisms for murine B cells. Journal of Immunology 159:3119–3125

Sun S, Kishimoto H, Sprent J (1998) DNA as an adjuvant: capacity of insect DNA and synthetic oligodeoxynucleotides to augment T cell responses to specific antigen. Journal of Experimental Medicine 187:1145–1150

Winzler C, Rovere P, Rescigno M, Granucci F, Penna G, Adorini L, Zimmermann VS, Davoust J, Ricciardi-Castagnoli (1997) Maturation stages of mouse dendritic cells in growth factor-dependent long-term cultures. Journal of Experimental Medicine 185:317–328

Yamamoto S, Yamamoto T, Shimada S, Kuramoto E, Yano O, Kataoka T, Tokunaga T (1992) DNA from bacteria, but not from vertebrates, induces interferons, activates natural killer cells and inhibits tumor growth. Microbiol Immunol 36:983–997

Yi A-K, Krieg AM (1998) Rapid induction of mitogen-activated protein kinases by immune stimulatory CpG DNA. Journal of Immunology 161:4493–4497

Zimmerman S, Egeter O, Hausmann S, Lipford GB, Rocken M, Wagner H, Heeg K (1998) CpG oligodeoxynucleotides trigger protective and curative Th1 responses in lethal murine leishmaniasis. Journal of Immunology 160:3627–3630

# Effects of CpG DNA on Th1/Th2 Balance in Asthma

J.N. Kline



## 1 Introduction

Asthma is a disease whose prevalence, morbidity, mortality, and cost have increased in the past two decades worldwide. These changes have led to increased attention to the causes of asthma and asthma exacerbations; it is now appreciated that airway inflammation is central to the problem of asthma. Recent international guidelines for the management of asthma emphasize that anti-inflammatory therapies are necessary for most asthma patients, and this focus is slowly improving compliance from asthmatics and their health-care providers. Unfortunately, current anti-inflammatory treatment in asthma is non-specific. In this chapter, the Th1/Th2 paradigm of inflammation, which may govern the eosinophilic inflammation found in asthma, is described. Data are presented that demonstrate that

---

Division of Pulmonary Medicine, University of Iowa College of Medicine, C33GH UIHC, 200 Newton Drive, Iowa City, IA 52242, USA
E-mail: joel-kline@uiowa.edu

CpG oligodeoxynucleotides (ODN) are capable of modulating the Th2-mediated inflammatory response in a murine model of asthma. These studies suggest that CpG DNA may be useful in the prevention and treatment of asthma.

## 2 Background

### 2.1 Asthma: Statement of the Problem

Asthma has attracted a great deal of attention from both the public and from the medical community in the past few years. It has been termed an "epidemic" and has been the subject of cover stories in major newspapers and magazines. This is primarily due to the observation that the disease is worsening, particularly in Western, industrialized nations. In the past three decades, the prevalence, severity, and mortality of asthma have increased significantly (EVANS et al. 1987). A recent study estimated the total annual cost in the United States at almost $6 billion (SMITH et al. 1997). Asthma is a disease characterized by eosinophilic airway inflammation, bronchospasm, and bronchial hyper-reactivity (National Asthma Education and Prevention Program 1997). Once thought to be due to airway muscle spasm, asthma is now known to be an inflammatory disorder; during an asthma exacerbation, inflammation precedes bronchospasm. In acute asthma, eosinophils may form up to half of the cellular infiltrate (ROBBINS and COTRAN 1979), and bronchoalveolar eosinophilia invariably follows allergen inhalation in asthma attacks (METZGER et al. 1985, 1986, 1987). Eosinophils cause inflammation and bronchial hyper-reactivity through release of mediators, such as leukotrienes, major basic protein, eosinophilic cationic protein, and eosinophilic peroxidase (BRUIJNZEEL 1994).

### 2.2 Th1 and Th2 Balance in Asthma

The number and activity of eosinophils are controlled by cytokines released from activated T cells, especially interleukin 4 (IL-4), IL-5, and IL-13. T-lymphocytes can be divided, on the basis of cytokine production, into T helper 1 (Th1) and Th2 (MOSMANN et al. 1986). Th1 cells produce IL-2 and interferon $\gamma$ (IFN-$\gamma$) but not IL-4 or IL-5, and Th2 cells produce IL-4, IL-5, IL-6, IL-10, and IL-13 but not IL-2 or IFN-$\gamma$. Th1 and Th2 cells interact in a counter-regulatory fashion: IL-4 and IL-10 promote Th2 development (SWAIN et al. 1990; PARRONCHI et al. 1992) and inhibit Th1 cell and cytokine production (MOORE et al. 1990), and IFN-$\gamma$ inhibits the proliferation of Th2 cells (GAJEWSKI and FITCH 1988) and promotes the development of Th1 cells (PARRONCHI et al. 1992). IL-12, mainly a product of activated macrophages, is also a strong promoter of Th1 responses (BLISS et al. 1996) and is often considered a Th1 cytokine; many of the activities ascribed to IL-12 are due to induction of IFN-$\gamma$ (Micallef et al. 1996). Th1 and Th2 cells have been identified in humans in vivo as well as in vitro (Wierenga et al. 1990; Del Prete et al. 1991; Romagnani 1991; Salgame et al. 1991; Field et al. 1993).

The Th2 cytokines, IL-4, IL-5, and IL-13 (Robinson et al. 1992; Grunig et al. 1998; Wills-Karp et al. 1998), have been increasingly implicated in the inflammation of asthma. IL-4 amplifies allergic responses by inducing immunoglobulin E (IgE) production by uncommitted B cells (Del Prete et al. 1988) and is a growth factor for mast cells (Saito et al. 1988). IL-5 also stimulates Ig secretion (Takatsu et al. 1980; Swain et al. 1990) as well as stimulating the proliferation and activation of eosinophils (Clutterbuck et al. 1988; Lopez et al. 1988; Walsh et al. 1990) and basophils (Hirai et al. 1990). IL-13 has recently been shown to cause airway hyperresponsiveness and inflammation independently of eosinophils or IL-4 (Grunig et al. 1998; Wills-Karp et al. 1998). In vitro, allergen-specific T-cell clones from atopic donors release Th2 cytokines after stimulation by specific allergens (Parronchi et al. 1991). During asthma exacerbations, peripheral T-cell activation and increased serum IL-5 correlate with eosinophilia and asthma symptoms (Corrigan et al. 1993), and bronchoalveolar lavage (BAL) T-cells release cytokines in a Th2-like pattern (Robinson et al. 1992, 1993a,b). Non-atopic asthmatics also have increased levels of Th2-like cytokines; increased IL-5 release from BAL T lymphocytes is characteristic of both atopic and non-atopic asthmatics (Walker et al. 1992), and expression of peripheral T-cell IL-5 mRNA from non-atopic asthmatic subjects correlates with increased BAL (Marini et al. 1992) and peripheral blood (Walker et al. 1992) eosinophilia.

## 2.3 Inflammation and the Treatment of Asthma

Because of these observations, the focus of treatment in asthma has shifted from primarily addressing bronchospasm to modulating inflammation. Recent guidelines for the management of asthma recommend that anti-inflammatory therapy be used for all but the most intermittent and benign cases of the disease (National Asthma Education and Prevention Program 1997). Current anti-inflammatory therapy, however, remains disappointingly broad; corticosteroids are the "gold standard" for asthma treatment, and inhaled corticosteroids are only incrementally better than they were 25 years ago. The much-touted leukotriene pathway antagonists, which have been released in the last 5 years, have been helpful only in a subset of asthmatics. Thus, the need remains for novel, effective treatments of inflammation in asthma.

## 2.4 DNA and Th1 Responses

The observation that ODN containing CpG motifs (the dinucleotide CpG in specific base-sequence contexts) have remarkable immunostimulatory properties, including induction of Th1 cytokines, has given rise to a series of studies examining how these agents may be useful in the modulation or abrogation of Th2-mediated inflammation in asthma. Bacterial, but not vertebrate, DNA causes immune activation (KRIEG et al. 1995); the dinucleotide CpG is markedly under-represented in vertebrate DNA and, when present, is almost invariably methylated (KRIEG et al. 1995). ODN containing unmethylated CpG motifs reproduce the effect of bacterial DNA strongly and rapidly; CpG DNA induces cytokine mRNA within 30min of administration (YI and KRIEG 1998), and IL-6 and IL-12 protein secretion is increased within 4h after in vivo treatment with CpG DNA (KLINMAN et al. 1996). CpG-DNA-induced IL-12 production precedes and is responsible for the subsequent natural killer cell IFN-$\gamma$ secretion (BALLAS et al. 1996; HALPERN et al. 1996; KLINMAN et al. 1996).

## 2.5 Early-Life Infections and Th1 Responses

The possibility that CpG DNA of prokaryotic origin may be protective against asthmatic inflammation is counterintuitive; indeed, it has been suggested that the pathogenesis of asthma may be related to poor control of airway infections. However, a number of studies have examined the association between childhood infection (resulting in early exposure to viral or prokaryotic DNA) and asthma, bolstering the hypothesis that this exposure may result in the conversion of a propensity for a Th2 response to allergen into a Th1 or non-atopic response. Von Mutius and colleagues hypothesized that worsening air pollution may be responsible for the increased morbidity of asthma in Western nations and compared the prevalence of asthma and atopy between children of Leipzig, in polluted former East Germany, and Munich, in less polluted former West Germany (VON MUTIUS et al. 1992). Surprisingly, a lower rate of asthma and allergic disorders was found among children from Leipzig than in Munich children. The investigators postulated that this finding might result from differences in living conditions rather than environmental exposures. In a follow-up study, they examined the relationship between atopy and number of siblings and found that children with increased numbers of siblings were less likely to express atopy (VON MUTIUS et al. 1994). As children with multiple older siblings were more likely to develop childhood illnesses earlier in life than those without siblings (the older siblings bring home infections from school or daycare), it is possible that the early-life infections prevent the later development of atopy. These results suggested an interpretation of their earlier study: the Leipzig children (who were more likely to be in out-of-home daycare than the Munich children) may also have been protected against atopy by an increase in early-life childhood infections.

A more direct examination of the relationship between infection and atopy was carried out by Shaheen and colleagues (SHAHEEN et al. 1996). A measles epidemic occurred in Guinea-Bissau in 1979. Subsequently, children were evaluated for the development of atopy, which was defined by reactivity to skin testing for common environmental allergens. After controlling for potentially confounding variables, such as numbers of older siblings, history of breast feeding, and in-home exposure to livestock, a history of measles infection was found to be associated with a significant reduction in the risk of atopy (relative risk 0.20, $P < 0.01$) (SHAHEEN et al. 1996). Finally, in a much-quoted recent study, Shirakawa and colleagues found that a positive response to tuberculin predicted a lower incidence of asthma and atopy in a cohort study of Japanese schoolchildren (SHIRAKAWA et al. 1997). Moreover, the tuberculin responsiveness correlated with induction of Th1 (IFN-γ) and suppression of Th2 (IL-4, IL-13) serum cytokine levels. This exciting study, while not definitive evidence, strongly supports a link between childhood infection and protection against the development of atopy.

# 3 Results

## 3.1 CpG ODN Prevent Eosinophilic Airway Inflammation in Asthma

Based on these epidemiological studies and observations that CpG ODN were potent inducers of the Th1 cytokines IL-12 and IFN-γ, we elected to examine the effects of CpG DNA in an animal model of asthma. For the initial studies, we adapted a murine model in which C57BL/6 mice were sensitized to *Schistosoma mansoni* proteins by intraperitoneal injection of schistosome eggs (5,000 eggs). The mice subsequently received two airway challenges with soluble egg antigen from the schistosome eggs (SEA; 10μg) (LUKACS et al. 1994; KLINE et al. 1998). Following exposure to the second airway challenge, mice were studied. Some mice received ODN (30μg, either CpG or control, on a phosphothioate backbone) by intraperitoneal injection at the time of sensitization. For these studies, the CpG ODN used was TCCATGACGTTCCTGACGTT, and the control ODN was TCCATGA*GC*TTCCTGA*GT*C*T*. In comparison with saline control mice, we found that significant eosinophilic pulmonary inflammation developed following airway challenge in the sensitized mice (Fig. 1). Although all cell types increased in lung lavage samples, a majority of the cells in the influx were eosinophils. This pleocytosis and eosinophilia were almost completely abrogated in the mice who received CpG ODN; interestingly, mice who received control ODN at the time of sensitization demonstrated a slight but consistent diminution of airway inflam-

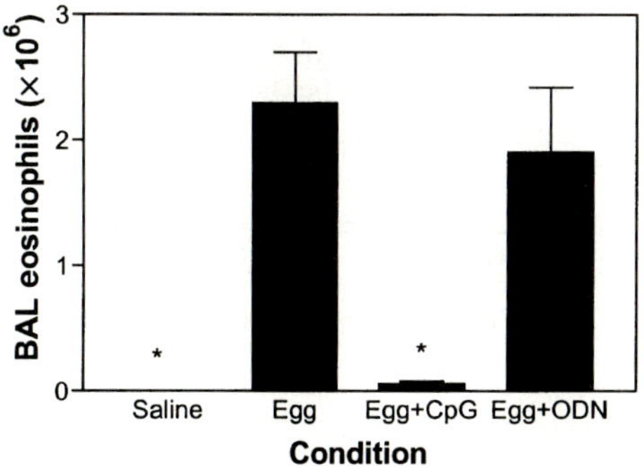


**Fig. 1.** CpG oligodeoxynucleotides (ODN) prevent the development of airway eosinophilia in a murine model of asthma. C57BL/6 mice who received CpG ODN at the time of sensitization, but not those who received control ODN, develop significantly less airway eosinophilia than mice sensitized to schistosome eggs in the absence of ODN. $^*P < 0.001$ vs the schistosome-egg-sensitized group

mation, suggesting the existence of a non-specific ODN effect. Histopathologic examination of the lungs of the mice confirmed the pronounced eosinophilic inflammation induced by schistosome sensitization and airway challenge; a marked peribronchial and perivascular inflammatory exudate included eosinophils, mast cells, macrophages, and lymphocytes, and the airway epithelium was thickened, with prominent mucus production (Fig. 2). These inflammatory changes were markedly reduced in those mice who received pretreatment with CpG ODN.

## 3.2 CpG ODN Prevent Bronchial Hyper-Reactivity

As asthma is defined by the presence of bronchial hyper-reactivity to non-specific stimuli in addition to eosinophilic airway inflammation, we next wanted to evaluate the effect of CpG ODN on the airway pathophysiologic responses found in this model. For these studies, we evaluated the response of the mice to inhaled methacholine and used a computer-interfaced whole-body plethysmograph (Buxco, Troy, NY, USA), which non-invasively analyzes the response to increasing doses of inhaled methacholine. "Enhanced pause" (Penh) is calculated as a function of expiratory time, relaxation time, and peak inspiratory and expiratory flow (Penh $= [(T_e/0.3T_r)-1] \times [2P_{ef}/3P_{if}]$); Penh correlates with directly measured

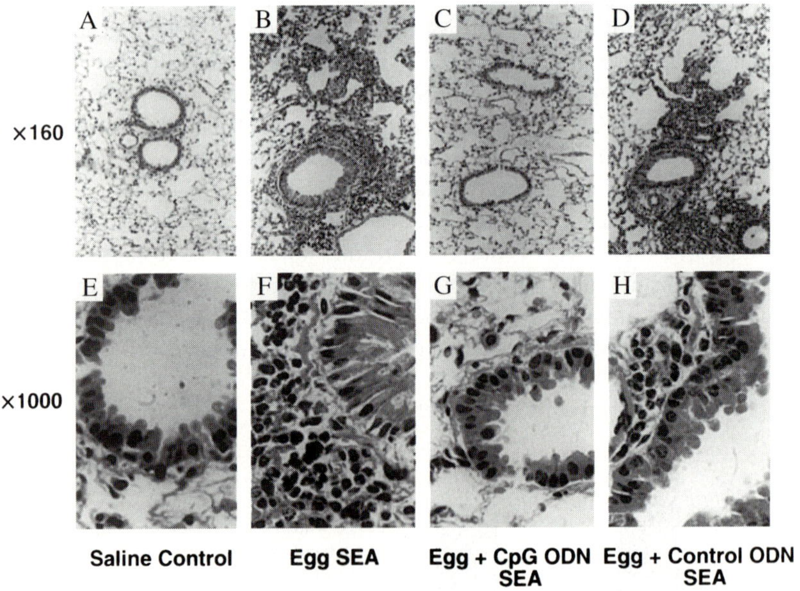


**Fig. 2A–H.** CpG oligodeoxynucleotides (ODN) prevent the development of peribronchial eosinophilia in a murine model of asthma. Compared with saline control mice (**A, E**), those mice who are sensitized to schistosome eggs and then are challenged with soluble egg antigen in the airway (**B, F**) develop marked peribronchial-eosinophilic inflammation and epithelial activation, which is markedly reduced in CpG-ODN-treated (**C, G**) but not control-ODN-treated (**D, H**) mice bronchospasm (Hamelmann et al. 1997; Kline et al. 1998). Although C57BL/6 mice are relatively insensitive to inhaled methacholine, those mice who are sensitized and later challenged with schistosome proteins have a marked increase in response to methacholine (Fig. 3); pretreatment with CpG ODN, but not control ODN, returned the bronchial reactivity nearly to baseline. These studies demonstrated that CpG ODN could prevent the physiologic sequellae of inflammation and the inflammatory response itself.

### 3.3 CpG ODN Suppress Pulmonary and Systemic Th2 Responses

Because induction by CpG ODN of Th1 cytokines, such as IFN-γ and IL-12, had previously been demonstrated, we next examined the effects that CpG ODN have on the regulation of cytokines in the lung. For these studies, we measured the concentration of IL-4, IFN-γ, and IL-12 in lung lavage fluid. We found (Table 1) that IL-4 was strongly induced in the sensitized/challenged mice. This induction was greatly diminished in mice that were treated with CpG ODN. Conversely, both IFN-γ and IL-12 were greatly upregulated in the lavage fluid of mice treated with CpG

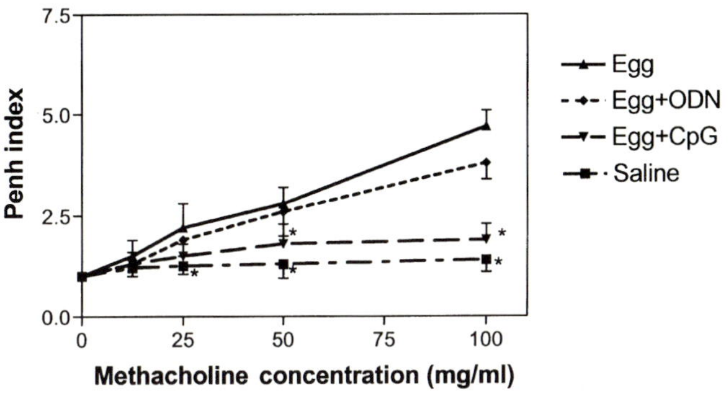


**Fig. 3.** CpG oligodeoxynucleotides (ODN) inhibit the development of bronchial hyper-reactivity to inhaled methacholine. Mice that are sensitized to schistosome eggs and challenged with soluble egg antigen in the airway develop bronchial hyper-reactivity (demonstrated as increased Penh index or fold increase in Penh, a measure of airway resistance) to inhaled methacholine, compared with saline-control mice. This response is markedly blunted in mice treated with CpG ODN but not in those mice treated with control ODN. $*P < 0.05$ vs the schistosome-egg-sensitized group

**Table 1.** Th1 and Th2 cytokine levels in bronchoalveolar lavage fluid

| Condition of sensitization | IL-4 (pg/ml) | IFN-γ (pg/ml) | IL-12 (pg/ml) |
|---|---|---|---|
| Saline | 5.3 ± 2.1* | 8.4 ± 3.1 | 6.9 ± 3.4 |
| CpG ODN + egg | 23.6 ± 4.7* | 403.2 ± 107.5** | 57.6 ± 19.7** |
| Control ODN + egg | 82.3 ± 17.6 | 107.6 ± 82.4 | 20.4 ± 7.1 |
| Schistosome egg | 115.8 ± 24.9 | 15.2 ± 9.9 | 18.3 ± 4.6 |

IFN, interferon; IL, interleukin; ODN, oligodeoxynucleotide.
$*P < 0.05$ vs schistosome-egg-sensitized mice.
$**P < 0.01$ vs schistosome-egg-sensitized mice.

ODN (Table 1). These data suggested that induction of the Th1 cytokines, IFN-γ and IL-12, were responsible for the anti-inflammatory effects of CpG ODN in the murine model of asthma and that their effect was downregulation of Th2 responses.

Humoral Th1 and Th2 responses also differ from one another. Humoral Th2 responses are characterized by induction of IgE antibodies, and Th1 responses are characterized by induction of IgG2a and IgG2b antibodies. In the schistosome murine model of asthma, total IgE is induced, and there is no change in the total IgG2a or IgG2b. Treatment with CpG ODN significantly reduced the induction of IgE (from 4.10 ± 0.54μg/ml to 1.04 ± 0.32μg/ml, $P < 0.05$) but had no effect on IgG subtypes.

## 3.4 The Effect of CpG ODN in Murine Asthma is Neither Strain nor Model Dependent

Because we wished to ascertain whether the effect of CpG ODN on atopic airway inflammation was strain-specific, we next examined the effect of the ODN on the development of airway inflammation and bronchial hyper-reactivity using the schistosome egg/SEA protocol on BALB/c mice. These mice developed less airway eosinophilia than C57BL/6 mice do, but their eosinophilia was completely abrogated by the CpG ODN, demonstrating that the protection offered by CpG ODN is not strain specific ("asthmatic" BALB/c mice developed $1.2 \pm 0.4 \times 10^6$ eosinophils/lavage; CpG-treated mice developed $0.06 \pm 0.04 \times 10^6$ eosinophils/lavage, $P < 0.001$). Moreover, using an ovalbumin murine model of asthma, we also examined whether this protection was model specific. For these studies, C57BL/6 mice were sensitized to ovalbumin with a single intraperitoneal injection of ovalbumin (10μg) heat-precipitated with alum; this was followed by a period of ovalbumin inhalation (6% solution aerosolized in a chamber for 30 minutes daily on days 7–11 and 14–18). Using this model, we found that the eosinophilia induced by ovalbumin in sensitized mice was significantly reduced by CpG ODN, but not by control ODN (ovalbumin-sensitized mice: $2.6 \pm 0.7 \times 10^6$ eosinophils; CpG ODN-treated mice: $0.03 \pm 0.02 \times 10^6$ eosinophils, $P < 0.001$; control ODN-treated mice: $1.9 \pm 0.8 \times 10^6$ eosinophils, $P$ = no significance).

## 3.5 The Effects of CpG ODN are Persistent

To examine whether the effects of CpG ODN are long-lasting, we next sensitized a cohort of mice to schistosome eggs in the presence or absence of CpG or control ODN and challenged them as per our protocol at days 7 and 14; 4 weeks later, we repeated the process, but this time all the mice received schistosome eggs without ODN. Lavage of the mice after the second series of SEA inhalation revealed that those mice who received CpG ODN along with the sensitizing antigen received long-lasting protection against sensitization (Fig. 4). These mice were also protected against the development of bronchial hyper-reactivity (data not shown).

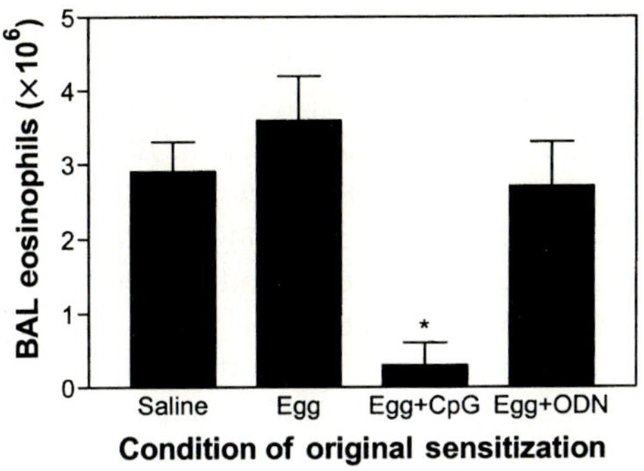


**Fig. 4.** Mice that receive CpG oligodeoxynucleotides (ODN) at the time of sensitization are protected against subsequent sensitization. C57BL/6 mice were sensitized to schistosome eggs in the presence or absence of CpG ODN or control ODN or received saline alone as a control, and were then challenged with soluble egg antigen (SEA) in the airway. Four weeks later, all mice received a second intraperitoneal injection of schistosome eggs (no ODN) followed by two weekly airway challenges with SEA. Those mice that received the CpG ODN (but not those that received the control ODN) were initially protected against the development of eosinophilic inflammation even when challenged after the second exposure to schistosome eggs. $*P < 0.01$ vs the schistosome-egg-sensitized group

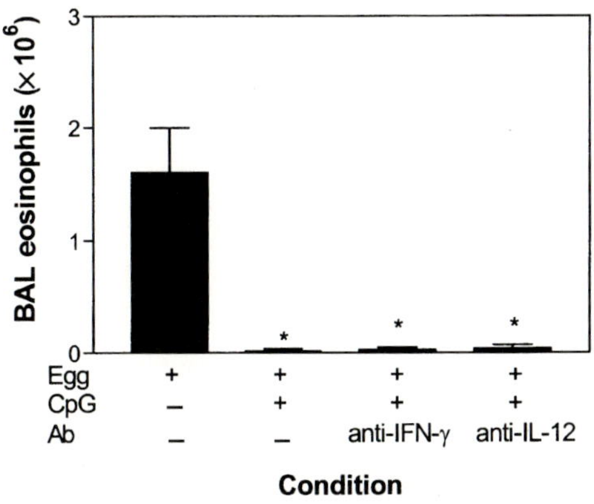


**Fig. 5.** CpG oligodeoxynucleotides (ODN) are effective in prevention of eosinophilic airway inflammation and bronchial hyper-reactivity in the presence of antibodies to interferon γ (IFN-γ) or interleukin 12 (IL-12). Administration of anti-IFN-γ or anti-IL-12 blocking antibodies does not significantly impair the effectiveness of CpG ODN in prevention of eosinophilic airway inflammation. $*P < 0.005$ vs the egg group. There is no significant difference between the three CpG groups

## 3.6 Neither IFN-γ nor IL-12 is Required for the Protective Effects of CpG ODN

To evaluate whether the Th1 cytokines, IFN-γ, and IL-12 are required for the protective effects of CpG ODN, we first examined whether they would prevent the development of airway eosinophilia and bronchial hyper-reactivity in the absence of IFN-γ. For these studies, we administered anti-IFN-γ antibodies (R4–6A2) at the time of the administration of the ODN and the schistosome eggs. These antibodies are known to specifically block the effects of IFN-γ in vivo (Spitalny and Havell 1984). To our surprise, we found that CpG ODN, in the absence of IFN-γ activity, were still effective at preventing airway eosinophilia (Fig. 5) and bronchial hyper-reactivity (not shown). To definitively evaluate whether CpG ODN could prevent the development of asthma-like inflammation in the absence of IFN-γ, we next used IFN-γ knockout (KO) mice (Dalton et al. 1993) in our asthma model. We found that these mice were also protected from the development of eosinophilic inflammation (Fig. 6A) and airway hyper-reactivity (Fig. 6B).

We performed similar studies to evaluate the role of IL-12 in the mediation of the CpG effect. These studies used the blocking antibody TOSH [a rat anti-murine IL-12 IgG1 antibody (Wenner et al. 1996)] or an isotype-specific control antibody; systemic blockade of IL-12 only modestly reduced the protective effects of CpG ODN on the prevention of airway eosinophilia (Fig. 5). Using IL-12 KO mice (Magram et al. 1996), we found that these mice were also protected from the development of schistosome/SEA-induced airway eosinophilia by administration of

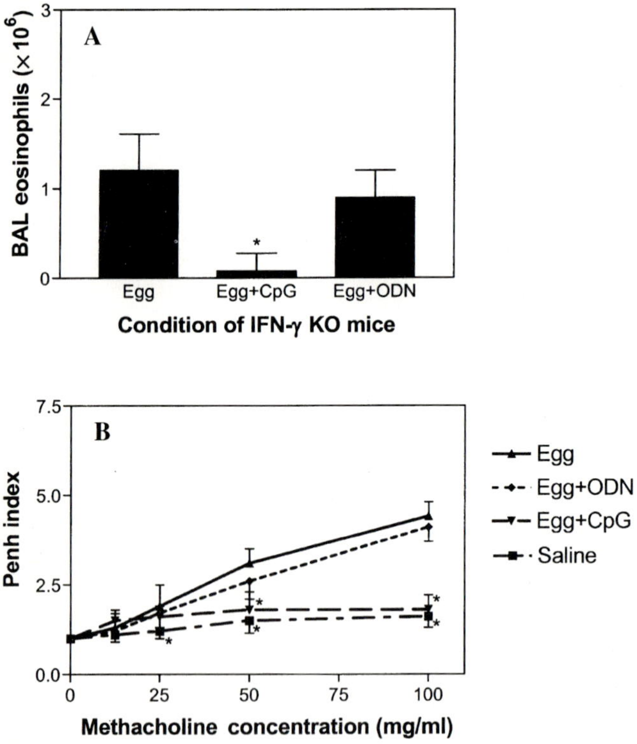


**Fig. 6A,B.** CpG oligodeoxynucleotides (ODN) are effective in prevention of eosinophilic airway inflammation and bronchial hyper-reactivity in interferon γ (IFN-γ) knockout (KO) mice. IFN-γ KO mice were sensitized to schistosome eggs in the presence or absence of CpG or control ODN. KO mice that received CpG, but not those who received control ODN at the time of sensitization, were protected against the development of (**A**) airway eosinophilia and (**B**) hyper-reactivity to inhaled methacholine. *$P < 0.01$ vs the schistosome-egg-sensitized group

CpG ODN but not control ODN (Fig. 7A). The CpG-treated IL-12 KO mice exposed to schistosome egg and then SEA also had diminished bronchial hyper-reactivity to inhaled methacholine compared with sensitized and challenged IL-12 KO mice who did not receive CpG ODN (Fig. 7B).

Since the protection conferred by CpG ODN on both IFN-γ KO and IL-12 KO mice was qualitatively similar to the protection offered to wild-type C57BL/6 mice, we next examined whether it was quantitatively identical. In the preceding studies, we used CpG ODN in an amount previously found to be well within the protective range, 30μg (Kline et al. 1998). Therefore, we proceeded to examine whether the dose-response relationship between the amount of CpG ODN used and protection against eosinophilia was altered in either or both of the cytokine KO mice. For these studies, we used the same protocol described above, but we administered varying amounts of CpG ODN (between 0.3μg and 30μg) at the time of sensitization to schistosome eggs (Fig. 8). We found that, at lower doses of CpG ODN ( < 10μg), both of the KO mice were offered significantly less protection against the development of airway eosinophilia than wild-type C57BL/6 mice.

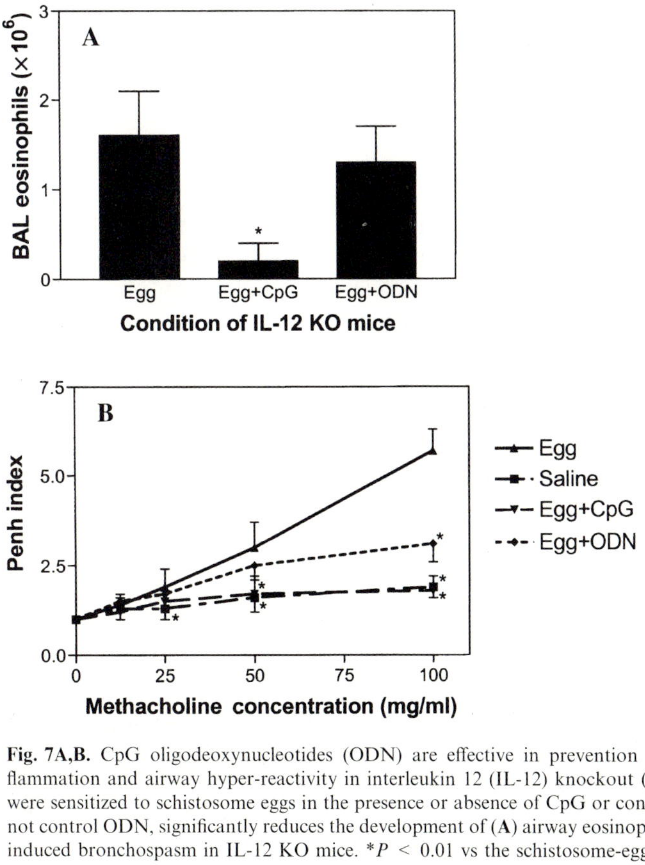


**Fig. 7A,B.** CpG oligodeoxynucleotides (ODN) are effective in prevention of eosinophilic airway inflammation and airway hyper-reactivity in interleukin 12 (IL-12) knockout (KO) mice. IL-12 KO mice were sensitized to schistosome eggs in the presence or absence of CpG or control ODN. CpG ODN, but not control ODN, significantly reduces the development of (**A**) airway eosinophilia and (**B**) methacholine-induced bronchospasm in IL-12 KO mice. *$P < 0.01$ vs the schistosome-egg-sensitized group

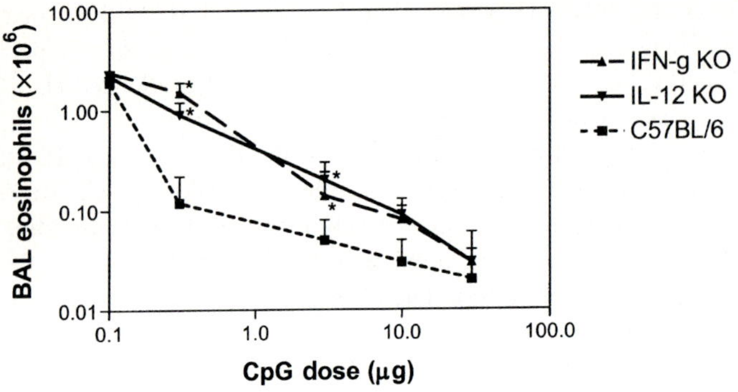


**Fig. 8.** In the absence of interferon γ (IFN-γ) or interleukin 12 (IL-12), mice are relatively deficient in response to CpG oligodeoxynucleotide (ODN). Varying amounts of CpG ODN (0, 0.3, 3, 10, or 30μg) were administered to wild-type C57BL/6, IFN-γ knockout (KO), and IL-12 KO mice (each on a C57BL/6 background) at the time of sensitization to schistosome eggs. All mice who received 30μg and the wild-type mice who received 10μg of CpG ODN were fully protected against airway eosinophilia. There was no difference in airway eosinophilia between any of the groups that received no CpG ODN. A significant difference in the development of airway eosinophilia was seen between the wild-type and KO mice groups that received between 0.3μg and 10μg of CpG ODN. $^{**}P < 0.005$ vs the egg group; $n = 4$ mice in each group

# 4 Summary

Thus, in our studies, we demonstrated that CpG ODN are effective in preventing the development of eosinophilic airway inflammation and bronchial hyper-reactivity in a murine model of asthma. Antigen-associated elevation of serum IgE levels is also suppressed. CpG ODN, administered in conjunction with antigen, is also effective in down-regulation of established Th2 responses. This protection is neither murine strain-dependent nor model-dependent. Although these effects of CpG ODN are associated with the induction of the Th1 cytokines IFN-γ and IL-12, neither cytokine is absolutely required for the protection. These results suggest that CpG ODN may be an effective immunomodulatory agent in the treatment, and possibly prevention, of asthma.

# References


Ballas ZK, Rasmussen WL, Krieg AM (1996) Induction of NK activity in murine and human cells by CpG motifs in oligodeoxynucleotides and bacterial DNA. J Immunol 157:1840–1845

Bliss J, Van Cleave V, Murray K, Wiencis A, Ketchum M, Maylor R, Haire T, Resmini C, Abbas AK, Wolf SF (1996) IL-12, as an adjuvant, promotes a T helper 1 cell, but does not suppress a T helper 2 cell recall response. J Immunol 156:887–894

Bruijnzeel PL (1994) Eosinophil tissue mobilization in allergic disorders. Ann N Y Acad Sci 725:259–267

Clutterbuck EJ, Hirst EMA, Sanderson CJ (1988) Human interleukin 5 (IL-5) regulates the production of eosinophils in human bone marrow cultures: comparison and interaction with IL-1, IL-3, IL-6, and GMCSF. Blood 73:1504–1513

Corrigan CJ, Haczku A, Gemou-Engesaeth V, Doi S, Kikuchi Y, Takatsu K, Durham SR, Kay AB (1993) CD4 T-lymphocyte activation in asthma is accompanied by increased serum concentrations of interleukin 5. Effect of glucocorticoid therapy. Am Rev Respir Dis 147:540–547

Dalton D, Pitts-Meek S, Keshav S, Figari I, Bradley A, Stewart T (1993) Multiple defects of immune cell function in mice with disrupted interferon-gamma genes. Science 259:1739–1745

Del Prete G, Maggi E, Parronchi P, Chretien I, Tiri A, Macchia D, Ricci M, Banchereau J, De VJ, Romagnani S (1988) IL-4 is an essential factor for the IgE synthesis induced in vitro by human T cell clones and their supernatants. J Immunol 140:4193–4198

Del Prete GF, De Carli M, Mastromauro C, Biagiotti R, Macchia D, Falagiani P, Ricci M, Romagnani S (1991) Purified protein derivative of Mycobacterium tuberculosis and excretory-secretory antigen(s) of *Toxocara canis* expand in vitro human T cells with stable and opposite (type 1T helper or type 2T helper) profile of cytokine production. J Clin Invest 88:346–350

Evans R, Mullally DI, Wilson RW, Gergen PJ, Rosenberg HM, Grauman JS, Chevarley FM, Feinleib M (1987) National trends in the morbidity and mortality of asthma in the U.S. Chest 91:65S–74S

Field EH, Noelle RJ, Rouse T, Goeken J, Waldschmidt T (1993) Evidence for excessive Th2 CD4+ subset activitity in vivo. J Immunol 151:48–59

Gajewski TF, Fitch FW (1988) Anti proliferative effect of IFN-$\gamma$ in immune regulation. I. IFN-$\gamma$ inhibits the proliferation of Th2 but not Th1 murine helper T lymphocyte clones. J Immunol 140:4245–4252

Grunig G, Warnock M, Wakil AE, Venkayya R, Brombacher F, Rennick DM, Sheppard D, Mohrs M, Donaldson DD, Locksley RM, Corry DB (1998) Requirement for IL-13 independently of IL-4 in experimental asthma. Science 282:2261–2263

Halpern MD, Kurlander RJ, Pisetsky DS (1996) Bacterial DNA induces murine interferon-gamma production by stimulation of interleukin-12 and tumor necrosis factor-alpha. Cell Immunol 167:72–78

Hamelmann E, Schwarze J, Takeda K, Oshiba A, Larsen GL, Irvin CG, Gelfand EW (1997) Noninvasive measurement of airway responsiveness in allergic mice using barometric plethysmography. Am J Resp Crit Care Med 156:766–775

Hirai K, Yamaguchi M, Misaki Y, Takaishi T, Ohta K, Morita Y, Ito K, Miyamoto T (1990) Enhancement of human basophil histamine release by interleukin 5. J Exp Med 172:1525–1528

Kline JN, Waldschmidt TJ, Businga TR, Lemish JE, Weinstock JV, Thorne PS, Krieg AM (1998) Modulation of airway inflammation by CpG oligodeoxynucleotides in a murine model of asthma. J Immunol 160:2555–2559

Klinman DM, Yi AK, Beaucage SL, Conover J, Krieg AM (1996) CpG motifs present in bacteria DNA rapidly induce lymphocytes to secrete interleukin 6, interleukin 12, and interferon gamma. Proceedings of the National Academy of Sciences of the United States of America 93:2879–2883

Krieg AM, Yi AK, Matson S, Waldschmidt TJ, Bishop GA, Teasdale R, Koretzky GA, Klinman DM (1995) CpG motifs in bacterial DNA trigger direct B-cell activation. Nature 374:546–549

Lopez AF, Sanderson CJ, Gamble JR, Campbell HR, Young IG, Vadas MA (1988) Recombinant human interleukin 5 is a selective activator of human eosinophil function. J Exp Med 167:219–223

Lukacs NW, Strieter RM, Chensue SW, Kunkel SL (1994) Interleukin-4-dependent pulmonary eosinophil infiltration in a murine model of asthma. Am J Respir Cell Mol Biol 10:526–532

Magram J, Connaughton SE, Warrier RR, Carvajal DM, Wu CY, Ferrante J, Stewart C, Sarmiento U, Faherty DA, Gately MK (1996) IL-12-deficient mice are defective in IFN gamma production and type 1 cytokine responses. Immunity 4:471–481

Marini M, Avoni E, Hollemborg J, Nattoli S (1992) Cytokine mRNA profile and cell activation in bronchoalveolar lavage fluid from nonatopic patients with symptomatic asthma. Chest 102:661–669

Metzger WJ, Hunninghake GW, Richerson HB (1985) Late asthmatic responses: inquiry into mechanisms and significance. Clin Rev All 3:145–165

Metzger WJ, Richerson HB, Worden K, Monick M, Hunninghake GW (1986) Bronchoalveolar lavage of allergic asthmatic patients following allergen bronchoprovocation. Chest 89:477–483

Metzger WJ, Zavala D, Richerson HB, Mosely P, Iwamota P, Monick MM, Sjoerdsma K, Hunninghake GW (1987) Local allergen challenge and bronchoalveolar lavage of allergic asthmatic lungs: description of the model and local airway inflammation. Am Rev Respir Dis 135:433–440

Micallef MJ, Ohtsuki T, Kohno K, Tanabe F, Ushio S, Namba M, Tanimoto T, Torigoe K, Fujii M, Ikeda M, Fukuda S, Kurimoto M (1996) Interferon-gamma-inducing factor enhances T helper 1 cytokine production by stimulated human T cells: synergism with interleukin-12 for interferon-gamma production. Eur J Immunol 26:1647–1651

Moore KW, Vieira P, Fiorenton DF, Trounstine ML, Khan TA, Mossmann TR (1990) Homology of cytokine synthesis inhibitory factor (IL-10) to the Epstein-Barr virus gene BCRF1. Science 248: 1230–1234

Mosmann TR, Cherwinski H, Bond MW, Giedlin MA, Coffman RL (1986) Two types of murine helper T cell clones. I. Definition according to profiles of lymphokine activities and secreted proteins. J Immunol 136:2348–2357

National Asthma Education and Prevention Program. (1997) Expert panel report 2: guidelines for the diagnosis and management of asthma. National Institutes of Health, Bethesda, MD.

Parronchi P, De Carli M, Manetti R, Simonelli C, Sampognaro S, Piccinni M-P, Macchia D, Maggi E, Del Prete G, Romagnani S (1992) IL-4 and IFN (α and γ) exert opposite regulatory effects on the development of cytolytic potential by Th1 or Th2 human T cell clones. J Immunol 149:2977–2983

Parronchi P, Macchia D, Piccinni M-P, Biswas P, Simonelli C, Maggi E, Ricci M, Ansari AA, Romagnani S (1991) Allergen- and bacterial antigen-specific T-cell clones established from atopic donors show a different profile of cytokine production. Proc Natl Acad Sci, USA 88:4538–4542

Robbins SL, Cotran RS (1979) Pathologic basis of disease. Saunders, Philadelphia.

Robinson DS, Hamid Q, Bentley A, Ying S, Kay AB, Durham SR (1993a) Activation of CD4+ T cells, increased TH2-type cytokine mRNA expression, and eosinophil recruitment in bronchoalveolar lavage after allergen inhalation challenge in patients with atopic asthma. J All Clin Immunol 92: 313–324

Robinson DS, Hamid Q, Ying S, Tsicopoulos A, Barkans J, Bentley AM, Corrigan C, Durham SR, Kay AB (1992) Predominant TH2-like bronchoalveolar T-lymphocyte population in atopic asthma. N Engl J Med 326:298–304

Robinson DS, Ying S, Bentley AM, Meng Q, North J, Durham SR, Kay AB, Hamid Q (1993b) Relationships among numbers of bronchoalveolar lavage cells expressing messenger ribonucleic acid for cytokines, asthma symptoms, and airway methacholine responsiveness in atopic asthma. J All Clin Immunol 92:397–403

Romagnani S (1991) Human Th1 and Th2 subsets: doubt no more. Immunol Today 12:256–257

Saito H, Hatake K, Dvorak AM, Leiferman KM, Donnenberg AD, Arai N, Ishizaka K, Ishizaka T (1988) Selective differentiation and proliferation of hematopoietic cells induced by recombinant human interleukins. Proc Natl Acad Sci, USA 85:2288–2292

Salgame P, Abrams JS, Clayberger C, Goldstein H, Convit J, Modlin RL, Bloom BR (1991) Differing lymphokine profiles of functional subsets of human CD4 and CD8 T cell clones. Science 254:279–282

Shaheen SO, Aaby P, Hall AJ, Barker DJ, Heyes CB, Shiell AW, Goudiaby A (1996) Measles and atopy in Guinea-Bissau. Lancet 347:1792–1796

Shirakawa T, Enomoto T, Shimazu S, Hopkin JM (1997) The inverse association between tuberculin responses and atopic disorder. Science 275:77–79

Smith DH, Malone DC, Lawson KA, Okamoto LJ, Battista C, Saunders WB (1997) A national estimate of the economic costs of asthma. Am J Respir Crit Care Med 156:787–793

Spitalny G, Havell E (1984) Monoclonal antibody to murine gamma interferon inhibits lymphokine-induced antiviral and macrophage tumoricidal activities. J Exp Med 159:1560–1565

Swain SL, Weinberg AD, English M, Huston G (1990) IL-4 directs the development of Th2-like helper effectors. J Immunol 145:3796–3806

Takatsu K, Tanaka K, Tuminaga A, Hamaoka T (1980) Antigen-induced T cell-replacing factor. J Immunol 124:2646–2653

von Mutius E, Fritzsch C, Weiland SK, Roll G, Magnussen H (1992) Prevalence of asthma and allergic disorders among children in united Germany: a descriptive comparison. Br Med J 305:1395–1399

von Mutius E, Martinez FD, Fritzsch C, Nicolai T, Reitmeir P, Thiemann HH (1994) Skin test reactivity and number of siblings. Br Med J 308:692–695

Walker C, Bode E, Boer L, Hansel TT, Blaser K, Virchow J-C (1992) Allergic and nonallergic asthmatics have distinct patterns of T-cell activation and cytokine production in peripheral blood and bronchoalveolar lavage. Am Rev Respir Dis 146:109–115

Walsh GM, Hartnell A, Wardlaw AJ, Kurihara K, Sanderson CJ, Kay AB (1990) IL-5 enhances the in vitro adhesion of human eosinophils but not neutrophils, in a leucocyte integrin (CD11/18)-dependent manner. Immunol 71:258–265

Wenner CA, Guler ML, Macatonia SE, O'Garra A, Murphy KM (1996) Roles of IFN-gamma and IFN-alpha in IL-12-induced T helper cell-1 development. J Immunol 156:1442–1447

Wierenga EA, Snoek M, de Groot C, Chretien I, Bos JD, Jansen HM, Kapsenberg ML (1990) Evidence for compartmentalization of functional subsets of CD4+ T lymphocytes in atopic patients. J Immunol 144:4651–4656

Wills-Karp M, Luyimbazi J, Xu X, Schofield B, Neben TY, Karp CL, Donaldson DD (1998) Interleukin-13: central mediator of allergic asthma [see comments]. Science 282:2258–2261
Yi AK, Krieg AM (1998) CpG DNA rescue from anti-IgM-induced WEHI-231 B lymphoma apoptosis via modulation of I kappa B alpha and I kappa B beta and sustained activation of nuclear factor-kappa B/c-Rel. J Immunol 160:1240–1245

# Responses of Human B Cells to DNA and Phosphorothioate Oligodeoxynucleotides

H. LIANG and P.E. LIPSKY



## 1 Introduction

Until recently, DNA was thought to be neither immunogenic nor immunostimulatory, but emerging information has documented that specific forms of DNA can be both. It has been found that certain DNA structures (Z-DNA) can induce significant antibody responses when administered to normal mice (STOLLAR 1975, 1986; RICH et al. 1984; ZARLING et al. 1984; FRAPPIER et al. 1989). In addition, studies in mice demonstrated that DNA from various micro-organisms (bacterial *Escherichia coli* DNA) induced B-cell proliferation, immunoglobulin (Ig) production and cytokine secretion both in vitro and in vivo. In contrast, mammalian DNA did not stimulate murine B-cell responses (MESSINA et al. 1991, 1993; KRIEG 1995; KRIEG et al. 1995; PISETSKY 1995, 1996; KLINMAN et al. 1996b; SUN et al. 1996, 1997). Comparison of the sequences of mammalian and microbial genomes indicated that CpG dinucleotides are ten times more frequent in bacterial than in vertebrate DNA, and 80% of the cytosines within CpG dinucleotides in vertebrate genomes are methylated, whereas those in microbes and synthetic ODNs are non-

University of Texas Southwestern Medical Center at Dallas, 5323 Harry Hines Blvd., Dallas, TX 75235-8884, USA methylated (Bird 1986, 1987; Cardon et al. 1994). After testing a battery of ODNs for their immunostimulatory activities, Krieg and his colleagues found that the stimulatory activity of microbial DNA and sODNs could be attributed to a CpG motif, in which an unmethylated CpG dinucleotide is flanked by two 5′ purines and two 3′ pyrimidines (Krieg et al. 1995; Krieg 1996a,b).

The immunostimulatory activity of DNA has also been demonstrated with certain polynucleotides, especially those synthesized from phosphorothioate derivatives of nucleotides. Phosphorothioate oligodeoxynucleotides (sODNs) have a sulfur substitution for a non-bridging oxygen in the backbone (Fig. 1) and differ from natural phosphodiester-linked polynucleotides in a number of properties, such as nuclease resistance and melting temperature (Stein and Cohen 1988; Stein and Cheng 1993; Tonkinson and Stein 1996). Among 18–27-bp sODN antisense constructs designed to inhibit the expression of specific target genes, some induced activation of murine B lymphocytes both in vivo and in vitro (Krieg et al. 1989; Branda et al. 1993b, 1996b; McIntyre et al. 1993, Mojcik et al. 1993; Pisetsky and Reich 1993; Monteith et al. 1997). B-cell stimulation by the antisense sODNs did not appear to result from inhibition of specific genes by the compounds but rather from direct stimulation of B cells. Although some stimulatory sODNs contained the CpG motifs, murine B-cell activation could also be induced by sODNs containing no CpG at all (Branda et al. 1996b, Monteith et al. 1997).

In addition to the activation of murine B cells, bacterial DNA and ODNs containing the CpG motif also directly activate murine macrophages and dendritic cells and induce the secretion of a variety of cytokines, which subsequently stimulate natural killer (NK) cells to secrete interferon γ (IFN-γ) and enhance NK lytic activity (Ballas et al. 1996; Cowdery et al. 1996; Halpern et al. 1996; Stacey et al. 1996; Chace et al. 1997; Lipford et al 1997b; Sparwasser et al. 1997, 1998; Jacob et al. 1998). Moreover, because of their ability to induce production of interleukin 12 (IL-12) and IFN-γ, bacterial DNA and ODNs containing the CpG motif can stimulate a selective T helper 1 (Th1) response both in vivo and in vitro (Raz et al. 1996; Carson and Raz 1997; Chu et al. 1997; Roman et al. 1997).

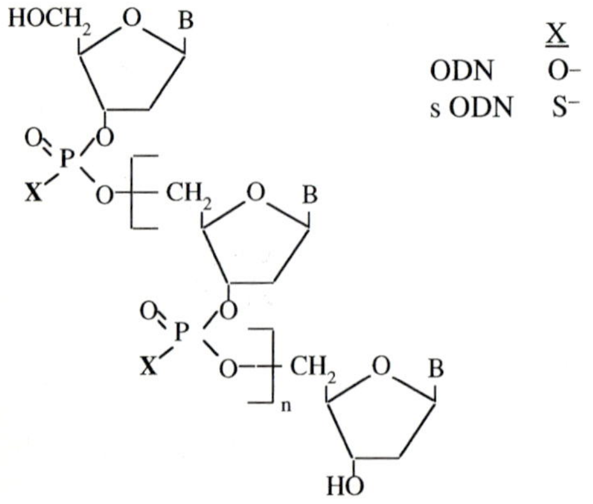


**Fig. 1.** Structure of oligodeoxynucleotides (*ODNs*) and phosphorothioate ODNs (sODNs)

Because of the variety of activities on the immune system leading to antibody production and Th1 responses, the immunostimulatory DNA or ODNs containing the CpG motif could be important adjuvants for conventional protein vaccination and necessary components for DNA vaccination against a variety of targets (Klinman et al. 1996a; Pisetsky 1996; Sato et al. 1996; Lipford et al. 1997a; Roman et al. 1997; Weiner et al. 1997; Davis et al. 1998; Krieg 1998; Moldoveanu et al. 1998; Sun et al. 1998). The immunostimulatory DNA or ODNs containing the CpG motif can also have important therapeutic effects, including an enhancement of the immune response to tumor cells and protection from allergic disease (Wooldridge et al. 1997; Kline et al. 1998).

These effects have been noted in mice and suggest that immunostimulatory DNA or sODNs may have important roles in vaccination, tumor resistance and modification of Th2-mediated responses in humans. However, there is little information available on the potential immunostimulatory effects of DNA or sODNs in humans. Because there are significant differences in the responsiveness of human and murine lymphocytes, it is not possible to extrapolate results obtained from mouse directly to man. Therefore, investigation of the capacity of immunostimulatory DNA and ODNs to induce activation of human lymphocytes is necessary to determine both the potential impact of these materials on human immune responsiveness and their potential role as immune modulators.

## 2 Polyclonal Activation of Human B Cells by DNA and sODNs

The first suggestion of an immunostimulatory activity of DNA in man came from studies on the binding of human sera to a variety of mammalian and non-mammalian DNA. Sera from normal subjects were noted to contain antibodies that bound DNA from two bacterial species, *Micrococcus lysodeikticus* and *Staphylococcus epidermidis*. These antibodies had high affinity and specificity for bacterial and non-mammalian DNA, which suggested that they bound to structural epitopes unique to these DNAs (Robertson et al. 1992; Bunyard and Pisetsky 1994; Pisetsky 1997; Wu et al. 1997; Korans et al. 1998; Pisetsky 1998). It was also noted that sera from normal subjects contained antibodies that bound to DNA from BK polyomavirus (Fredriksen et al. 1993). Moreover, it was noted that antisense sODNs complementary to parts of the hepatitis-B-virus genome stimulated proliferation of B cells from chronic hepatitis patients (Chen et al. 1996). Furthermore, it has been demonstrated that 27bp and 21bp antisense sODNs to the rev region of the human immunodeficiency virus (HIV) genome increased proliferation and Ig production of peripheral blood mononuclear cells (PBMCs) from normal subjects as well as those with common variable immunodeficiency and chronic lymphocytic leukemia (Branda et al. 1993a; Branda et al. 1996a). The stimulation indices of PBMCs from patients were similar to those from normal human subjects, suggesting that responsiveness did not require antecedent exposure to HIV (Branda et al. 1996a). Taken together, these reports suggest that certain DNA or sODNs can induce lymphocyte proliferation and antibody production from patients and normal human subjects. However, the mechanism of the human B-cell activation and the question of whether DNA and sODNs can directly activate normal human B cells remain unclear.

Many studies have been carried out recently to investigate the immunostimulatory activity of CpG-containing DNA and ODNs in mouse and other species, but minimal information on the responsiveness of human B cells is available. To address this issue, a series of experiments was carried out to determine whether bacterial DNA and various ODNs (Table 1) could activate human B cells directly (Liang et al. 1996). These studies employed highly purified human peripheral blood B cells so that a direct stimulatory activity on human B cells could be assessed. As shown in Fig. 2, three of the various ODNs tested consistently induced significant human B-cell proliferation. Each of these stimulatory polynucleotides was a phosphorothioate. Of note, neither *E. coli* DNA nor calf thymus DNA was stimulatory. Other sODNs (poly G, poly C, poly T, randomers) and ODNs induced

**Table 1.** Sequence, length and description of representative oligodeoxynucleotides (ODNs). The categories were defined according to their ability to activate human B cells. There was no significant difference in the magnitude of activation by any pair of synthetic ODNs (sODNs) within each category, although there were significant differences between activation induced by sODNs from different categories ($P < 0.05$) and between activation induced by any sODN and the response of B cells alone. *HIVas*, antisense human immunodeficiency virus; *HSVas*, antisense herpes simplex virus; *NK*, natural killer

| Name | Sequence | Length | Description |
|---|---|---|---|
| Most active | | | |
| HSVas | 5′-GCCGAGGTCCATGTCGT ACGA-3′ | 21bp | Antisense to a translation initiation region of HSV |
| 20-mer | 5′-TTGCTTCCATCTTCCTC GTC-3′ | 20bp | |
| HIVas | 5′-TCGTCGCTGTCTCCGCTTC TTCTTGCC-3′ | 27bp | Antisense to *rev* of HIV |
| $TCG_4$ | 5′-TCGTCGTCGTCG-3′ | 12bp | |
| $MCMT^+$ | 5′-TGACGTTTGACGTTT-GACGTT-3′ | 21bp | Murine CpG motif |
| $MCMT^+C\Delta T$ | 5′-TGATGTCTGATGTCT-GATGTC-3′ | 21bp | Murine CpG motif with CΔT |
| $MCMT^+C\Delta G$ | 5′-TGAGGTCTGAGGTCT-GAGGTC-3′ | 21bp | Murine CpG motif with CΔG |
| Moderately active | | | |
| NKNSO | 5′-GGGGGGGGGGGGACCG GTGGGGGGGGGGGG-3′ | 30bp | NK nonstimulatory sODN |
| NKSO | 5′-GGGGGGGGGGGGAACGT TGGGGGGGGGGG-3′ | 30bp | NK stimulatory sODN |
| Minimally active | | | |
| $G_{20}$ | 5′-GGGGGGGGGGGGGGGG GGGG-3′ | 20bp | Poly G |
| $C_{20}$ | 5′-CCCCCCCCCCCCCCCCC CCC-3′ | 20bp | Poly C |
| $CG_{7.5}$ | 5′-CGCGCGCGCGCGCGC-3′ | 15bp | Poly CG |

| Name | Sequence of oligodeoxynucleotide | Size |
|---|---|---|
| **Phosphorothioates:** | | |
| HSVas | 5′-GCC GAG GTC CAT GTC GTACGC-3′ | 21bp |
| 20mer | 5′-TTG CTT CCATCT TCC TCG TC-3′ | 20bp |
| HIVas | 5′-TCG TCG CTG TCT CCG CTT CTT CTT GCC-3′ | 27bp |
| $T_{20}$ | 5′-TTT TTT TTT TTT TTT TTT TT-3′ | 20bp |
| $A_{20}$ | 5′-AAA AAA AAA AAA AAA AAA AA-3′ | 20bp |
| $G_{20}$ | 5′-GGG GGG GGG GGG GGG GGG GG-3′ | 20bp |
| $C_{20}$ | 5′-CCC CCC CCC CCC CCC CCC CC-3′ | 20bp |
| $G_{10}$ | 5′-GGG GGG GGG G-3′ | 10bp |
| $G_5$ | 5′-GGG GG-3′ | 5bp |
| $pGG_{19}$ | 5′p-GGG GGG GGG GGG GGG GGGGG-3′ | 20bp |
| $Rmer_{10}$ | 5′-NNN NNN NNN N-3′ | 10bp |
| $Rmer_{20}$ | 5′-NNN NNN NNN NNN NNN NNN NN-3′ | 20bp |
| $A_6G_2$ | 5′-AAA AAA GG-3′ | 8bp |
| $A_7G_1$ | 5′-AAA AAA AG-3′ | 8bp |
| $C_6G_2$ | 5′-CCC CCC GG-3′ | 8bp |
| **Diesters:** | | |
| HSVaso | 5′-GCC GAG GTC CAT GTC GTACGC-3′ | 21bp |
| $Rmer_{20}o$ | 5′-NNN NNN NNN NNN NNN NNN NN-3′ | 20bp |
| NKNSOo | 5′-GGG GGG GGG GGG ACC GGT GGG GGG GGG GGG-3′ | 30bp |
| NKSOo | 5′-GGG GGG GGG GGG AAC GTT GGG GGG GGG GGG-3′ | 30bp |
| $G_{20}o$ | 5′-GGG GGG GGG GGG GGG GGG GG-3′ | 20bp |
| $C_{20}o$ | 5′-CCC CCC CCC CCC CCC CCC CC-3′ | 20bp |
| No stimulus | | |

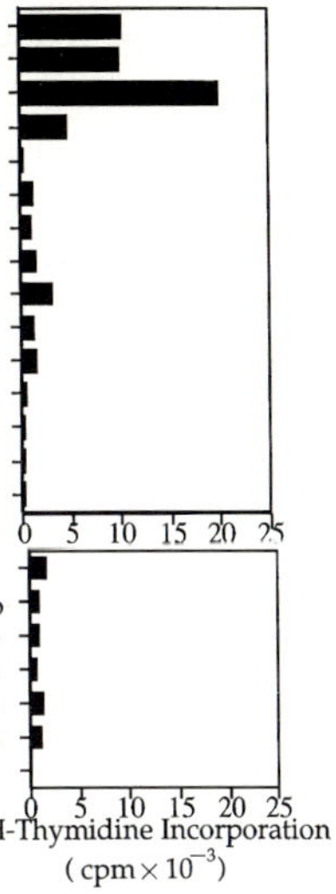


**Fig. 2.** B-cell proliferation induced by phosphorothioate oligodeoxynucleotides (sODNs). Highly purified B cells (50 × $10^3$/well) were cultured with various concentrations of ODNs (1µg/ml–50µg/ml) for 4 days, and proliferation was assessed by $^3$H-thymidine incorporation. The data indicate results with the optimal stimulatory concentrations, usually 5µg/ml (except 1µg/ml for antisense human immunodeficiency virus and 25µg/ml for Rmer10). Data with herpes simplex virus antisense oligodeoxynucleotide (HSVaso) were obtained by incubation of B cells with 5µg/ml antisense herpes simplex virus and then adding 5µg/ml of HSVaso to culture daily. A single addition of HSVaso induced no response. The data are the means of triplicate cultures with a standard error of the mean of less than 10%, and are representative of three separate experiments, each carried out with B cells from a different donor

minimal, but nonetheless detectable, responses that were statistically different from the $^3$H-thymidine incorporation manifested by B cells cultured alone. Poly A and three 8bp sODNs did not stimulate B cells at all. Moreover, the diester form of the herpes simplex virus (HSV) antisense (HSVas) was much less stimulatory than its phosphorothioate counterpart within this concentration range, even when added to culture repetitively (the mean maximum response to the diester form of HSVas was 615 ± 22cpm vs. 3113 ± 93cpm for the phosphorothioate form of HSVas). Although modest, the response induced by the diester form of HSVas was significantly greater than the tritiated thymidine incorporation of control B cells (99 ± 7cpm, $P < 0.01$) (Liang et al. 1996). These results indicated that certain sODNs induced marked human B-cell proliferation far more efficiently than their phosphodiester counterparts did.

In addition, the sODNs that induced maximal proliferation also induced production of IgM, IgG and IgA from purified B cells in the absence of exogenous cytokines or T cells. The other ODNs and DNA induced little or no Ig production (Liang et al. 1996).

IgM anti-DNA antibody was also induced by the sODNs that stimulated maximal proliferation, and the anti-DNA antibodies secreted bound all sODNs tested. Therefore, maximal stimulatory sODNs activated B cells committed to the production of anti-DNA antibodies. However, specificity of the anti-DNA antibody for the activating sODNs was not apparent (LIANG et al. 1996).

Of note, certain cytokines (IL-2) enhanced responses of B cells activated by the most stimulatory sODNs (LIANG et al. 1996). Other cytokines tested, including IL-4 and IL-10, had no augmenting effects. Purified human T cells also enhanced the responses of human B cells to sODNs (LIANG et al. 1996), although T cells did not bind sODNs (Fig. 3B) and did not proliferate when cultured with sODNs (LIANG et al. 1996). These results indicated that IL-2 and intact T cells enhanced B-cell responses to the most stimulatory sODNs, but they were not required for the stimulatory activity of sODNs.

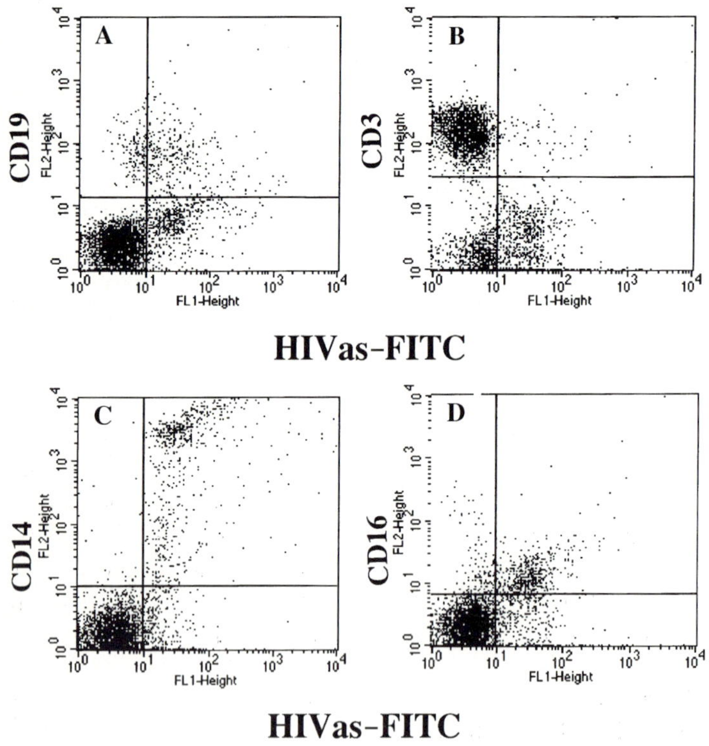


**Fig. 3A–D.** Antisense human immunodeficiency virus (HIVas) binds to B cells, natural killer cells and CD14-positive myeloid cells but not T cells. Peripheral blood mononuclear cells were incubated with HIVas–fluorescein isothiocyanate (FITC; 4μg/sample) at 37°C for 30min. After washing to remove unbound HIVas–FITC, cells were stained with monoclonal antibodies (mAbs) against CD19 (**A**), CD3 (**B**), CD14 (**C**) and CD16 (**D**) or their isotype-matched controls. The FACscan plots show mAb staining on the y-axis and HIVas-FITC staining on the x-axis and are representatives of one of three experiments, each using cells from a different donor. HIVas–FITC is bound by 67% of CD19-positive cells, 3% of CD3-positive cells, 96% of CD14-positive cells and 76% of CD16-positive cells sODNs also induced the expression of activation markers by highly purified B cells. Of interest, a much larger number of B cells expressed activation markers than might have been anticipated from the moderate degree of B-cell proliferation noted. Thus, the sODNs that induced maximal proliferation activated more than 80% of B cells to express activation markers, including CD69, CD86 and CD25. In contrast, sODNs that stimulated minimal proliferation also induced either minimal or no increased expression of activation markers. The finding that sODNs induced activation-marker expression by more than 80% of B cells implies that the response is polyclonal. This was confirmed by documenting that B cells expressing all six major $V_H$ families were activated by stimulatory sODNs (LIANG et al. 1996).

The large number of B cells induced to express activation markers, compared to the modest degree of proliferation induced by sODNs, implied that many B cells were initially activated, but few underwent clonal expansion in response to sODNs. Additional experiments directly analyzing cell-cycle entry and progression confirmed that active sODNs stimulated a few cells to enter the cell cycle, and these cells underwent multiple rounds of proliferation.

Taken together, certain sODNs induced T-cell-independent polyclonal activation of human B cells, resulting in polyclonal expansion of a small number of activated B cells and differentiation of some stimulated B cells into Ig secreting cells. Thus, certain sODNs provide sufficient signals to induce human B-cell responses in vitro in the absence of T cells and other cells capable of facilitating B-cell activation.

# 3 Mechanism of Activation of Human B Cells by sODNs

## 3.1 sODNs Activate Human B Cells by Engaging Surface Receptors

The mechanism of human B-cell activation by sODNs has not been completely delineated. To examine whether sODNs activated human B cells by engaging surface receptors or whether cellular entry of sODNs was necessary for the stimulatory activities, active sODNs were covalently coupled to Sepharose beads and tested for their ability to activate B cells. The results showed that Sepharose-bound sODNs stimulated B cells as effectively as soluble sODNs (LIANG et al. 1996). Additional experiments documented that the stimulatory activity did not result from soluble sODNs released from the beads. These results indicate that human B-cell stimulation results from engagement of sODN-binding surface receptors.

Examination of binding of fluorescein-isothiocyanate-labeled sODNs indicated that binding was rapid, saturable, specific and initially temperature independent, suggesting that the binding is likely to be receptor mediated. Most sODNs tested appeared to bind to the same or to a limited set of receptor(s). However, the nature of the sODN-binding stimulatory receptor remains unknown. These data indicate that sODNs directly bind to surface receptors on human B cells and suggest that binding to specific receptors initiates polyclonal B-cell activation. Although a number of DNA or sODN-binding receptors have been found on a variety of cells, and binding of sODNs to some of these receptors may have functional consequences (BENNETT et al. 1985, 1987, 1991; LOKE et al. 1989; YAKUBOV et al. 1989; HEFENEIDER et al. 1990, 1992; GESELOWITZ and NECKERS 1992; PEARSON et al. 1993; KIMURA et al. 1994; BELTINGER et al. 1995; BENIMETSKAYA et al. 1997), the nature of the sODN-binding receptor on human B cells that eventuates in polyclonal activation remains unclear.

## 3.2 The Nature of the Stimulatory sODNs

A number of possible characteristics could account for the stimulatory properties of sODNs. First, there might be unique sequence motifs accounting for the stimulatory activity of sODNs. Alternatively, sODNs might assume unique secondary or tertiary structures that are necessary to activate human B cells.

To identify potential sODN sequences necessary for B-cell responses, one of the most stimulatory sODN, antisense HIV (HIVas), was examined in detail. Three 15bp sODNs that contained the sequences corresponding to the 5′, central, and 3′ regions of HIVas were synthesized. The 5′ 15bp of HIVas stimulated B-cell activation to a degree comparabe to that induced by the full-length 27-mer, whereas the central and 3′ 15-mers were less stimulatory (LIANG et al. 1996). Comparison of the sequences indicated that TCGTCG was present only in the 5′ 15bp of HIVas, but not in the others. To determine whether this motif was stimulatory, the activity of additional 15bp sODNs was tested. Of these compounds, $(TCG)_5$ induced B-cell activation equal to or greater than that induced by HIVas, whereas a 5′ TCG doublet followed by a 3′ G nonamer failed to stimulate B cells. Additional sODNs, in which the TCG repeat was altered to ACG, TCC or TGG, had no stimulatory activity (LIANG et al. 1996). In order to examine the minimal length of active sODNs, additional sODNs that contained two to four repeats of TCG were synthesized. sODNs containing three TCG repeats were stimulatory, but a tandem TCG had minimal stimulatory activity. Taken together, these data indicate that $(TCG)_n$, where $n > 3$, is a minimal stimulatory element. In addition to TCG repeats, repeats of the entire CpG motif that stimulated murine B cells also activated human B cells (LIANG et al. 1996). However, these motifs are not necessary for human B-cell activation, since sODNs containing no CpG also induce maximal human B-cell activation. Thus, after testing 62 ODNs and sODNs, a unique sequence accounting for the stimulatory activity of sODNs could not be identified.

To examine the second hypothesis that secondary or tertiary structure rather than a specific sequence might be involved in stimulating human B cells, $^{32}$P-labeled sODNs were resolved by polyacrylamide gel electrophoresis (PAGE) and detected by autoradiography. The results showed that HIVas, HSVas and MCMT$^{+}$ form tertiary structures, as indicated by their aberrant mobilities in non-denaturing gel. This finding suggested that structure rather than primary sequence might be an important determinant of the stimulatory capacity of sODNs, but the structures that could be necessary for stimulating B cells remain unknown.

## 4 Human Monocytes Enhance Response of Human B Cells to sODNs

In addition to binding human B cells, most stimulatory sODNs also bound to almost all the monocytes and most NK cells in the peripheral blood, but they did not bind to T cells (Fig. 3). Moreover, highly purified human monocytes augmented B-cell Ig production induced by most stimulatory sODNs. These results suggested that, in addition to a direct stimulatory effect on human B cells, sODNs can also co-stimulate human B cells by an indirect effect mediated by monocytes.

## 5 Differences Between Human and Murine B Cells in the Responsiveness to DNA and sODNs

The current data suggest that human B-cell activation induced by sODNs is governed by different principles than those observed in the mouse (Table 2). First, the response of human and murine B cells to bacterial DNA is different. It has been demonstrated in mice that *E. coli* genomic DNA activated B cells both in vivo and in vitro. In contrast, human B cells failed to respond to *E. coli* DNA in vitro (LIANG et al. 1996). Whereas sera from normal subjects contained antibodies that bound DNA from two bacterial species, *M. lysodeikticus* and *S. epidermidis*, it is not known whether DNA from these micro-organisms directly stimulate human B cells.

**Table 2.** Differences between human and murine B cells in responsiveness to DNA and synthetic oligodeoxynucleotides. *IgM*, immunoglobulin M; *IL*, interleukin

| | Murine B cells | Human B cells |
|---|---|---|
| Stimulation by bacterial DNA | *E. coli* | None |
| Activation motifs | CpG motif (necessary) | CpG motif (not necessary); TCG repeats |
| Cellular entry of stimulus | Necessary | Not necessary; activation is receptor mediated |
| Cellular activation | 95% of B cells in cell cycle | <20% of B cells in cell cycle |
| Mechanism | IgM production requires IL-6 | IgM production is independent of stimulation of IL-6 |

Secondly, the basic activation motif appears to be different in man and mouse. The CpG motif has been shown to be sufficient and necessary for inducing murine B-cell activation (Krieg et al. 1995). In contrast, for stimulation of human B cells, the CpG motif is sufficient but not necessary. In addition, $(CpG)_n$, which is a potent stimulator of murine B cells, does not appear to be capable of inducing maximal human B-cell activation, although it did stimulate human B cells minimally (Liang et al. 1996). Of note, TCG repeats activate human B cells comparably to CpG motifs (Liang et al. 1996), but this sequence has not been tested in murine systems. Of importance, sODNs containing no CpG dinucleotides are also capable of activating human B cells. It should be noted that sODNs containing no CpG motifs also have been found to activate murine B cells (Branda et al. 1996b; Monteith et al. 1997), suggesting that the difference between mouse and human in this regard may be more relative than absolute.

Thirdly, the mechanism of B-cell activation also appears to be different between mice and humans. It has been suggested that binding to surface receptors is not involved in murine B-cell activation, whereas cellular uptake of the sODNs seemed to be required (Krieg et al. 1995). In contrast, sODNs immobilized so as to prevent uptake induced maximal activation of human B cells (Liang et al. 1996). Binding of sODNs to human B cells appears to be receptor mediated, indicating that sODNs stimulate human B-cell activation by binding to surface receptors. The nature of these receptors remains to be delineated.

Fourthly, the character of B-cell activation seems to be different in man and mouse. It has been reported that DNA or sODNs containing the CpG motif induced more than 95% of murine B cells to enter the cell cycle and also rescued murine B cells from apoptosis (Krieg et al. 1995; Yi et al. 1996b, 1998; Macfarlane et al. 1997; Wang et al. 1997; Yi and Krieg 1998). In contrast, less than 20% of human B cells were in the cell cycle after stimulation with active sODNs. Whether this represents a species difference or is related to the responsiveness of human peripheral blood B cells relative to murine splenic B cells remains to be determined.

Finally, the cellular mechanism of B-cell activation induced by sODNs seems to be different between mice and humans. It has been shown that DNA or sODNs containing the CpG motif induced marked IL-6 secretion from murine B cells and that IL-6 was necessary for IgM production induced by sODNs (Klinman et al. 1996b, Yi et al. 1996a,c). In contrast, limited IL-6 secretion was detected in the supernatant of human B cells cultured with stimulatory sODNs, and neutralizing antibodies to IL-6 only modestly inhibited the IgM production of human B cells. These results suggest that sODNs directly stimulate human B-cell proliferation and differentiation and that IL-6 production plays only a modest role. Taken together, human and murine B cells appear to differ significantly in their responses to DNA and sODNs, implying that results derived from studies of murine B cells can not be directly extrapolated to human B-cell responses.

# 6 Summary and Conclusion

Emerging information has documented that certain DNA and sODNs can be both immunogenic and immunostimulatory. sODNs, but not DNA, induce T-cell-independent polyclonal activation of human B cells by engaging cell-surface receptors. Manifestations of sODN-induced human B-cell activation include expression of activation markers, proliferation, Ig production and anti-DNA antibody production. IL-2 and intact T cells enhanced B-cell responses to sODNs but were not required. Monocytes also provided a modest enhancement of human B-cell responses induced by sODNs. The chemical nature of sODNs capable of stimulating human B cells and the specific cell-surface receptors involved have not been completely delineated. Further studies will be necessary to elucidate the potential role of stimulatory sODNs in disease pathogenesis and to develop a means to employ ODNs as therapeutic agents in humans.

*Acknowledgement.* The authors would like to thank Dr. David S. Pisetsky for providing the oligonucleotide reagents and for many helpful discussions. The work described in this review was supported by National Institutes of Health grant 2-P01-AI31229-8.

# References

Ballas ZK, Rasmussn WL, Krieg AM (1996) Induction of NK activity in murine and human cells by CpG motifs in oligodeoxynucleotides and bacterial DNA. J Immunol 157:1840–1845

Beltinger C, Saragovi HU, Smith RM, LeSauteur L, Shah N, DeDionisio L, Christensen L, Raible A, Jarett L, Gewirtz AM (1995) Binding, uptake, and intracellular trafficking of phosphorothioate-modified oligodeozynucleotides. J Clin Invest 95:1814–1823

Benimetskaya L, Loike JD, Khaled Z, Loike G, Silverstein SC, Cao L, Khoury JE, Cai T, Stein CA (1997) Mac-1 (CD11b/CD18) is an oligoseoxynucleotide-binding protein. Nat Med 3:414–420

Bennett RM, Gabor GT, Merritt MM (1985) DNA binding to human leukocytes: Evidence for a receptor-mediated association, internalization, and degradation of DNA. J Clin Invest 76:2182–2190

Bennett RM, Kotzin BL, Merritt MJ (1987) DNA receptor dysfunction in systemic lupus erythematosus and kindred disorders: induction by anti-DNA antibodies, antihistone antibodies and antireceptor antibodies. J Exp Med 166:850–863

Bennett RM, Cornell KA, Merritt MJ, Bakke AC, Hsu PH (1991) Autoimmunity to a 28–30kD cell membrane DNA binding protein: occurrence in selected sera from patients with SLE and mixed connective tissue disease (MCTD). Clin Exp Immunol 86:374–379

Bird AP (1986) CpG-rich islands and the function of DNA methylation. Nature 321:209–213

Bird AP (1987) CpG islands as gene markers in the vertebrate nucleus. Trends genet 3(12):342–347

Branda RF, Moore AL, Mathews L, McCormack JJ, Zon G (1993a) Stimulation of peripheral blood mononuclear cells (PBMNCs) from chronic lymphatic leukemia (CLL) patients by an antisense phosphorothioate oligomer to the rev gene of HIV. (Abstract) Proceeding American Association of Cancer Research 34:455

Branda RF, Moore AL, Mathews L, McCormarck JJ, Zon G (1993b) Immune stimulation by an antisense oligomer complementary to the rev gene of HIV-1. Biochem Pharmacol 45:2037–2043

Branda RF, Moore AL, Hong R, McCormack JJ, Zon G, Cunningham-Rundles C (1996a) Cell proliferation and differentiation in common variable immunodeficiency patients produced by an antisense oligomer to the rev gene of HIV-1. Clin Immunol Immunopatho 79(2):115–121

Branda RF, Moore AL, Lafayette AR, Mathews L, Hong R, Zon G, Brown T, McCormack JJ (1996b) Amplification of antibody production by phosphorothioate oligodeoxynucleotides. J Lab Clin Med 128:329–338

Bunyard MP, Pisetsky DS (1994) Characterization of antibodies to bacterial double-stranded DNA in the sera of normal human subjects. Int Arch Allergy Immunol 105:122–127

Cardon LR, Burge C, Clayton DA, Karlin S (1994) Pervasive CpG suppression in animal mitochondrial genomes. Proc Natl Acad Sci USA 91:3799–3803

Carson DA, Raz E (1997) Oligonucleotide adjuvants for T helper 1 (Th1)-specific vaccination. J Exp Med 186:1621–1622

Chace JH, Hooker NA, Mildenstein KL, Krieg AM, Cowdery JS (1997) Bacterial DNA-induced NK cell IFN-gamma production is dependent on macrophage secretion of IL-12. Clin Immunol Immunopathol 84:185–193

Chen C, Zhou Y, Yao Z, Zhang Y, Feng Z (1996) Stimulation of human lymphocyte proliferation and CD40 Ag expression by phosphorothioate oligodeoxynucleotides complementary to hepatitis B virus genome. J Viral Hepat 3:167–172

Chu RS, Targoni OS, Krieg AM, Lehmann PV, Harding CV (1997) CpG oligodeoxynucleotides act as adjuvants that switch on T-helper (Th1) immunity. J Exp Med 186:1623–1631

Cowdery JS, Chace JH, Yi A, Krieg AM (1996) Bacterial DNA induced NK cells to produce IFN-γ in vivo and increases the toxicity of lipopolysaccharides. J Immunol 156:4570–4575

Davis HL, Weeranta R, Waldschmidt TJ, Tygreett L, Schorr J, Krieg AM (1998) CpG DNA is a potent enhancer of specific immunity in mice immunized with recombinant Hepatitis B surface antigen. J Immunol 160:870–876

Frappier L, Price GB, Martin RG, Zannis-Hadjopoulos M (1989) Characterization of the binding specificity of two anticruciform DNA monoclonal antibodies. J Biol Chem 264:334–341

Fredriksen K, Skogsholm T, Flaegstad T, Traavik T, Rekvig OP (1993) Antibodies to ds DNA are produced during primary BK virus infection in man, indicating that anti-dsDNA antibodies may be related to virus replication in vivo. Scand J Immunol 38:401–406

Geselowitz DA, Neckers LM (1992) Analysis of oligonucleotide binding, internalization, and intracellular trafficking utilizing a novel radiolabeled crosslinker. Antisense Res Dev 2:17–25

Halpern MD, Jurlander RJ, Pisetsky DS (1996) Bacterial DNA induces murine interferon-γ production by stimulation of interleukin-12 and tumor necrosis factor-α. Cell Immunol 167:72–78

Hefeneider SH, Bennett RM, Pham TQ, Cornell K, McCoy SL, Heinrich MC (1990) Identification of a cell-surface DNA receptor and its association with systemic lupus erythematosus. J Invest Dermatol 94:29S–84S

Hefeneider SH, McCoy SL, Morton JI, Bakke AC, Cornell KA, Brown LE, Bennett RM (1992) DNA binding to mouse cells is mediated by cell-surface molecules: the role of these DNA-binding molecules as target antigens in murine lupus. Lupus 1:167–173

Jacob T, Walker PS, Krieg AM, Udey MC, Vogel JC (1998) Activation of cutaneous dendritic cells by CpG-containing oligodeoxynucleotides: a role for dendritic cells in the augmentation of Th1 responses by immunostimulatory DNA. J Immunol 161:3042–3049

Karounos DG, Grudier JP, Pisetsky DS (1988) Spontaneous expression of antibodies toDNA of various species origin in sera of normal subjects and patients with systemic lupus erythematosus. J Immunol 140:451–455

Kimura Y, Sonehara K, Kuramoto E, Makino T, Yamamoto S, Yamamoto T, Kataoka T, Tokunaga T (1994) Binding of oligoguanylate to scavenger receptors is required for oligonucleotides to augment NK cell activity and induce IFN. J Biochem 116:991–994

Kline JN, Waldschmidt TJ, Businga TR, Lemish JE, Weinstock JV, Thorne PS, Krieg AM (1998) Cutting edge: modulation of airway inflammation by CpG oligodeoxynucleotides in a murine model of asthma. J Immunol 160:2555–2559

Klinman DM, Yamshchikov G, Ishigatsubo Y (1996a) Contribution of CpG motifs to the immunogenicity of DNA vaccines. J Immunol 158:3635–3639

Klinman DM, Yi A, Beaucage SL, Conover J, Krieg AM (1996b) CpG motifs present in bacterial DNA rapidly induce lymphocytes to secret interleukin-6, interleukin-12, and interferon-γ. Proc Natl Acad Sci USA 93:2879–2883

Krieg AM (1995) CpG DNA: a pathogenic factors in systemic lupus erythematosus. J Clin Immuno 15:284–292

Krieg AM (1996a) An innate immune defense mechanism based on the recognition of CpG motifs in microbial DNA. J Lab Clin Med 128:128–133

Krieg AM (1996b) Lymphocytes activation by CpG dinucleotides motifs in prokaryotic DNA. Trends Microbiol 4:73–76

Krieg AM (1998) The role of CpG dinucleotides in DNA vaccines. Trends Microbiol 6:23–27

Krieg AM, Gause WC, Gourley MF, Steinberg AD (1989) A role for endogenous retroviral sequences in the regulation of lymphocyte activation. J Immunol 143:2448–2451
Krieg AM, Yi A, Matson S, Waldschmidt TJ, Bishop GA, Teasdale R, Koretzky GA, Klinman DM (1995) CpG motifs in bacterial DNA trigger direct B cell activation. Nature 371:546–549
Liang H, Nishioka Y, Reich CF, Pisetsky DS, Lipsky PE (1996) Activation of human B cells by phosphorothioate oligodeoxynucleotides. J Clin Invest 98:1119–1129
Lipford GB, Bauer M, Blank C, Reiter R, Wagner H, Heeg K (1997a) CpG-containing synthetic oligonucleotides promote B and cytotoxic T cell responses to protein antigen: a new class of vaccine adjuvants. J Exp Med 27:2340–2344
Lipford GB, Sparwasser T, Bauer M, Zimmermann S, Koch E, Heeg K, Wagner H (1997b) Immunostimulatory DNA: sequence-dependent production of potentially harmful or useful cytokines. Eur J Immunol. 27:3420–3426
Loke SL, Stein CA, Zhang XH, Mori K, Nakanishi M, Subasinghe C, Cohen JS, Neckers LM (1989) Characterization of oligonucleotide transport into living cells. Proc Natl Acad Sci USA 86:3474–3478
Macfarlane DE, Manzel L, Krieg AM (1997) Unmethylated CpG-containing oligodeoxynucleotides inhibit apoptosis in WEHI 231 B lymphocytes induced by several agents: evidence for blockade of apoptosis at a distal signalling step. Immunology 91:586–593
McIntyre KW, Lombard-Gillooly K, Perez JR, Kunsch C, Sarmiento UM, Larigan JD, Landreth KT, Narayanan R (1993) A sense phosphorothioate oligonucleotide directed to the initiation codon of transcription factor NF-κB p65 causes sequence-specific immune stimulation. Antisense Res Dev 3:309–322
Messina JP, Pisetsky DS (1991) Stimulation of in vitro murine lymphocyte proliferation by bacterial DNA. J Immunol 147:1759–1764
Messina JP, Gilkeson GS, Pisetsky DS (1993) The influence of DNA structure on the in vitro stimulation of murine lymphocytes by natural and synthetic polynucleotide antigens. Cell Immunol 147:148–157
Mojcik CF, Gourley MF, Klinman DM, Krieg AM, Gmelig-Meyling F, Steinberg AD (1993) Admission of a phosphorothioate oligonucelotide antisense to murine endogeous retroviral MCF env causes immune effects in vivo in a sequence-specific manner. Clin Immunol Immunopathol 67:130–136
Moldoveanu Z, Love-Homan L, Huang WQ, Krieg AM (1998) CpG DNA, a novel immune enhancer for systemic and mucosal immunization with influenza virus. Vaccine 16:1216–1224
Monteith DK, Henry SP, Howard RB, Flournoy S, Levin AA, Bennett CF, Crooke ST (1997) Immune stimulation-a class effect of phosphorothioate oligodeoxynucleotides in rodents. Anticancer Drug Des 12:421–432
Pearson AM, Rich A, Krieger M (1993) Polynucleotide binding to macrophage scavenger receptors depends on the formation of base-quartet-stabilized four-stranded helices. J Biol Chem 268: 3546–3554
Pisetsky DS (1995) Immunological properties of bacterial DNA. Ann NY Acad Sci 772:152–163
Pisetsky DS (1996) Immune activation by bacterial DNA: a new genetic code. Immunity 5:303–310
Pisetsky DS (1997) Specificity and immunochemical properties of antibodies to bacterial DNA. Methods: a companion to methods in enzymology 11:55–61
Pisetsky DS (1998) Antibody responses to DNA in normal immunity and aberrant immunity. Clin Diagn Lab Immunol 5:1–6
Pisetsky DS, Reich CF (1993) Stimulation of murine lymphocyte proliferation by a phosphorothioate oligonucleotide with antisense avtivity for herpes simplex virus. Life Sci 54:101–107
Raz E, Tighe H, Sato Y, Corr M, Daudler JA, Roman M, Swain S, Spiegelberg HL, Carson DA (1996) Preferential induction of a Th1 immune response and inhibition of specific IgE antibody formation by plasmid DNA immunization. Proc Natl Acad Sci USA 93:5141–5145
Rich A, Nordheim A, Wang AH (1984) The chemistry and biology of left-handed Z-DNA. Annu Rev Biochem 53:791–846
Robertson CR, Gilkeson GS, Ward M.M, Pisetsky DS (1992) Patterns of heavy and light chains utilization in the antibody response to single-stranded bacterial DNA in normal human subjects and patients with systemic lupus erythematosus. Clin Immunol Immunopathol 62:25–32
Roman M, Martin-Orozco E, Goodman JS, Nguyen M, Sato Y, Ronaghy A, Kornbluth RS, Richman DD, Carson DA, Raz E (1997) Immunostimulatory DNA sequences function as T helper-1-promoting adjuvants. Nat Med 3:849–854
Sato Y, Roman M, Tighe H.m Lee D, Corr M, Nguyen M, Silverman GJ, Lotz M, Carson DA, Raz E (1996) Immunostimulatory DNA sequences necessary for effective intradermal gene immunization. Science 273:352–354

Sparwasser T, Miethke T, Lipford G, Erdmann A, Hacker H, Heeg K, Wagner H (1997) Macrophages sense pathogen via DNA motifs: induction of tumor necrosis factor-α-mediated shock. Eur J Immunol 27:1671–1679

Sparwasser T, Koch E, Vabulas RM, Heeg K, Lipford GB, Ellwart JW, Wagner H (1998) Bacterial DNA and immunostimulatory CpG oligonucleotides trigger maturation and activation of murine dendritic cells. Eur J Immunol 28:2045–2054

Stacey KJ, Sweet MJ, Hume DA (1996) Macrophage ingest and are activated by bacterial DNA. J Immunol 157:2116–2122

Stein CA, Cohen JS (1988) Oligodeoxynucleotides as inhibitors of gene expression: a review. Cancer Res 48:2659–2668

Stein CA, Cheng YC (1993) Antisense oligonucleotides as therapeutic agents-is the bullet really magical? Science 261:1004–1012

Stollar BD (1975) The specificity and applications of antibodies to helical nucleic acids. CRC Crit Rev Biochem 3:45–69

Stollar BD (1986) Antibodies to DNA. CRC Crit Rev Biochem 20:1–36

Sun S, Cai Z, Langlade-Demoyen P, Kosaka H, Brunmark A, Jackson MR, Peterson PA, Sprent J (1996) Dual function of *Drosophila* cells as APCs for na $CD8^+$ T cells: implication for tumor immunotherapy. Immunity 4:555–564

Sun S, Beard C, Jaenisch R, Jones P, Sprent J (1997) Mitogenicity of DNA from different organisms for murine B cells. J Immunol 159:3119–3125

Sun S, Kishimoto H, Sprent J (1998) DNA as an adjuvant: capacity of insect DNA and synthetic oligodeoxynucleotides to augment T cell responses to specifc antigen. J Exp Med 187:1145–1150

Tonkinson JL, Stein CA (1996) Antisense oligodeoxynucleotides as clinical therapeutic agents. Cancer Invest 14:54–65

Wang Z, Karras JG, Colarusso TP, Foote LC, Rothstein TL (1997) Unmethylated CpG motifs protect murine B lymphocytes against FAS-mediated apoptosis. Cell Immunol 180:162–167

Weiner GJ, Liu H, Wooldridge JE, Dahle CE, Krieg AM (1997) Immunostimulatory oligodeoxynucleotides containing the CpG motifs are effective as immune adjuvants in tumor antigen immunization. Proc Natl Acad Sci USA 94:10833–10837

Wooldridge JE, Ballas Z, Krieg AM, Weiner GJ (1997) Immunostimulatory oligodeoxynucleotides containing CpG motifs enhance the efficacy of monoclonal antibody therapy of lymphoma. Blood 89:2994–2998

Wu Z, Drayton D, Pisetsky DS (1997) Specificity and immunochemical properties of antibodies to bacterial DNA in sera of normal human subjected and patients with systemic lupus erythematosus (SLE). Clin Exp Immunol 109:27–31

Yakubov LA, Deeva EA, Zarytova VF, Ivanova EM, Ryte AS, Yurchenko LV, Vlassov VV (1989) Mechanism of oligonucleotide uptake by cells: Involvement of specific receptors? Proc Natl Acad Sci USA 86:6454–6458

Yamamoto S, Yamamoto T, Kataoka T, Kuramoto E, Yana O, Tokunaga T (1992) Unique palindromic sequences in synthetic oligonucleotides are required to induce IFN and augment IFN-mediated natural killer activity. J Immunol 148:4072–4076

Yi A, Chace JH, Cowdery JH, Krieg AM (1996a) IFN-γ promotes IL-6 and IgM secretion in response to CpG motifs in bacterial DNA and oligodeoxynucleotides. J Immunol 156:558–564

Yi A, Hornbeck P, Lafrenz DE, Krieg AM (1996b) CpG DNA rescue of murine B lymphoma cells from anti-IgM-induced growth arrest and programmed cell death is associated with increased expression of *c-myc* and *bcl-xL*. J Immunol 157:4918–4925

Yi A, Klinman DM, Martin TL, Matson S, Krieg AM (1996c) Rapid immune activation by CpG motifs in bacterial DNA: systemic induction of IL-6 transcription through an antioxidant-sensitive pathway. J Immunol 157:5494–5402

Yi A, Krieg AM (1998) CpG DNA rescue from anti-IgM-induced WEHI-231 B lymphoma apoptosis via modulation of IκBα and IκBβ and sustained activation of nuclear factor-κB/c-Rel. J Immunol 160:1240–1245

Yi A, Chang M, Peckham DW, Krieg AM, Ashman RF (1998) CpG oligodeoxynucleotides rescue mature spleen B cells from spontaneous apoptosis and promote cell cycle entry. J Immunol 160: 5898–5906

Zarling DA, Arndt-Jovin DJ, Robert-Nicoud M, McIntosh LP, Thomae R, Jovin TM (1984) Immunoglobulin recognition of synthetic and natural left-handed Z DNA conformations and sequences. J Mol Biol 176:369–415

# Subject Index

# Current Topics in Microbiology and Immunology

Volumes published since 1989 (and still available)

Vol. 206: **Chisari, Francis V.; Oldstone, Michael B. A. (Eds.):** Transgenic Models of Human Viral and Immunological Disease. 1995. 53 figs. XI, 345 pp. ISBN 3-540-59341-1

Vol. 207: **Prusiner, Stanley B. (Ed.):** Prions Prions Prions. 1995. 42 figs. VII, 163 pp. ISBN 3-540-59343-8

Vol. 208: **Farnham, Peggy J. (Ed.):** Transcriptional Control of Cell Growth. 1995. 17 figs. IX, 141 pp. ISBN 3-540-60113-9

Vol. 209: **Miller, Virginia L. (Ed.):** Bacterial Invasiveness. 1996. 16 figs. IX, 115 pp. ISBN 3-540-60065-5

Vol. 210: **Potter, Michael; Rose, Noel R. (Eds.):** Immunology of Silicones. 1996. 136 figs. XX, 430 pp. ISBN 3-540-60272-0

Vol. 211: **Wolff, Linda; Perkins, Archibald S. (Eds.):** Molecular Aspects of Myeloid Stem Cell Development. 1996. 98 figs. XIV, 298 pp. ISBN 3-540-60414-6

Vol. 212: **Vainio, Olli; Imhof, Beat A. (Eds.):** Immunology and Developmental Biology of the Chicken. 1996. 43 figs. IX, 281 pp. ISBN 3-540-60585-1

Vol. 213/I: **Günthert, Ursula; Birchmeier, Walter (Eds.):** Attempts to Understand Metastasis Formation I. 1996. 35 figs. XV, 293 pp. ISBN 3-540-60680-7

Vol. 213/II: **Günthert, Ursula; Birchmeier, Walter (Eds.):** Attempts to Understand Metastasis Formation II. 1996. 33 figs. XV, 288 pp. ISBN 3-540-60681-5

Vol. 213/III: **Günthert, Ursula; Schlag, Peter M.; Birchmeier, Walter (Eds.):** Attempts to Understand Metastasis Formation III. 1996. 14 figs. XV, 262 pp. ISBN 3-540-60682-3

Vol. 214: **Kräusslich, Hans-Georg (Ed.):** Morphogenesis and Maturation of Retroviruses. 1996. 34 figs. XI, 344 pp. ISBN 3-540-60928-8

Vol. 215: **Shinnick, Thomas M. (Ed.):** Tuberculosis. 1996. 46 figs. XI, 307 pp. ISBN 3-540-60985-7

Vol. 216: **Rietschel, Ernst Th.; Wagner, Hermann (Eds.):** Pathology of Septic Shock. 1996. 34 figs. X, 321 pp. ISBN 3-540-61026-X

Vol. 217: **Jessberger, Rolf; Lieber, Michael R. (Eds.):** Molecular Analysis of DNA Rearrangements in the Immune System. 1996. 43 figs. IX, 224 pp. ISBN 3-540-61037-5

Vol. 218: **Berns, Kenneth I.; Giraud, Catherine (Eds.):** Adeno-Associated Virus (AAV) Vectors in Gene Therapy. 1996. 38 figs. IX,173 pp. ISBN 3-540-61076-6

Vol. 219: **Gross, Uwe (Ed.):** Toxoplasma gondii. 1996. 31 figs. XI, 274 pp. ISBN 3-540-61300-5

Vol. 220: **Rauscher, Frank J. III; Vogt, Peter K. (Eds.):** Chromosomal Translocations and Oncogenic Transcription Factors. 1997. 28 figs. XI, 166 pp. ISBN 3-540-61402-8

Vol. 221: **Kastan, Michael B. (Ed.):** Genetic Instability and Tumorigenesis. 1997. 12 figs.VII, 180 pp. ISBN 3-540-61518-0

Vol. 222: **Olding, Lars B. (Ed.):** Reproductive Immunology. 1997. 17 figs. XII, 219 pp. ISBN 3-540-61888-0

Vol. 223: **Tracy, S.; Chapman, N. M.; Mahy, B. W. J. (Eds.):** The Coxsackie B Viruses. 1997. 37 figs. VIII, 336 pp. ISBN 3-540-62390-6

Vol. 224: **Potter, Michael; Melchers, Fritz (Eds.):** C-Myc in B-Cell Neoplasia. 1997. 94 figs. XII, 291 pp. ISBN 3-540-62892-4

Vol. 225: **Vogt, Peter K.; Mahan, Michael J. (Eds.):** Bacterial Infection: Close Encounters at the Host Pathogen Interface. 1998. 15 figs. IX, 169 pp. ISBN 3-540-63260-3

Vol. 226: **Koprowski, Hilary; Weiner, David B. (Eds.):** DNA Vaccination/Genetic Vaccination. 1998. 31 figs. XVIII, 198 pp. ISBN 3-540-63392-8

Vol. 227: **Vogt, Peter K.; Reed, Steven I. (Eds.):** Cyclin Dependent Kinase (CDK) Inhibitors. 1998. 15 figs. XII, 169 pp. ISBN 3-540-63429-0

Vol. 228: **Pawson, Anthony I. (Ed.):** Protein Modules in Signal Transduction. 1998. 42 figs. IX, 368 pp. ISBN 3-540-63396-0

Vol. 229: **Kelsoe, Garnett; Flajnik, Martin (Eds.):** Somatic Diversification of Immune Responses. 1998. 38 figs. IX, 221 pp. ISBN 3-540-63608-0

Vol. 230: **Kärre, Klas; Colonna, Marco (Eds.):** Specificity, Function, and Development of NK Cells. 1998. 22 figs. IX, 248 pp. ISBN 3-540-63941-1

Vol. 231: **Holzmann, Bernhard; Wagner, Hermann (Eds.):** Leukocyte Integrins in the Immune System and Malignant Disease. 1998. 40 figs. XIII, 189 pp. ISBN 3-540-63609-9

Vol. 232: **Whitton, J. Lindsay (Ed.):** Antigen Presentation. 1998. 11 figs. IX, 244 pp. ISBN 3-540-63813-X

Vol. 233/I: **Tyler, Kenneth L.; Oldstone, Michael B. A. (Eds.):** Reoviruses I. 1998. 29 figs. XVIII, 223 pp. ISBN 3-540-63946-2

Vol. 233/II: **Tyler, Kenneth L.; Oldstone, Michael B. A. (Eds.):** Reoviruses II. 1998. 45 figs. XVI, 187 pp. ISBN 3-540-63947-0

Vol. 234: **Frankel, Arthur E. (Ed.):** Clinical Applications of Immunotoxins. 1999. 16 figs. IX, 122 pp. ISBN 3-540-64097-5

Vol. 235: **Klenk, Hans-Dieter (Ed.):** Marburg and Ebola Viruses. 1999. 34 figs. XI, 225 pp. ISBN 3-540-64729-5

Vol. 236: **Kraehenbuhl, Jean-Pierre; Neutra, Marian R. (Eds.):** Defense of Mucosal Surfaces: Pathogenesis, Immunity and Vaccines. 1999. 30 figs. IX, 296 pp. ISBN 3-540-64730-9

Vol. 237: **Claesson-Welsh, Lena (Ed.):** Vascular Growth Factors and Angiogenesis. 1999. 36 figs. X, 189 pp. ISBN 3-540-64731-7

Vol. 238: **Coffman, Robert L.; Romagnani, Sergio (Eds.):** Redirection of Th1 and Th2 Responses. 1999. 6 figs. IX, 148 pp. ISBN 3-540-65048-2

Vol. 239: **Vogt, Peter K.; Jackson, Andrew O. (Eds.):** Satellites and Defective Viral RNAs. 1999. 39 figs. XVI, 179 pp. ISBN 3-540-65049-0

Vol. 240: **Hammond, John; McGarvey, Peter; Yusibov, Vidadi (Eds.):** Plant Biotechnology. 1999. 12 figs. XII, 196 pp. ISBN 3-540-65104-7

Vol. 241: **Westblom, Tore U.; Czinn, Steven J.; Nedrud, John G. (Eds.):** Gastroduodenal Disease and Helicobacter pylori. 1999. 35 figs. XI, 313 pp. ISBN 3-540-65084-9

Vol. 242: **Hagedorn, Curt H.; Rice, Charles M. (Eds.):** The Hepatitis C Viruses. 2000. 47 figs. IX, 379 pp. ISBN 3-540-65358-9

Vol. 243: **Famulok, Michael; Winnacker, Ernst-L.; Wong, Chi-Huey (Eds.):** Combinatorial Chemistry in Biology. 1999. 48 figs. IX, 189 pp. ISBN 3-540-65704-5

Vol. 244: **Daëron, Marc; Vivier, Eric (Eds.):** Immunoreceptor Tyrosine-Based Inhibition Motifs. 1999. 20 figs. VIII, 179 pp. ISBN 3-540-65789-4

Vol. 245/I: **Justement, Louis B.; Siminovitch, Katherine A. (Eds.):** Signal Transduction and the Coordination of B Lymphocyte Development and Function I. 2000. 22 figs. XVI, 274 pp. ISBN 3-540-66002-X

Vol. 245/II: **Justement, Louis B.; Siminovitch, Katherine A. (Eds.):** Signal Transduction on the Coordination of B Lymphocyte Development and Function II. 2000. 13 figs. XV, 172 pp. ISBN 3-540-66003-8

Vol. 246: **Melchers, Fritz; Potter, Michael (Eds.):** Mechanisms of B Cell Neoplasia 1998. 1999. 111 figs. XXIX, 415 pp. ISBN 3-540-65759-2

Printing: Saladruck, Berlin
Binding: H. Stürtz AG, Würzburg